APPROPRIATE TECHNOLOGY SOURCEBOOK
VOLUME II

by Ken Darrow
Kent Keller
Rick Pam

January 1981

The second in a two volume set
of guides to practical books and plans
for village and small community technology

A Volunteers in Asia Publication

ABSTRACT

Appropriate Technology Sourcebook, Volume Two, January 1981, 496 pages.

This is the second volume in a set of guides to practical books and plans for village and small community technology. The first volume, with over 30,000 copies in print, is being used in more than 100 countries, to find a wide range of published technical information that can be used by individuals and small groups. In volume two 500 more publications from international and U.S. sources are reviewed, covering small water supply systems, renewable energy devices such as windmills and solar dryers, agricultural tools and implements, intensive gardening, workshop tools and equipment, crop preservation, housing, health care and other topics. New topic areas include forestry, aquaculture, nonformal education, small enterprises, transportation and others. Price and ordering address are provided for each publication. 300 illustrations. Extensive index.

ISBN 0-917704-06-1 (Volume 2, paperback)
ISBN 0-917704-09-6 (2 volume set, paperback)
ISBN 0-917704-11-8 (Volume 2, clothbound)
ISBN 0-917704-13-4 (2 volume set, clothbound)

First Printing, January 1981
Printed in U.S.A.

Front cover design by Louis Saekow
20 drawings in text by Greg Shaw

Additional copies of this book can be ordered from: Appropriate Technology Project, Volunteers in Asia, Box 4543, Stanford, California 94305, USA. The regular price of the paperback edition is $6.50 (plus $1.38 postage for a single copy). For local groups in developing countries, the price is $3.25 (plus $1.38 postage for a single copy). For either group, discounts are available for purchases of 10 copies or more, and postage is less per copy. A special clothbound edition with a sewn binding is available for $11.50 ($8.25 to local groups in developing countries), plus postage.

Contents

ACKNOWLEDGEMENTS

Volume Two of the Appropriate Technology Sourcebook was made possible in large part due to the financial support of the following organizations:

United Methodist Committee on Relief
United Church Board for World Ministries
United Presbyterian Church
Maryknoll Fathers
Church World Service
Bishop's Fund for World Relief, Episcopal Church
Catholic Relief Services
CODEL

To these organizations we extend our warmest thanks for their enthusiasm, confidence in the worth of this effort, and support in carrying it out.

Many people contributed to this book. Rick Pam wrote most of the reviews in the Water Supply and Sanitation chapter, and many others in the Solar energy chapter and elsewhere. Thomas Fricke wrote the majority of the reviews in the Agriculture and Biogas chapters. Jim Kalin helped to acquire many books and wrote drafts of a number of reviews. Mike Connor wrote the Aquaculture reviews, with some input from Cliff Halverson. Bill Bower and David Werner made recommendations and observations on the materials reviewed in the Health Care chapter. Mary Pat O'Connell and Bob Huppe offered helpful advice on the Science Teaching chapter. Jim Bateson, Marcus Kauffman, Charles Kolstad, Martha Lewis, Maryanna Maloney, and Michael Saxenian each contributed one or more reviews. Mahesh Shrestha and Alison Davis helped with typesetting. Other members of the Volunteers in Asia staff who offered great support, encouragement, and understanding during the long process of compiling and producing this book include Carol Benedict, Sukie Jackson, Dwight Clark, Matt Lippert, Linda Shaw, Terry George and John Doll. We send a special thanks to all of you.

—K.D., K.K.

HOW TO USE THIS BOOK

This set of books was written for use by individuals and groups around the world that are looking for clearly presented, basic technical books and plans, on village technology topics. An updated edition of *Appropriate Technology Sourcebook Volume One* is now available. *Volume Two* (this book) is a guide to additional practical publications in print in 1980/81. Each review contains the price and source address. (In some cases, acronyms [abbreviations] have been used; these addresses can be found on pages 354-356.)

The tools and techniques described by these books and plans should be *adapted* to fit local materials, skills, and conditions, and then carefully tested to be sure that they are genuinely valuable in a new setting. We recommend reading the introduction and the chapters on background reading, strategies for local self-reliance and appropriate technology, and local communications before undertaking technology promotion activities of any kind.

Introduction

Since we wrote volume one of this book in 1976, the tide of interest in appropriate technology has continued to rise. The number of active individuals and groups has multiplied, as have the programs supported by the international community and governments of both rich and poor countries. With so much activity, there appear to be both favorable and unfavorable trends in the way appropriate technology is being thought about and pursued. This introductory section reflects our understanding of some of these trends, and suggests some of the issues that appropriate technology people should be thinking about: changes in institutional structures, research and development strategies, the politics of technologies, and the relationships between community organizing and appropriate technology. We have tried to evaluate where the appropriate technology movement now stands, and what steps seem to be needed to carry us a little further.

It is increasingly evident that appropriate technology workers should be forging relationships and making alliances with people active in (potentially) complementary development activities—such as community organizers, transport planners, university students and faculty, the staff of small business and cooperatives promotion programs, personnel of non-formal education programs, members of unions and peasant organizations, teachers in technical high schools, and librarians. All have much to share with technologists. For this reason we have decided to add new categories of readings to volume two of this book. These materials seem increasingly important to appropriate technology activities; they range from the very thought-provoking to the highly practical. One new section is concerned with strategies for appropriate technology and local self-reliance; discussed here are the advantages of small scale in community-based efforts, and how government policies can support those efforts. Another new section reviews some of the literature on the successful operation of small enterprises and cooperatives—important vehicles for the application of appropriate technologies and the equitable distribution of the benefits. Educational strategies that support local problem-solving are necessary if local knowledge of needs and the power of community action are to be tapped. We have therefore included chapters on science teaching to support local technical innovation, and nonformal education approaches and training techniques.

What is appropriate technology? It is a way of thinking about technological change, recognizing that tools and techniques can evolve along different paths toward different ends. It includes the belief that human communities can have a hand in deciding what their future will be like, and that the choice of tools and techniques is an important part of this. It also includes the recognition that technologies can embody cultural biases and sometimes have political and distributional effects that go far beyond a strictly economic evaluation. "A.T." therefore involves a search for technologies that have, for example, beneficial effects on income distribution, human development, environmental quality, and the distribution of political power—in the context of particular communities and nations.

Different origins

The appropriate technology movement in rich countries like the United States got started due to a convergence of a wide variety of concerns. These include the need to find a more harmonious and sustainable relationship with the environment, identify a way out of the accelerating energy and resource crises, reduce alienating work disconnected from its products and goals, develop more democratic workplaces, bring local economies back to health with diverse locally owned and operated enterprises, and revitalize local culture to counter the increasingly homogeneous and sterile mass culture channeled through the electronic media. Thoughtful, careful social choices are needed to correct the excesses and imbalances of an industrial culture driven by endless materialism. An essential quality of the appropriate technology movement in the United States can therefore be expressed by the word "**restraint**."

The appropriate technology movement in poor countries has, on the other hand, developed in a very different fashion. In the rich countries a job created in manufacturing typically requires an investment in the range of $20,000-$150,000, and in heavy industry this figure is higher still. In the poor countries a fundamental obstacle to widespread industrialization has been the large amounts of capital required; as a consequence the small amounts of capital available have usually been concentrated in a small industrial sector, creating very few jobs. The appropriate technology movement in poor countries has come out of the recognition that industrialization strategies have not been successfully solving the problems of poverty and inequality, and that in many cases "modernization" efforts have been essentially massive assaults on local culture. The result for hundreds of millions of people has been the modernization of poverty—the neglect or destruction of traditional crafts occupations, the consolidation of farm lands into fewer and fewer hands, and the division of communities, leaving these people to eke out an existence on the fringe of economic activity. The appropriate technology movement in the Third World has developed as "**the art of the possible**" among the world's poor, seeking ways to solve pressing basic problems and create jobs with resources consisting of local skills and materials but little surplus cash.

From these different origins, the appropriate technology movements in rich and poor countries have been moving towards each other. The development of renewable energy technologies, long a chief area of activity among U.S. appropriate technology groups, has moved high on the list of priorities in oil-importing poor countries as well, as they have faced high prices and scarcity of fuel for buses, tractors, and irrigation pumps. Similarly, environmental protection has gained increased attention in poor countries as multinational firms have sold pesticides banned in the United States, and energy-related deforestation has reached a critical level. Meanwhile, A.T. proponents in rich countries have begun to select and use technologies from poor countries (such as methods for making tempeh, the Indonesian fermented soybean food) in their own efforts.

This book, though primarily oriented towards appropriate technology activities in poor countries, contains much of relevance to the North American appropriate technology movement, and reflects the increasingly shared goals of both. The published technical materials reviewed here describe tools and techniques that:[1]

1) require only small amounts of capital;

2) emphasize the use of locally available materials, in order to lower costs and reduce supply problems;

3) are relatively labor-intensive but more productive than many traditional technologies;

---

1. This list is much the same as that used in volume one; see pages 10-12.

4) are small enough in scale to be affordable to individual families or small groups of families;

5) can be understood, controlled and maintained by villagers whenever possible, without a high level of special training;

6) can be produced in villages or small metal working shops;

7) suppose that people can and will work together to bring improvements to communities;

8) offer opportunities for local people to become involved in the modification and innovation process;

9) are flexible, can be adapted to different places and changing circumstances;

10) can be used in productive ways without doing harm to the environment.

Part of the appropriate technology strategy has been to start with and build on locally available skills and materials, based on the initiative and full participation of local people. This should mean that local needs will be met more effectively, that mistakes will be on a scale that is understandable and correctable, and that technological and social changes that follow are more likely to harmonize with evolving local traditions and culture.

For those who subscribe to this approach to technological change, a key question has been and continues to be: Where do we start? The evidence so far suggests that there is little cause for believing that we are about to make major steps forward. One way of summarizing this metaphorically is to say that "there is nobody as the helm"—charting the world's technological course, directing the international economy, or managing the ecological systems of the planet. There is no global Solomon who will mediate the claims and needs of rich vs. poor countries, present vs. future generations, or homo sapiens vs. other species vanishing before the destructive tide of our uncontrolled tools and habits. The problems that will continue to face all of us for the rest of this century and the beginning of the next have become increasingly evident over the past decade, but the political commitment to respond imaginatively and on a necessary scale is simply nowhere in sight.

How far have we come?

Judging by the number of programs and organizations involved in appropriate technology work, we might have hoped to be able to say that the purely technological reasons for poverty (inadequate tools and techniques and therefore inefficient use of labor and resources) were well on their way to being eliminated. Unfortunately, this is far from being the case. The achievement of quasi-respectable status for appropriate technology within both the academic community and the still relatively ineffectual development institutions over the past five years does not mean that a major effort to apply A.T. concepts is underway. By contrast, during that same five-year period, the running shoe industry has experienced extremely rapid growth, reaching $1.5-2.0 billion in annual sales in the United States alone. The evolution in shoe designs has been equally dramatic, as large sums of money have been devoted to research and development in the race to make a better shoe. In comparison, progress in appropriate technology has been probably less noticeable and certainly not as well funded.

Lest this seem unduly negative, let us remember that much can and is being accomplished in small efforts around the world. That the principles of appropriate technology can be applied and the goals can begin to be visibly achieved in small activities is cause for optimism—there are many people around the world who can and will make a difference. They need wait for no one. This is the fundamental message of hope that we can read.

People in both small and large organizations are affected, though to different degrees, by the relative shortage of funds to support appropriate technology work. But the largest single factor preventing more rapid progress in the development and application of relevant improved village technologies does not seem to be a shortage of money. It is instead the lack of a coherent set of ideas about how funds and human resources might be combined in workable strategies to apply the concepts of appropriate technology, both within and between countries. There appear to be, for example, a number of communications tasks or "overhead" functions that include smoothing the flow of technical documentation, disseminating information on successful policy measures, supporting dialog between partner institutions, and channeling small amounts of funds to grassroots technology research and development efforts. Some of these might be tasks that existing international agencies or networks should take on or support others to take on (see list below under "Tasks for International Efforts in Appropriate Technology"). Other tasks are perhaps best initiated and financed by local networks or individual groups. Yet differing languages and cultural attitudes are barriers to building systems that work; and the differences in organizational forms among cultures make the challenge all the more difficult.

Currently organizations of all types and sizes are stabbing at ideas, without a sense of their role within a larger strategy to support appropriate technology. Feeling an imperative to be involved in some work called "A.T.", they are committing resources to technology activities that seem promising in an immediate sense, but may be a poor use of those still scarce resources. There have been, for example, perhaps as many as 20 international directories of A.T. groups produced, at least 4 of them well-funded efforts by major international organizations. There continue to be million-dollar pilot projects demonstrating too-expensive "village" technology. Genuine grassroots practitioners are scarce at conferences, where sympathetic non-practitioners, ill-versed in the daily obstacles to technology improvement in the field, have difficulty identifying the truly useful "overhead functions" that coalitions and aid agencies could perform. Well-heeled expatriates running projects heavily funded by bilateral and international aid agencies find that the daily obstacles to technology development and dissemination have dropped below their horizon; they cannot see where judicious use of funds could make a difference.

INSTITUTIONAL CHANGE

Programs and projects

For many of the institutions suffering from these problems, "appropriate technology" means something different than it did when it was obscure and unlikely to ever be generally accepted. At that time it did not have to be uncontroversial and tailored to match the strengths and weaknesses of international and governmental aid agencies. It could be honest, radical (going to the root to re-examine), visionary—and not necessarily politic, incremental, or pragmatic.

Now that "appropriate technology" is being added to the activities of national and international organizations, it is coming under the critical scrutiny of those who would like to use elements from it to repair the creaking, battered bridges of development aid. Armies of these development professionals are now chopping away at the lush growth of ideas that have come to represent A.T. They have been charged with "operationalizing" A.T., and to do so they are busily cutting back the undergrowth to get at what they see as a few good pieces of timber.

A.T. is a difficult approach to incorporate into large agency planning efforts. The concept of "local self-reliance," for example, is difficult to define or quantify and will vary from place to place. Furthermore, it is a quality that can be either nurtured or destroyed from the outside, but never created. "Self-reliance" also sounds vaguely utopian or ideologically-tainted. To many planners it looks unnecessary, and out it goes.

Equally difficult is the concept of "people's participation." "Participation" is probably the most often invoked and least often attempted aspect of rural development programs. "Participation" is often interpreted to mean carrying out instructions. This kind of interpretation makes "participation" simply a measure of the degree of local acceptance of a project, not a strategy for success and human development. Nor does this approach to "participation" contribute at all to an on-going process of community problem-solving. A higher stage of "participation" may be reached when community reaction to the activities planned by intervening agencies is sought. Such an approach is sometimes taken in the hope that local enthusiasm will be increased and gross mistakes (due to factors unknown to the planners but evident to the community) can be avoided. Yet even this leaves the community in essentially a passive role, as Denis Goulet aptly points out:

> "One may plausibly argue that to structure feedback is merely to assure that any participation elicited will be a mere 'reaction' to what is proposed. To **propose** is thus, in effect, to **impose**, inasmuch as those who plan initial arrangements do not provide for a **feed-in** at early moments of problem-definition. Feedback prevents non-experts from gaining access to essential parameters of the decision process **before** these are congealed."[2]

To better understand how a planner tends to be forced to deal with the concept of "participation," let us consider a hypothetical set of choices. Suppose we arbitrarily define different degrees of participation as follows: 1) The people will interact with the new technology in some active way. 2) The people will use the technology and determine themselves just what they want to accomplish with it. 3) The people will make and repair the new tool or machine themselves, in addition to doing what they want with it. 4) The people will be involved in the design and development of the technology to address needs they feel to be most pressing. To the planner, the fourth option (beneficiary involvement in design and development of technology) is likely to seem inefficient, uncertain, and slow. The third option, villager involvement in production, maintenance, and repair, may be thought to restrict too severely the kinds of materials and processes used. The second option would probably seem to introduce an irrelevant issue, as beneficiaries must use a new technology as originally planned if a project is to meet its production targets. That leaves option number one, "the people will interact with the technology in some active way," a choice that neglects both the essence and the advantages of genuine community participation.

Another important concept in the A.T. approach is that development should increase (or at least not reduce) the ability of the poor to cope with their problems. Yet the project designer is hard-pressed to ensure this; no project's consequences can be fully known beforehand. We must, however, retain a healthy respect for the complexity of the world and the unpredictability of

---

2. Denis Goulet, **The Uncertain Promise: Value Conflicts in Technology Transfer**, (1977), emphasis in original. See review on page 360.

human events. We should keep unpredictability in mind when we evaluate the kinds of activities that are chosen. We can ask ourselves: Given that certainty is impossible, what new risks come with a proposed activity? Will this activity make the poor more or less capable of dealing with unforeseen problems and crises which may arise? These questions are particularly important when we consider projects that tie the poor into the world economy. Planners and decision-makers have pointed to international markets and invoked the benefits of "interdependence", but have neglected the question of who stands to lose out when times are difficult. For it is the poor, in the end, who will be hit by famine when the crop fails, left with their equipment idle when fuel is unavailable, and find themselves on the street without work when fashions change and their exported handicraft product is no longer in vogue. The oil crisis illustrates, in perhaps the most dramatic way, the consequences of this increased vulnerability. Citizens of fuel-importing poor countries are finding that they simply will not be able to stay in the bidding for fuel as the prices climb. Current price levels are leading to sharp drops in consumption in the Third World, while the industrialized world, much ballyhoo to the contrary, has shown relatively little real change in its own consumption. It thus appears that the brunt of oil shortfalls are felt by the oil-importing countries of the Third World, a circumstance which can only get worse in the future.

These, then, are some of the ways in which it is difficult to incorporate A.T. principles into large agency programs of action. Concepts like "local self-reliance" and "people's participation" are difficult to "operationalize" and they will therefore not be popular with planners. The need to integrate the poor into the world economy is taken as given; "the little guy" is subjected to the roller coaster price cycles of international markets, and his chances for self-reliant advancement are undercut. But unless thoughtful attention is given to these difficult issues, projects and programs created will have little about them that is really "appropriate".

Policies

Policy-makers responsible for the national directives which influence technological change also need to carefully examine whether their strategies are supportive of appropriate technologies. People who work with appropriate technologies at the grassroots levels in many countries consider governmental measures out of touch and ineffectual due to the national level at which they originate. Thus independent-minded A.T. workers, generally isolated from each other and receiving no support from the national policy-making level, have become thoroughly skeptical of government and international agency activities labeled "A.T.". If the world were made up of autonomous communities, this would merely mean that people would be free to make their own choices about technological change, unhampered by external influences. But the fact is that the image of the completely isolated village is an illusion. In a poor country like Indonesia, for example, where more than 70% of the population lives in rural areas, the majority of that 70% has watched television and knows of Honda motorcycles and Levi's jeans. The economic culture of the West has driven literally to the farthest corners of the earth, and profoundly affected the aspirations and behavior of much of humanity. Thus it is that in a village in which the largest killer is infant diarrhea, people save from tiny disposable incomes to buy quartz-crystal watches, cameras, and motorcycles while the need for clean water goes unaddressed. Thus it is that not only the landless but also the young, the bright, and the enterprising people are lured to the city. They leave behind village economies drained of resources by international soft drinks, cigarettes, and consumer goods, and drained of talent by the magnetic pull of economic opportunity and "modernization."

Clearly local A.T. workers must consider these powerful forces influencing the communities in which they work. Community-based development strategies will be up against great odds where the vitality of the community has been sapped. It is just as clear that neglect at the national policy level will greatly increase the odds against A.T. development. There are simply too many policy questions—in communications, education, and economics—that directly affect the chances of appropriate technologies to evolve and survive.

However, we should not assume that appropriate technology must therefore emerge out of top-down directives. Rather, the challenge is to devise policy measures and government programs which will support local A.T. efforts—instead of replacing or directing them. Nicolas Jequier, a development strategist himself, notes: The A.T. movement does

> "...not really know how to handle A.T. on the scale that is required by the needs of hundreds of millions of poor people...**The problem in effect is not simply to develop more appropriate types of technology, important as this may be, but to start redesigning the existing system of planning, investment, and development...**"[3]

The designing of more and more appropriate technology projects is not enough. Changes are needed not just in the content of research and development programs, but more importantly in the ways this content is determined. For planners and policy-makers, this means that emphasis should shift from project design and implementation to creating policies that will foster decentralized initiative and innovation.

RESEARCH AND DEVELOPMENT OF APPROPRIATE TECHNOLOGIES

We noted earlier that the running shoe industry in the United States has experienced enormous growth in sales over the past five years. With this growth have come some dramatic improvements in the shoes. They have evolved from padded foot coverings into sophisticated combinations of cushions and supports to protect the long distance runner from foot, ankle, and knee injuries that can follow long hours of steady pounding. These design achievements have come within an approach of mass-marketing to an affluent, rapidly expanding group of consumers. Buyers have kept informed about the performance of different shoes through annual extensive comparative testing programs sponsored by popular magazines. Manufacturers that have succeeded in designing and producing the better shoes have been rewarded with dramatically increased sales. This is, in fact, a textbook case of market demand linked technical innovation. It is not, however, a model that is very relevant in village technology development. The meager financial resources of the poor, the geographically scattered nature of the "market", and technology needs that vary from place to place make conditions and thus strategies substantially different for development and dissemination of appropriate technologies.

Who are the innovators?

Who originates successful technological innovations? The answer to this question should significantly affect any strategy for village technology research

---

3. Emphasis added. See "Appropriate Technology: The Second Generation", paper by Nicolas Jequier, (1978).

and development. It has been commonly recognized that professionally-trained engineers, scientists, and foreigners have made significant innovations in a variety of settings. Each of these sometimes overlapping groups has its own strengths: engineers are versed in the fundamentals of design and methods of presentation; scientists have powerful conceptual and methodological skills; foreigners bring ideas from outside and insights from a different way of looking at problems.

It is not as commonly recognized that craftsmen, farmers, and other villagers have been contributing to the village technology innovation process for much longer than the professionals and outsiders. The assumption that poor farmers and craftsmen are not inventive is frequently built into technology improvement strategies. This kind of assumption is an unfortunate misconception held (often unconsciously) by people who have had little direct contact with villagers in their own or other countries, or who do not bear in mind the fact that human beings are inherently creative. In fact, the poor in both rural and urban areas around the world show considerable ingenuity in using the materials available to them to solve their problems: recycling industrial materials into shoes and oil lamps, imitating natural ecological interactions in small farming systems, keeping vehicles running for decades without the proper spare parts, crafting windmills and watermills in a multitude of different but durable designs within a single community.

A person convinced of the existence of lively inventive activity among the people of the rural areas and urban slums of the Third World may still be inclined to dismiss this as too small in magnitude to be relevant to appropriate technology efforts. But is it? Suppose we assume that roughly 2% of any human population acts in ways that might earn the label "inventor". (This is the figure used for the San Francisco bay area of California by a regional inventor's council.[4]) If this is accurate, there would be 1.7 million "inventors" among the 85 million inhabitants of Java! Even if the figure of 2% is reduced by a factor of ten, that would leave 170,000 people informally involved in day-to-day village technology adaptation—a figure that simply dwarfs the number of researchers working in institutionalized programs on the same tasks. Appropriate technology efforts should be designed to take advantage of this large and creative group of people and support them, with technical assistance when necessary and with formal and nonformal educational programs, to put such inventive activity on a firmer technical footing and accelerate it.

Clearly we must keep in mind the fact that craftsmen and farmers commonly have a knack for devising tools. They also have a firm grasp of acceptability, affordability, and usefulness that is sorely needed in institutional research and development (R&D) programs. The magnitude of the potential contribution by craftsmen and farmers in an innovation process is at least as great as that of the professionals and outsiders. This suggests that programs for the indigenous development of appropriate technologies should draw on the different perspectives and innovative talents of each of these groups, by giving special attention to imaginative ways to directly include the poor.

---

4. Obviously these population groups are quite different. The San Francisco bay area (100,000 "inventors" in a population of 5 million) has a large number of highly educated people, and is the heart of the rapidly growing electronics industry. On the other hand, the mostly rural population of Java has greater daily involvement with tools and materials, and a great range of traditional technologies.

Structure of research and development efforts

Most research and development institutions in the Third World today are structured in ways that work against the development of appropriate technologies. This continues to be true even when the content of R&D activities is changed to "appropriate technology". Many of these institutions have foreigners in top administrative positions; the local staff tend to be from the urban elite, are highly educated (often in the industrialized countries), and look to their peers for recognition. Facilities are located far from villages and trials of new technologies are conducted in artificial environments. There is generally little place in this kind of institutional setting for input from farmers and other villagers.

It is certainly not new to suggest that this problem should be corrected. For some years now observers around the world have been pointing to the need for direct involvement of farmers and other villagers in any technological research intended to benefit them:

"The farmers in a land parcelization project complained of little or no corn response to fertilizer even though it was required in the complete credit package...(Subsequent) results from Farm Trials indicated response in some cases, especially in some of the hybrids tested, but in none was it profitable; conventional wisdom, coupled with the natural tendency to consider fertilizer necessary in any complete recommendation, had created a situation in which the farmers were being forced into unprofitable investments..."

—Peter Hildebrand, "Generating Small Farm Technology", (1977)

"The philosophical foundation of...research and development centers ought to be based on the active participation and agricultural wisdom of the small farmer...The fact is that the farmer is the best judge of what he does and what he hopes to do better. The farmer will improve his methods of production and accept new technologies only if he has been an active participant in the generation of the new technologies."

—E.G. Vallianatos, **Fear in the Countryside**, (1976) [5]

"The challenge before us is to establish a system which will produce machines that will make poor people more productive—machines that will work, will last, and are affordable. In developing this system, we must ensure that the villager becomes an active member of the research team. For it is the villager...who is the focal point of all this activity, and ultimately it is the villager who will judge if we are making a serious effort to solve his problems, or if we are merely continuing to tinker with his future."

—David Henry, "Designing for Development:
What is A.T. for Rural Water Supply and Sanitation?", (1978) [6]

It is commonly assumed that poor people should not be involved in their own research because they do not have the necessary surplus to afford the false starts and mistakes that inevitably are part of experimentation and adaptation. To some extent this is true, particularly when it comes to expensive tech-

---

5. See review on page 426.

6. In **The Social and Ecological Effects of Water Development in Developing Countries**, (1978), see review on page 528.

nologies that require heavy R&D expenditures and can only be afforded after adaptation is complete (and credit is made available). In these cases, it may make sense to involve farmers and craftsmen in technology development efforts funded by governments and aid agencies. The same is true when technically complex topics require investigation. However, there is a wide range of very low cost village technologies that require little more than local materials and labor for adaptation work, which could be carried out largely by local people themselves. Included here are such vitally important ones as improved cookstoves and grain storage bins. Moreover, it is precisely the technological improvements involving the lowest cash investments which hold the greatest potential for quickly reaching widespread use in poor communities.

Topics for R&D

"To what extent is 'good' research thought to be research which is methodologically sound, designed to refine a paradigm, related to earlier respectable research, requiring sophisticated equipment and measurement, and enabling the researcher to enhance his reputation with a tidy, citable, footnoted paper with tabulations to two places of decimals, published in a hard international journal?"

— Robert Chambers, "Identifying Research Priorities in
Water Development", (1978)[7]

Most applied scientific investigations take a very thin slice of the world and explore it in depth. Scientists are skilled at pushing back the frontiers of knowledge by identifying and exploring nooks and crannies and bringing to them the light of day. This pattern of activity, however, often serves to divert attention from obvious, everyday problems which also deserve investigation. For example, most of the world's trees have already disappeared, in part a consequence of inefficient open cooking fires. Yet until recently scarcely a handful of research efforts to devise more efficient cooking stoves had been conducted. This is a response that is in no way proportional to the magnitude of the problem. Similarly, with the cart still serving as the primary mode of transport of goods in many countries around the world, virtually no systematic design work to improve these carts had taken place until the last few years.[8] Such neglect is not the result of oversight but part of a pattern in which the research content and application of science has been heavily biased toward the needs of industrialized countries and the exotic. The everyday pressing problems in poor countries have rarely been sufficiently unique to attract the attention of the scientist eager to investigate some phenomenon never before researched. This means that the conduct of science and its applications tends toward that which is exciting and new, rather than what is practical or relevant to the concerns of poor people.

How might research become more responsive and related to everyday problems? One way, mentioned above, is to involve people who will use the fruits of research in the research process itself, and in decisions about research

---

7. In **The Social and Ecological Effects of Water Development in Developing Countries,** (1978), see review on page 528.

8. Recently there has been some interesting work on bullock cart design in India; see **The Management of Animal Energy Resources and the Modernization of the Bullock Cart System,** by N.S. Ramaswamy, (1979), reviewed on page 566.

content. An alternative approach to research, already evidenced by some committed scientists and technical people in the Third World, also offers promise. Whereas commercial research and development has responded to the promise of economic gain for the innovative firm (as in the case of the running shoe industry), the investigator who is a "social entrepreneur" identifies needs and organizes responses not in hopes of capturing large profits, but rather in hopes of contributing to a workable solution. These investigators will opt to study new roofing materials and productive, ecologically sound small farming systems, while their counterparts in the rich countries study such things as the mating behavior of exotic fish.

"The prime criterion for good research should be that it is likely to mitigate poverty and hardship among rural people, especially the poorer rural people, and to enhance the quality of their lives in ways which they will welcome; that in short, priorities should be arrived at less by an overview than by an underview, grounded in the reality of the rural situation. Starting with rural people, their world view, their problems and their opportunities, will give a different perspective. To be able to capture that perspective requires a revolution in professional values and in working styles; it requires that scientists should learn the skills and approaches of anthropologists; it requires humility and a readiness to innovate which may not come easily in many research establishments."

—Robert Chambers [9]

Science teaching

We have identified the experience and native inventiveness of local people as key elements in relevant research and development efforts. Learning about science could also bolster and broaden local capability to do research and design work, harnessing the systematic methods of scientific inquiry to the creativity and experience which people already possess. Yet science education as it is conducted in developing countries is rarely concerned with fostering a basic understanding of scientific approaches to problem-solving, nor does it offer students skills that are relevant to their daily lives. This is true in the secondary schools, as well as the primary schools which provide the only years of schooling for most rural people. Major problems include the lack of affordable texts and lab equipment, the lack of written or printed materials in the local language, the failure of curricula to show connections between science (with its odd lab apparatus) and the natural world, and meager science background among teachers. With little or no equipment, students do not learn to take systematic steps in identifying a natural phenomenon or solving a mystery; instead science is presented to them as a set of abstract concepts to be memorized. Educational systems geared to the needs of the few students who pass to subsequent levels (instead of the larger numbers who leave school at the end of each level) make science a topic for quizzing and screening students rather than a useful approach to asking questions.

A way out of the dilemma may be found by relating science more directly to the natural processes going on around students in their daily lives, by making low cost lab equipment [10], and by using devices and materials that are

---

9. From "Identifying Reseach Priorities in Water Development", in **The Social and Ecological Effects of Water Development in Developing Countries**, (1978); see review on page 528.

10. See the materials reviewed in the SCIENCE TEACHING chapter.

normally found in the community (such as bicycle pumps and market scales). Students could then become directly involved in the systematic procedures of science, learning skills invaluable in problem-solving. They could begin to escape the deadening effect of rote schooling where memorization rather than skill development and understanding has become the goal. Special courses on simple machines and agriculture, directly related to farm activities, could be included. Some curriculum development and teacher training would be crucial for the success of such efforts.

This kind of shift in science education may be some of what Albert Baez (director of UNESCO's division of science teaching from 1961-67) had in mind when he observed that "The inquiry mode of science and the design mode of technology should both infuse the science education of the future." [11] In that 1979 paper he went on to note that both Einstein and Edison were stifled and powerfully alienated by their early contact with rote schooling. An essential part of a new approach to science education, he argues, is the fostering of creativity. He cites a study which indicates that creative people "challenge assumptions, recognize patterns, see in new ways, make connections, take risks, take advantage of change, and construct networks."

It may be possible to create a corps of people who can use both "the inquiry mode of science and the design mode of technology" to help solve the technological problems of their communities. These people would receive special practical training in addition to the new science courses. They would play a role analogous to that of the "barefoot doctors" that have been successful in China and an increasing number of other developing countries. These "barefoot engineers" would not replace other engineers, but would greatly increase the availability of technical skills for problem-solving at the grassroots level. In rural Colombia, the FUNDAEC program has been training such a corps of "barefoot engineers".[12] These young people, coming from the rural communities with a sixth grade education, go through a three year training program. A university-based group distills and combines concepts from a variety of technical fields, to arm students with a set of skills relevant to the problems of their communities.

Basic steps for R&D

To summarize the observations on research and development in these pages, we can identify at least four basic steps that are likely to increase the relevance and productivity of appropriate technology R&D efforts:

1) Change the criteria for "good" research. Good research should be that which is likely to reduce poverty.

2) Seek to understand the viewpoint of the poor: their perceptions of problems and opportunities.

3) Actively include the poor, especially small farmers and craftsmen, in both decisions about research content, and in the research itself.

4) Offer basic relevant science education geared to the challenges of local design problems, with curricula adapted to employ available materials and common devices to illustrate principles, and to provide young people and farmer-inventors with a more scientifically sound basis for their innovation efforts.

---

11. "Curiosity, Creativity, Competence and Compassion—Guidelines for Science Education in the Year 2000," by Albert V. Baez, June 1979.

12. See review of some materials describing this program, on page 740.

EDUCATION AND TRAINING

Relevant science education might accelerate the process of generating useful and affordable village technology adaptations. There are other potential links between educational efforts and improved tools and techiques as well.

Increases in the standard of living come with advances in productivity. This is accomplished through upgraded technologies—which embody human knowledge—and increases in the skills of the people. When better technologies and skill development opportunities are widely available, increases may be seen in the standard of living of the poor majority. Yet conventional development strategies and programs instead concentrate resources on a narrow range of activities. Scarce capital and R&D funds are channeled to industry, which employs only a very small part of the work force. Costly training programs yield small numbers of graduates for a few sectors of the economy. A high rate of economic growth may follow in these sectors, but the benefits reach only a few people. At the same time, the vast majority of economic activities remain stagnant, experiencing no or only slow increases in productivity. The appropriate technology approach therefore requires a broader view, and asks the question, "How can we create conditions in which productivity in all activities will increase?" In a sense it is the abandonment of the great leap forward for the few, in favor of regular steps forward for the many. With the resources used to train one engineer, it would be possible to train 10-50 or more farmer/inventors who would have incentive to focus their efforts on raising productivity and earnings in the activities of the poor majority. Such a shift in the emphasis of technical training could thus mean that the mass of a population would no longer be forced to simply wait for industrialization to make them either prosperous or destitute. Instead, they could take over the development of new tools and skills that emerge from the old, setting into motion a dynamic process of productivity increases that actually involves the whole society.

Educational and training opportunities are also more than a strategy for raising the productivity of ordinary people. They have intrinsic value too, in that they broaden intellectual and technical perspectives, widen the horizons of general knowledge, and help liberate individuals from the oppression of meaningless tasks, poverty, and political domination. This "humanistic" perspective, which asserts that relevant learning opportunities are inherently worthwhile, can be seen at work in nonformal education efforts offering reading and vocational skills to groups of out-of-school adults. Such an orientation would result in decisions favoring technologies and programs which have educational consequences for the poor majority. Too often, however, development planners are not interested in widespread development of skills and knowledge that comes with decentralized technology and participatory community development programs. And, in fact, any extra effort required to initiate an educational process would be seen as an obstacle to achieving the more measurable goals—the number of wells and hand pumps installed or the number of patients examined, for example. The arbitrary disregard of informal training effects when pursuing narrow goals can be seen regularly in the way large programs are organized. John Turner observes:

> "Modern administrations, from the World Bank to the smallest and poorest national agencies, are trapped in the syndrome symptomized by categorical programmes. As long as thought, planning and institutions demand the classification of demand and supply and their combination in

fixed categories around which goals, instruments, and strategies are organized, no real or direct progress will be made towards the **liberation of resources, the realization of their potential and the regeneration of culture.''** [13]

In other words, the benefits of informal training, community organization, and an increased level of local experimentation and problem-solving do not show up in the calculations of categorical programs (which focus on one kind of development objective, such as improved health or housing). These benefits are invisible, and they are not taken into account—except by the few categorical programs that focus on these as their own particular objectives!

The case of the 2800 small scale cement plants operating in the People's Republic of China illustrates how widespread training benefits can follow directly from choice of technology. These cement plants provide cheap cement for local infrastructure construction projects such as irrigation canals; they employ ten times as many people (per unit of output) as the conventional larger rotary kilns; and they provide workers with a range of practical skills which will serve as a valuable foundation for future small industry activities:

> "Firstly, a large number of people are being trained in the process of industrial technology. Secondly, a sizable number of people inside production units have received training in organizational skills. A smaller but still sizable number have been trained in administrative skills relating to the procurement of machinery and raw materials, distribution of products and coordination with other industrial units."
>
> — Jon Sigurdson, **Small Scale Cement Plants: A Study in Economics,** (1977) [14]

In this case, the skills of the rural population have broadened significantly— more than would have been the case had the centralized "higher technology" rotary kiln been chosen. Clearly the choice of technology affects how much local people will learn and how much control they will have over production.

Valuable informal training effects can also be seen in a wide variety of private, day-to-day activities. For example, while in most poor countries managerial and entrepreneurial talent is thought to be in short supply, children of the ethnic group controlling some particular economic activity (e.g. shipping or commerce) become exceptionally talented and successful business-people. They do not, for the most part, graduate from schools with degrees in business. Rather, they participate in daily business activities, and have a high motivation to learn. Informal "learning by doing" can also be seen when communities undertake their own development projects. Mistakes are more evident and more likely to occur on a scale which is correctable by the participants. Appropriate technology advocates should be aware of these natural informal educational processes, and should think about ways to open them up to more people.

---

13. John Turner, **Housing by People**, (1976), emphasis added. See review on page 658.

14. See review on page 388.

LOCAL RESOURCES

> "Personal and local resources are imagination, initiative, commitment and responsibility, skill and muscle-power; the capability for using specific and often irregular areas of land or locally available materials and tools; the ability to organize enterprises and local institutions; constructive competitiveness and the capacity to co-operate."
>
> — John Turner [15]

Formal and informal learning opportunities—whether science education offered in the classroom or the chance to acquire managerial skills in a factory or business—are crucial in mobilizing these local human resources. Skilled, creative local people will, in turn, be able to better use local material resources—often the only alternative in poor countries:

> "To construct using renewable resources is not a sentimental fad in an area without exportable products to pay for imports...In a low cash economy it is the interactions of human resources with the immediate materials of the land that provide for the richness and fullness of life."
>
> —Peter van Dresser, **Homegrown Sundwellings**, (1977) [16]

We noted above that learning opportunities are often overlooked in the calculations of development planners. Likewise, local labor and materials are frequently ignored when planners consider the assets of a poor community. The investment of cash has the effect of mobilizing the efforts of other, distant people; the money we pay for sheet metal, for example, pays for the efforts of the workers in the steel mill and the miners who extract ore and coal, along with the transport workers and the middlemen necessary to bring this material to us. If we can instead accomplish a task by mobilizing our **own** effort (e.g. building effective grain storage bins using our own labor and locally available basket materials and clay) we can avoid spending cash on sheet metal for bins. In other words, we can directly convert our labor and local materials into capital, without any cash input. Which we choose (or are forced) to do depends on whether we know about alternatives, whether opportunities to earn cash are available, what skills we have, and which use of time and effort will most easily accomplish the task at hand. Wherever jobs that pay cash are few but local materials and labor abundant, a reliance on cash investment poses an unnecessary obstacle to accomplishing tasks like building effective grain storage bins, basic houses, or simple waterpumping windmills.

The use of local materials not only offers opportunities for action where money is scarce, but has other advantages as well. Nicolas Jequier has pointed to a distinction between what he calls "systems-dependent" and "systems-independent" technologies. [17] "Systems-dependent" technologies are those which require, for example, a supply line of spare parts, fuel, maintenance and repair skills, and materials in order to be efficiently used. "Systems-independent" technologies, on the other hand, can be efficiently used without such supply lines; they obviously include tools, techniques, and structures that are made of local materials.

---

15. John Turner, **Housing by People**, (1976), see review on page 658.

16. See review on page 633.

17. See "Appropriate Technology: Some Criteria" by Jequier, in **Towards Global Action for Appropriate Technology**, (1979), reviewed on page 361.

Local materials are thus often the key to what the members of a poor community can afford to do. "Affordable alternative", in this context, has a very different meaning than "economically competitive alternative". A technology, for example a renewable energy technology, is termed "competitive" if it is cheaper than the conventional commercial equipment and fuel used over time. This does not mean, however, that it is cheap enough to be affordable and a good trade-off for the poor compared to what they currently use. A windmill designer once expressed to us his satisfaction that with the way oil prices were going, windmills were very soon going to be economically competitive with diesel-driven pumps for water lifting. The poor, however, who couldn't afford the engine-driven pumps when oil was cheap, will hardly be in a position to afford a windmill which requires an even greater initial investment. This is not to argue that windmills do not have a place in appropriate technology efforts; on the contrary, they do seem to have a significant role to play. But the particular designs that will be successful are likely to be those that use mainly local materials and labor, and thus are far cheaper than engine-driven pumps once were.[18] Similarly, a careful calculation of what constitutes an "affordable alternative" should be made when we look at investments in improved cooking technologies from the point of view of the poor. Because firewood used in cooking is the major form of energy consumed in rural areas, fuel-efficient stoves could be a major tool in the effort to increase available rural energy supplies (and stop deforestation). But since most firewood is gathered (not purchased) and the open fire is a no-cost technology, current cooking methods require very little if any cash investment. Therefore improved stoves, however fuel-efficient they may be, will also have to be very inexpensive to be attractive and affordable.

COMMUNICATIONS

The concept that improved village technologies should be based upon local human and material resources, and in harmony with local culture, has gained great acceptance within the appropriate technology movement. As a result, prevailing models of the role of communications in development have become increasingly inadequate for A.T. advocates. The "diffusion of innovations" theories, on which so many extension programs have been based, have assumed that it was to be centralized agencies that would determine which technologies to promote. For the most part the task of communications has been, therefore, to persuade the poor recipient population to accept these solutions. This process has not left much room for input from the poor as to what solutions they might be interested in, nor has it allowed for the fact that some needs are more pressing in one community, while entirely different needs may have priority in nearby communities. The extension agent has typically been assigned to a vast area due to funding shortages, and has found his/her impact seriously diluted. In addition, errors and misunderstandings have been compounded as information passes from trainer to extension agent to "opinion leader" to the rest of the farmers.

Two of the most serious problems with this kind of approach are: 1) there is little room for "participation" by the beneficiaries except in the most minor sense — carrying out instructions; and 2) information flows almost entirely one

---

18. More expensive forms of this and other technologies might be made available through innovative, realistic credit programs. Instead of collateral requirements, terms could provide for repossession of machines on which loans are defaulted; this would allow access to equipment for people who would not otherwise qualify for loans.

way, from central agency to the poor. It is therefore essentially impossible for the villager to get technical assistance with anything except what the extension agent is promoting. Not surprisingly, the technologies promoted through this process have not been notably "appropriate", nor have the majority of extension programs achieved the success rate (measured in numbers of people adopting the prescribed technologies) expected.

What are some of the elements that should be incorporated into communications strategies to make them more consistent with A.T. concepts and more supportive of the complementary parts of A.T. efforts? If we agree that an active level of participation in problem identification and solution by the members of poor communities is highly desirable, this requires that they have:

a) access to information in a form in which it can be of practical use;
b) the ability to initiate communications in search of relevant experience and information from other communities, including information on the successful technologies that have been developed nearby, within the region, and around the world;
c) support from those with more advanced scientific and technical skills, through technical assistance centers that respond to requests.

Communications techniques and strategies linked with appropriate technology should therefore be able to support locally relevant science teaching, a research and development strategy that includes local participation, the operation of community organizations as vehicles for problem-solving, and institutional changes that allow more horizontal and two-way flow of information and initiative in poor communities.

Networks

A key local informal communication mechanism is the network—not the deliberate product of a "networker", but created by the social ties of a community, that lead people to help each other with skills and pass on information. When new skills and information get into such a network they become available for all the members to tap. An approach based on this phenomenon has been used effectively in teaching people to build solar water heaters in the United States.[19] A weekend solar training workshop is offered to members of a social club—a natural network. Later these people are in a good position to help each other properly complete solar water heater installations for their homes.

On local, national, regional, and international levels, networks of appropriate technology people are exchanging ideas and information in a highly active, decentralized fashion. Mimeographed newsletters can be low cost vehicles for information exchange among groups across some distance. Grassroots radio programs produced on cassette tapes can be used as a forum for questions and ideas about common problems.[20] In this way, people at the grassroots can listen to each other. Some of the other low cost technologies for horizontal communications strategies, many of them affordable at a village level, are documented in the LOCAL COMMUNICATIONS chapter.

Catalogs

Organizations trying to help the poor find out more about technology options should consider the possibility of producing and distributing catalogs to

---

19. See **A Solar Water Heater Workshop Manual**, by Ecotope Group, (1978), reviewed on page 639.

20. See **Grass Roots Radio**, by Rex Keating, (1977), reviewed on page 761.

document widely-relevant technologies that are traditional and efficient, new, or from outside the country. [21] The cost of producing such a catalog is much less than the extremely high cost of operating a technical information data bank that gathers information from all over, stores it for retrieval, and responds to individual requests. Even in developing countries, when the total cost of running a conventional technical information data bank is divided by the number of requests, the cost per request comes out to $100-300—equal to the annual per capita income in many countries. [22] Catalogs, by contrast, can be produced at a cost of a few dollars each when a few thousand are printed. They should be designed to anticipate and offer answers to many commonly asked questions, in addition to stimulating new thinking. This will not eliminate the need for information data banks at some levels, but it should serve to reduce the number of databanks and the expense for skilled staff to handle routine questions.

Computerized information systems

Another extremely powerful tool for information collection, storage and retrieval is the computer. Now under preparation are several computerized "appropriate technology" information systems which should do a wonderful job of controlling a vast amount of information, but do not appear likely to succeed in supplying that information to "end-users", that is, appropriate technology practitioners with grassroots ties. The institutions scheduled to receive the computer terminals are portraying themselves as end-users of the information, yet they rarely have any grassroots connections or direct experience in village technology work that has successfully improved the lot of the poor. Computerized systems for controlling and providing access to information are ill-suited to a situation in which end-users are scattered about the globe in remote parts of poor countries. The other fundamental problem with such systems is that the information going into the computers has to be screened for relevance and accuracy. Unless this is done well by experienced and knowledgeable people, the computer system will become a processor of "garbage in, and garbage out."

Appropriate technology advocates should not rule out this powerful information handling tool; it may be one of the cards to be played in creating information exchange that genuinely aids in achieving the goals of the movement. But the problems of how to screen input and ensure broad access are going to be difficult to solve; we may well see the institutions possessing the computer terminals not allowing "uncredentialled" grassroots groups to use them. In any event, in the foreseeable future, the computerized information systems are not going to be the cornerstones of an effective international information sharing system; with luck they won't divert too many scarce resources away from networks using the more mundane, but far more accessible printed and other low cost communications technologies.

---

21. Many examples of such catalogs are contained within this book, which itself is such a catalog. Perhaps the best example of a catalog with information selected for a particular developing country is the **Liklik Buk** from Papua New Guinea, reviewed on page 50.

22. See Nicolas Jequier's discussion of this in **Appropriate Technology: Problems and Promises**, (1976), reviewed on page 31.

APPROPRIATE TECHNOLOGY AND INCOME DISTRIBUTION: SOME POLITICAL QUESTIONS

Saving labor

Saving labor serves no purpose if it is your own job that is "saved" and no other real employment—and thus income—opportunities exist. In China, in 19th century England, and in the Third World today, a central question regarding mechanization and industrialization has been: Whose labor will be saved? Who will benefit and who will simply lose any opportunity to earn a living?

In the industrialized countries, new jobs are theoretically created to replace those that are destroyed—the fact that historically this has been an awkward, unequal and demoralizing process is generally overlooked, while much is made of productivity increases leading to benefits for all. (The stubborn persistence of high unemployment and increasing inequality in wealth and income in industrialized countries like the United States suggest that this process may have finally come to a halt.)

Employment statistics from poor countries generally show that the industrial labor force is a very small fraction of the whole, with little or no growth in this percentage over time. Mechanization and industrialization appear to be destroying jobs much faster than they are creating new ones. This is due in part to great inequalities in power, wealth, and income and the availability of imported technologies that represent a great leap from what is locally available. The result is greatly increased income disparities, further concentration of economic and political power, and the impoverishment of a large percentage of the population.

Much mechanization and industrialization in the Third World has simply cut people out of the production process, changing who makes the goods (and thus who ends up with some purchasing power) without changing very much the amount of goods produced or their prices. For example, when industrially-produced aluminum pots take over the market in a rural area from clay pots made by villagers, the result is a) a marginally better product (perhaps), b) the same number of pots in use, c) perhaps a lower price when calculated over the life of the pots, d) a lot of unemployed potters who now have no purchasing power, e) a smaller number of new jobs in an urban-based industry, f) an enriched industrialist, and g) a shift away from the products of village industry previously bought by potters (e.g., tools produced by local blacksmiths) towards a smaller number of more highly processed foods, transportation services, and imported goods bought by urban residents. In this example "productivity" (as measured by the amount of product created per worker hour) has gone up, although total product has not. More people are unemployed, income inequality has increased, and village development has been set back. This is the process of the destruction of the crafts that has been repeated all over the world. Without a political commitment to full employment and the sharing of the benefits of technological change, the great technological gap created by the sudden arrival of imported technology can thus quickly increase poverty. It is in grappling with this kind of problem in societies shot through with inequalities that labor-intensive, decentralized, productivity-increasing technologies have a role to play. The need is for both national productivity increases and individual worker participation in production and consumption.

This need is acute in the agricultural sector of poor countries, where most of the people earn their livelihoods. To help the small farm family better use their resources, a program to adapt agricultural tools and equipment could be aimed at several categories of activities. One group of needed technologies are those that speed work at the points in the agricultural calendar when all available labor is fully employed, and production cannot be increased unless some manpower is released. Another tactic, one that the Chinese have used effectively in the commune system, is to mechanize the particularly low productivity activities in the rural sector and shift this labor to other activities, such as making threshers, pumps, and hand tractors. [23]

This kind of abstract process works relatively well when viewed from the perspective of a single farm, and where landholdings are equitably divided or shared. When land ownership is heavily concentrated in a few hands, this labor reallocation process is likely to be incomplete, leading to technology choices which may not benefit the majority of the population.

> "To maintain...a rational growth of capital use in a low-income economy, small farms are better suited than large ones, for the small farmers do not experience the same pressure to substitute capital for labor; **no one wants to mechanize himself out of his job.** Large farms are in fact the least economical, in social account, in the use of scarce capital and underemployed labor. Land reform countries generally exhibit a better record of a resource use that is rational in social account."
>
> —Folke Dovring, "Macro-Economics of Farm Mechanization", (1978) [24]

Farmers who hire labor to carry out low productivity activities are likely to substitute machines when possible, thereby replacing a number of farm labor jobs with a smaller number of machine operation and maintenance jobs. If farm labor is in short supply, wages may rise with productivity and everyone benefits. This is rarely the case in poor countries, however; more likely, the landowning farmer reaps the benefits and some laborers lose their jobs.

The tenant farm family is less likely to be hiring labor, and has less capital in any case to invest in machines. [25] These families are faced with the possibility that new technologies that raise the productivity of the land may lead to rent increases which capture the gains. There are of course many ways they can invest their own labor, through composting, tree planting, and general soil conservation work that pays off over a long term. But they probably will be unwilling to make this investment, fearing that they will not be around to reap the benefits.

These considerations make it unlikely that improved tools and methods for increased agricultural production can be designed to bring larger benefits **primarily** to the tenant farmer, and of course the landless laborer is even less

---

23. The production of agricultural machinery in the rural areas of China has meant that the equipment made is very responsive to technology needs and that mechanization does not necessarily replace agricultural jobs in urban industry. This situation is virtually unique in developing countries. See **Rural Small-Scale Industry in the People's Republic of China**, by Dwight Perkins, et al., (1977), reviewed on page 390.

24. In **Agricultural Technology for Developing Nations: Farm Mechanization Alternatives for 1-10 Acre Farms,** (1978), emphasis added. See review on page 460.

25. Whereas systems for renting machinery can be an important means of distributing costs and benefits of new equipment among a large group of farmers, such systems work best when competition keeps the charges reasonable. Otherwise, rental systems can be a means for owners to make monopoly profits; this is unfortunately quite common. Cooperative ownership, by contrast, usually involves more management problems but ensures broader distribution of benefits. Cooperatives too, however, are frequently taken over by elites.

likely to be able to directly increase his income through improved field crop technologies. [26]

Of course in many communities, most of the people own or have guaranteed access to some land, as individuals, through extended family ties, through legislated communal land ownership, or through secure tenant-landlord relationships. In these places there may be great scope for the application of small-scale agricultural technologies which would increase the viability of small farms and result in broadly distributed benefits. In many other communities, however, vigorous land reform efforts may be necessary before much can be accomplished through agricultural technology change.

How small is small?

Thus it seems clear that we cannot automatically assume that new or adapted low-cost technology will be ''accessible'' to the masses or equitable in its distributional effects. Many small-scale machines imported into poor rural areas (such as Japanese engine-driven rice mills brought to Java) are ''accessible'' in that they are ''affordable''—but they are affordable only to the wealthiest of rural people. Such machines may be accessible and small, but they are not accessible or small enough. Use of such machines can destroy jobs while providing benefits primarily to the owners.

There seems to be no reason why this same effect will not be observed with indigenously produced ''intermediate'' technology, especially that generated by formal R&D activities dominated by people who are not part of the poor rural majority. Under these circumstances, the tools, machines and techniques which receive the greatest attention are likely to be those considered most advantageous by the friends and relatives of the R&D staff. These friends and relatives are more likely to be wealthy farmers than poor farmers; they are more likely to employ labor than to sell labor; and when labor is overabundant, they alone will gain most of the benefits of productivity increases brought by new tools. In such cases, greater income stratification and increased concentration of land holdings are likely to follow. In Tanzania, for example, one observer warns that the successful development of the costlier range of intermediate technology is likely to be seized upon by the better-off private farmers, further improving their position and undermining the appeal of the collective ujamaa villages (which might not be able to afford the machines). [27]

Community organizing — a key step for appropriate technology

The foregoing discussion suggests that there are fundamental political qualities associated with scale and cost of technology. Large scale expensive technologies and centralized production systems tend to concentrate wealth, and can be vehicles that destroy the livelihoods of the poor majority in developing countries. Conversely, while examples mentioned above suggest that village elites are quite capable of consolidating their positions with ''intermediate'' technologies, truly accessible small-scale tools and techniques are still less compatible with exploitation and accelerating inequality. This is why technology policy—the set of codes, incentives, and restrictions affecting

---

26. More to the advantage of tenants and laborers would appear to be technologies for small scale crop processing, drying, and storage, home gardening, and household needs.

27. See **Technology for Ujamaa Village Development in Tanzania**, by Donald Vail, (1975), reviewed on page 389.

the direction of technological change—is such an important political matter. In almost all poor countries, government policies are determined primarily by a narrow group of urban-based elites; thus the resolution of this important matter will not necessarily be in the interests of rural people, much less the poor majority of rural people. Those who govern, it seems, may perceive little benefit for themselves in whole-heartedly instituting policy measures needed to support a small scale technology strategy.

An equally important related political question is whether—with or without substantial policy support—community organizations can exist or be created to serve as mechanisms for technological improvements that benefit everyone. The answer here seems to be a qualified "yes". Community organizers are becoming interested in A.T. because some technologies offer the opportunity for substantial benefits through community action, and thus encourage organization building. Technologists, on the other hand, have become increasingly aware of community organizing as a crucial activity (and even a requirement) for the success of their technology programs; a climate of community awakening, self-respect, and cooperation, and the chance to participate in a two-way dialog, can have a major effect on whether improved tools and techniques are applied successfully. Thus appropriate technology and community organizing work are seen by many as mutually supportive, each contributing to the progress and growth of the other.

Community organizers, while not seeking conflict for its own sake, recognize that the small communities of the Third World are often riddled with inequalities of wealth and power. An important step in community organizing is to awaken a community to its own political, economic and technological problems and opportunities. Then, the challenge is to find mechanisms which will allow progress for all, and prevent elites from taking over the benefits of community action. The promotion of small-scale technologies faces the same hurdle—either a strong community organization or a very careful technology group (or both) is needed to ensure that small scale equipment will not be monopolized by elites. [28] An important strategy for both appropriate technologists and community organizers may be to concentrate initially on technologies which will benefit all, regardless of differences in wealth and power. On this point, a training team in rural India observes:

> "Today there is much talk about 'total revolution' and radical transformation of society. But what really matters are the changes taking place in the socio-economic reality of the villages where poverty crushes the poor. In this stark reality of life the rural poor can hardly envisage more than creating for themselves some free space in society where they can breathe more freely and begin to stretch themselves. What is crucial at the moment is to create a base for joint action which is relatively free from control of the locally powerful. Wherever this has been achieved, people begin to move."

—**Moving Closer to the Rural Poor**, MOTT, (1979) [29]

Crucial "breathing space" for the poor may be created, without local political opposition, through the use of improved stoves which save 1/3 to 1/2 of the

---

28. Monopoly control is most likely with crop processing equipment, pumps and tillers, vehicles, and equipment for small industry; it is least likely with household technologies —improved stoves, home crop storage units, sanitation systems, and new construction materials.

29. See review on page 736.

fuel normally used in cooking, low cost grain storage units that can drama-tically reduce losses of grain stored in the home, or by water supply and sanitation systems that can markedly improve community health. Properly chosen and developed, such very low cost (especially household) technologies can provide a crucial entry point for community organizing efforts. At a later stage, existing great inequalities may have to be confronted directly or neither community organizing nor technological advance can proceed.

Those who have overlooked the importance of community organizing should bear in mind that in addition to helping prevent elites from controlling new technologies, community organizations open up other possibilities. Community technologies, such as water supply and sanitation systems, commonly fail when there is not a high degree of participation (i.e., a local committee for maintenance and repair). Technology adaptation and skills acquisition can also often be effectively pursued by community groups. In addition, coopera-tive community organizations offer members the chance to share resources and consolidate buying power. These intertwining functions can be seen at times in the more successful farm cooperatives:

> "It was the farm co-op that got America's farmer out from under the oppressive crop-lien system, which kept nineteenth-century farmers in hock [debt] to local merchants and distant brokers. Co-ops gave farmers an equal measure of bargaining power in the marketplace. With co-ops, farmers could market their crops directly, and could also do away with the hated middleman to purchase their supplies."

> — Jerry Hagstrom [30]

The links beginning to form between appropriate technology and community organizing should lead to ideas for organized cooperation on higher levels. Community organizing, appropriate technology, and other allied groups from different nations, for example, would find much in common and benefit greatly by learning about each other's experiences. How might this come about?

TASKS FOR INTERNATIONAL EFFORTS IN APPROPRIATE TECHNOLOGY

By their very nature, appropriate technology organizations working at the community level have few "disposable" resources to spend on anything but their immediate activities. Lower priority is therefore usually given to experi-mentation that is not linked to direct applications, careful preparation of documentation on successful and unsuccessful work, and searches for other groups with relevant experiences. But, as the preceeding pages indicate, the opportunity to innovate at the community level, and create a stock of useful information and experiences (including that from other groups), are important steps towards the decentralization of technology choice and the strengthening of community self-reliance.

There seem to be two categories of needs in international A.T. cooperation. One is for grassroots groups to know more about what other grassroots groups are doing, both in terms of technology research/adaptation and approach or strategy within the community. A second need is for large agencies to better understand the aims and activities of grassroots groups so that assistance to them will be of greater value. One tactic which has been used extensively in attempting to address both of these needs is the international conference.

---

30. Jerry Hagstrom, "Whose Co-op Bank?", Working Papers, (July/August 1980).

Unfortunately, few if any people with direct experience working at the community level are included in such meetings because they are below the horizons of funders and conference organizers. Also, in many cases community workers are understandably reluctant to attend meetings at which, all too often, little is accomplished; and they know that by attending they risk being vacuumed up into a planning, advisory, or administrative role which would take them away from work in their communities. The more common conference-goers are expatriates and university-based Third World people for whom international travel is (an expensive) part of a way of life.

Regional and international networks and coalitions often suffer from the same problem seen in international conferences. When those involved are a step or two away from the real grassroots A.T. practitioners, the connection to real problems and needs rapidly dwindles and may disappear altogether. Currently the greatest need appears to be for more decentralized local networks and coalitions, whose members may have more of immediate relevance to share with each other, and who are less likely to face language and cultural barriers. Once local networks are established and healthy, the need for more active regional and international networks should lead to sensible mechanisms and relationships.

An international A.T. effort might try to identify and support several kinds of people. One is the "social entrepreneur", a creative individual who can recognize social needs, overcome obstacles, and find ways to perform needed overhead functions. The business entrepreneur (and his counterparts in cooperatives and worker-owned enterprises) who can create and produce needed tools at affordable prices is clearly also important. Perhaps the most vigorous and enthusiastic group of people whose talents should be tapped are the young people all over the world who would like to put their minds and skills to work doing something socially useful. So far, few employment opportunities in A.T. have existed for such people. They often have little experience but some technical or organizational skills and most of all, great enthusiasm for the work, eagerness to learn, and openness to change. For the salary of one expert, it might be possible to provide a living for five or ten of these young people. Energetic and not trapped by outdated concepts of the nature of development problems, these people could become an invaluable source of insight and leadership in appropriate technology efforts. Volunteer programs, especially domestic volunteer programs, are one way to attempt this, yet they tend to generate experience and insight and then pull it away as the volunteer leaves. The development of experience among a group of local people who will stay in the area for a longer period seems a more fruitful goal.

The following are examples of specific functions in international cooperation which could be supported by regional, national and international groups.

Communications, documentation and reference functions

1. Compilations of documentation on successful traditional technologies within a country; of interest in the same country, in the region, and in the world; published in book form.

2. Catalog publishing for low cost dissemination of commonly relevant technical information, and to provide access to additional, more specific assistance (some examples of catalogs that have successfully done this are **Liklik Buk, Appropriate Technology Sourcebook, The Whole Earth Catalog,** and the **Sears and Roebuck Catalogs**; the last of these had wide circulation among North American farm families before World War II.)

ORDERING INFORMATION

Ordering Information

Prices

The prices listed in this sourcebook are approximately what you will have to pay. Fluctuations in the international currency exchange rates, rising postal charges, and inflation mean that it is impossible to provide price information that is completely accurate for long. For example, on January 1, 1981, postage on books mailed overseas from the United States was raised more than 50%. A large increase in domestic book mailing rates is also expected in early 1981. Consequently, many postage costs will be higher than those listed with the reviews.

Many of these publications are offered by small appropriate technology groups, who are often quite willing to trade publications with other groups, avoiding the problem of acquiring foreign currency, additional bank charges, and so forth.

The prices listed sometimes include surface postage. Airmail postage will often double the cost of a book; it may be necessary, however, when it is important to avoid the delay of 6 weeks to 4 months (depending on location) involved in surface mail.

When ordering publications that require airmail shipping and/or an awkward currency exchange process, we recommend sending an airmail letter of inquiry to obtain the most up to date price information in advance.

Addresses

Most of the nearly 500 publications reviewed in this book are from outside of the United States. For the majority of these, ordering information is listed with the reviews, including prices and addresses. There are, however, some organizations that either have produced a number of publications reviewed here, or act as distribution centers for some of these publications. To save space we have simply listed an acronym (abbreviation) in the text for many of these organizations. The full addresses are given below. (This is not a comprehensive list of publishers, nor is it a list of the most active appropriate technology groups.)

EARS—Environmental Action Reprint Service. This group stocks many U.S. publications on energy conservation and renewable energy technologies. EARS, Box 545, La Veta, Colorado 81055, USA.

FAO—Food and Agriculture Organization of the United Nations. Many of their publications are available through the UN Sales Section in New York, UNIPUB, and FAO book distributors in developing countries. To order directly, write to: Distribution and Sales Section, FAO, Via delle Terme di Caracalla, 00100 Rome, Italy.

GATE—German Appropriate Technology Exchange. Has a variety of English language publications issued between 1978 and 1980, documenting technologies of interest in developing countries. GATE, Dag-Hammarskjold-Weg 1, 6236 Eschborn 1, Federal Republic of Germany.

GRET—Groupe de Recherche sur les Techniques Rurales. This organization has an extensive set of French language publications on appropriate technologies. GRET, 34, rue Dumont d'Urville, 75116-Paris, France.

IDRC—International Development Research Center. This Canadian aid organization has consistently produced valuable and solidly researched books on subjects of interest to village technology workers. Unlike their counterparts to the south, IDRC has people from the developing countries on their board. Their excellent publications are free to local people in developing countries. IDRC, Box 8500, Ottawa, Canada K1G 3H9.

ITDG—Intermediate Technology Development Group. The organization founded by the late E.F. Schumacher and the present chairman George McRobie in the 1960's continues to publish the largest selection of books on appropriate technology, and has set a standard of quality equalled by few. High prices have recently been eased a bit with special editions supported by charitable organizations and an arrangement with a new U.S. distributor (Intermediate Technology Development Group of North America, Suite 1231, 60 East 42nd St., New York, New York 10017, USA). For a current publications list, write to Intermediate Technology Publications, 9 King St., London WC2E 8HN, England.

META—These people maintain a wide selection of appropriate technology publications, mostly from the U.S. and U.K. but some from elsewhere as well. We recommend working with META when setting up a library. META, P.O. Box 128, Marblemount, Washington 98267, USA.

NTIS—National Technical Information Service; a U.S. government organization that reprints many of the publications other government agencies have originally produced. This is where things can be found when they have gone out of print at the National Academy of Sciences, for example. A description of possibly relevant material held by the NTIS is contained in a book reviewed on page 377. Prices are very high, usually double outside the U.S. Some documents are now available free or at lower cost through U.S. Agency for International Development missions in Caribbean and Latin American nations, in a special arrangement with NTIS. You must include the Accession Number of each book when ordering (included with reviews). NTIS, Springfield, Virginia 22161, USA.

Popular Mechanics—A magazine aimed at the do-it-yourself handyman market, Popular Mechanics has over the years printed many plans for workshop power tools that can be made by the reader. A special service offers photocopies of these for about 25 to 50 cents per page. We have reviewed the most relevant ones in the WORKSHOP chapter. When ordering, write to Popular Mechanics Plans, Dept. 77, Box 1014, Radio City, New York 10019, USA.

RODALE—The publishing center of the North American organic gardening movement is the Rodale Press, which in recent years has become increasingly active in publishing books about small scale technology and renewable energy systems. Rodale also publishes magazines about health, organic gardening, small farming, and owner-building of homes. Rodale Press, 33 E. Minor St., Emmaus, Pennsylvania 18049, USA.

SWD—The Steering Committee on Wind-Energy for Developing Countries. This group is financed by the Netherlands Ministry for Development Cooperation, and is staffed by the Eindhoven University of Technology, the Twente University of Technology and DHV Consulting Engineers. The SWD tries to help governments, institutes and private parties in the Third World with their efforts to use wind energy. Research institutes in the Third world may ask for one copy of any of their publications free of charge, by writing directly to: SWD, c/o DHV Consulting Engineers, P.O. Box 85, 3800 AB Amersfoort, The Netherlands. Other groups wishing to order SWD publications should ask for a publications list with current prices and ordering information.

TALC—Teaching Aids at Low Cost. An excellent source of low-cost books and teaching slides in the health care field. Postage and packing charges are £1.00 for orders under £4.00 (US$9.00); add 25% of total price of order for postage and packing of orders over £4.00. If paying by check or money order in currency other than sterling drawn on a British bank, please add 50 pence. Teaching Aids at Low Cost, Institute of Child Health, 30 Guilford Street, London WC1N 1EH, England.

TPI—Tropical Products Institute. In addition to a variety of publications on tropical agricultural products, this institute has recently begun issuing a series of booklets (Rural Technology Guides) on simple processing tools. No charge is made for single copies of publications sent to governmental and educational establishments, research institutions and non-profit organizations working in countries eligible for British Aid. All publications are obtainable from: Publications, Publicity and Public Relations Section, Tropical Products Institute, 56/62 Gray's Inn Road, London WC1X 8LU, England.

UNIPUB—This U.S. company handles a large number of United Nations publications, along with those of other international organizations such as IDRC. UNIPUB, 345 Park Avenue South, New York, New York 10021, USA.

U.N. Sales Section—Many United Nations publications can be obtained from this distribution center in New York. Be sure to quote the identification number for any books ordered. U.N. Sales Section, Room LX 2300, United Nations, New York, New York 10017, USA.

VITA—Volunteers in Technical Assistance. This group does not send volunteers, but handles requests for technical information, which it forwards to a network of primarily U.S.-based volunteers for response. A large U.S.A.I.D. grant has recently increased their activity in renewable energy technologies. They have a long publications list, and accept UNESCO coupons in payment. VITA, 3706 Rhode Island Avenue, Mt. Rainier, Maryland 20822, USA.

BACKGROUND READING

Background Reading

The books reviewed in this chapter offer a variety of views on the cultural and economic aspects of technology choice, some of the political choices reflected in development strategies, and common technology needs in rural areas of the Third World.

For readers interested in the hard economic basis for appropriate technology, there is no better reference than **Technology and Underdevelopment**. A thoughtful analysis of the problems and issues involved in transnational "technology transfer" is contained in **The Uncertain Promise: Value Conflicts in Technology Transfer.** This volume is particularly concerned with the impact of alien technology on cultural value systems. **Appropriate Technology: Problems and Promises—Part One: The Major Policy Issues** (reviewed on page 31) gives an historical perspective on factors that have influenced the development of practical technologies in the United States and China, and explores a large number of policy issues that surround appropriate technology. This continues to rank as one of the most insightful books in the A.T. literature. **Towards Global Action for Appropriate Technology** presents the views of a number of observers on the nature of national and international mechanisms that could support appropriate technology research, development, and dissemination.

Paper Heroes, while favorably reviewing several particular tools and techniques, is critical of many of the basic assumptions and perceived benefits associated with appropriate technology. For the most part these are "the excessive claims and unsubstantiated promises of paper heroes," argues author Witold Rybczynski.

Stepping Stones is a collection of many of the best articles that have contributed to appropriate technology thinking in and for the United States. The next two books examine steps needed to establish a sustainable, environmentally-sound society. **Repairs, Reuse and Recycling** discusses the technological alternatives in reducing the flow of valuable materials to dumps and landfills. **Ecology and the Politics of Scarcity** summarizes the reasons for believing there are limits to industrial growth and resource consumption on the planet, and argues that political controls on wasteful consumption will be necessary in the future.

Questioning Development takes the position that development should be understood in terms of political and economic power within and among nations; at the local level in the Third World development must therefore mean "the more equal distribution of power among people." The author suggests that a critically important yardstick for evaluating the worth of development projects should be their anticipated effects on the distribution of power in the community, the nation, or the world.

The last five publications suggest what kinds of everyday activities in the rural Third World most urgently need improved technologies, and give many examples of tools and techniques that may be appropriate. **Appropriate Technology for African Women** and **Rural Women** are specifically concerned with the effects of technological change on women's lives, and discuss improved technologies that might particularly help women. "Technology for the Masses" brings together a number of articles by Indian scientists and A.T. practitioners, examining the kinds of technologies that show the most promise under Indian conditions. **Appropriate Technology: Technology with a Human Face** also summarizes many of the arguments that point to the intermediate technology approach as a better alternative than conventional technology dissemination strategies.

In Volume I:

Small is Beautiful, by E.F. Schumacher, 1973, page 30.

Appropriate Technology: Problems and Promises, edited by Nicolas Jequier, 1976, page 31.

A Handbook on Appropriate Technology, by Brace Research Institute and the Canadian Hunger Foundation, 1976, page 32.

Tools for Conviviality, by Ivan Illich, 1973, page 33.

Lectures on Socially Appropriate Technology, U.S. edition entitled **Introduction to Appropriate Technology**, edited by R. Congdon, 1975, page 34.

Alternative Technology and the Politics of Technical Change, by David Dickson, 1974, page 34.

Radical Technology, by editors of Undercurrents, 1976, page 35.

Design for the Real World, by Victor Papanek, 1972, page 35.

Sharing Smaller Pies, by Tom Bender, 1975, page 36.

Environmentally Appropriate Technology, by Bruce McCallum, 1975, page 37.

Technology and Employment in Industry, edited by A.S. Bhalla, 1975, page 38.

Technology and Underdevelopment, book, 320 pages, by Frances Stewart, 1977, $27.75 (hardback) from Westview Press, 5500 Central Ave., Boulder, Colorado 80301, USA; or £4.95 (paperback) from Macmillan Publishers Ltd., Houndmills Estate, Basingstoke, Hants., United Kingdom.

A necessary book for the person who says to him- or herself, "I sense that the appropriate intermediate technology strategy for development is correct, but how can it be justified in the context of traditional economic thinking?" Stewart leads the reader through the most comprehensive book to date dealing with the economic theory of appropriate technology and development in the Third World. The book is technical but can be absorbed by anyone with a modest background in economic jargon.

There are two parts: 1) a theoretical discussion of the nature of technology and the social consequences of its use, and 2) a set of case studies. The theoretical discussion is aimed at the reader familiar with conventional economic theory. The author points out precisely where the false assumptions and unwarranted extrapolations are found.

Stewart begins with a discussion of the nature of technology, and the kinds of technical choices open to a developing country. She defines "technological choice" both in terms of **product** (a glossy, refined consumer item vs. a plain but functional commodity) and the technology to create that product (a sophisticated tractor vs. an animal-drawn plow). Stewart states that the major reason developing countries are limited in their choice of technology is that technology has evolved to meet the different needs of developed countries. A new technology is more than a purely scientific achievement—it reflects a society's needs, standard of living, tastes, and relative scarcity of labor, capital and resources. Thus it would be purely coincidental that an ideal technology would exist for a specific application in a developing area.

The author notes the usefulness of some but not all technologies used in the evolution of developed countries. If a technology becomes obsolete in the West purely because of changes in the relative prices of capital and labor, or due to shifts in consumer tastes, then that technology may be usable in a developing country. However, technologies that became obsolete due to technical improvements may be obsolete in any context. (The failure to make this distinction underlies much of the disagreement on the relevance of older industrial technologies.)

Another of Stewart's major themes is that technical choice is not a narrow choice of a particular technique at a particular time. Rather, a national economic system is either oriented towards foreign advanced technology or towards more appropriate technology. If the "modern" approach is emphasized, consumers, infrastructure, and urban concentration tend to lock the country into "foreign" technology. An indigenous technology may not be able to compete in such an environment, despite its overall social desirability. The point is that the national choice of technology and life-style is a social choice which will dominate the narrow choice of technology for a specific application.

Highly recommended.

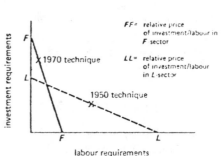

The different capital/labor substitution ratios in two sectors of the economy (or in two countries) should affect the optimal choice of technology

The Uncertain Promise: Value Conflicts in Technology Transfer, book, 324 pages, by Denis Goulet, 1977, $5.95 (10% discount to non-profit organizations, students and professors), from Overseas Development Council, 1717 Massachusetts Ave. N.W., Washington D.C. 20036, USA.

"This study of value conflicts in technology transfer has attempted to peel away the mystifications which veil the true impact of technology on societies nurturing diverse images of development. Technology is revealed herein as a two-edged sword, simultaneously bearer and destroyer of values. Yet technology is not static: it is a dynamic and expansionist social force which provides a 'competitive' edge enabling its possessors to conquer economic, political, and cultural power. Consequently, Third World efforts to harness technology to broader developmental goals are paradigmatic of a still greater task: to create a new world order founded not on elitism, privilege, or force but on effective solidarity in the face of human needs. The gestation of a new world order poses two troubling questions for all societies: Can technology be controlled, and will culture survive?''

''Technology is indispensable in struggles against the miseries of under-development and against the peculiar ills of overdevelopment. Technology can serve these noble purposes, however, only in those societies in which ideology, values, and decisional structures repudiate the tendency of technology to impose its own logic in striving after goals.''

''At least three values must now be internalized in any efficiency calculus: the abolition of mass misery, survival of the ecosystem, and defense of the entire human race against technological determinism...It is no longer correct to label some procedure efficient if it exacts intolerable social costs, proves grossly wasteful of resources, or imposes its mechanistic rhythms on its operator...Firm managers and designers of technology will need to explore ways of becoming integrally efficient—that is, of producing efficiently while optimizing social and human values.''

This book offers an insightful examination of the values implicit in tech-nological society; international mechanisms, and high financial and social costs, of transnational transfer of industrial technology; and basic strategies and policies in the Third World to channel technology to serve development goals. For those who can make their way through the difficult (even for native English speakers) language used, this will make excellent background reading for the discussion of appropriate technology policies.

Towards Global Action for Appropriate Technology, book, 220 pages, edited by A.S. Bhalla, 1979, $13.00 from Pergamon Press, Inc., Maxwell House, Fairview Park, Elmsford, New York 10523, USA.

A set of essays examining the need for and nature of national and international mechanisms to support appropriate technology research, development and dissemination. Nicolas Jequier writes about some of the non-economic criteria that should be considered in evaluating possible appro-priate technologies. Ajit Bhalla discusses the elements of a basic needs strategy for development, and policy choices to make appropriate technology part of that strategy. Amulya Reddy offers a framework for understanding why existing R&D institutions in developing countries do not generate appropriate technologies, and what shifts in policy and orientation are needed to ensure the development of ATs within some of these institutions. Willem Floor describes the activities of the UN agencies, and how they do or do not touch on appropriate technology. Frances Stewart summarizes and compares the functions of a variety of existing and proposed international bodies and programs in appropriate technology. Paul Marc Henry, Reddy and Stewart present a final 13-page proposal for a ''new international mechanism for appropriate technology'', a non-governmental organization to be associated with, but outside of, the United Nations.

National policy initiatives and programs, and international institutions can all play a major role in improving the climate for A.T. work. In reality, however, many of the most effective A.T. efforts around the world were initiated without government and international agency support. This volume unfortunately does not discuss the possibilities for in-country and international cooperation among grass-roots appropriate technology groups themselves, often forced to operate without policy support or funding.

Paper Heroes: A Review of Appropriate Technology, paperback book, 181 pages, by Witold Rybczynski, ¶1980, $4.95 from Doubleday and Company, 245 Park Ave., New York 10017, USA.

Paper Heroes is an attack on both romantic myths and basic assumptions of the "appropriate technology movement." The author has himself done some very important work on low cost technologies, co-authoring, for example, the instant classic **Low Cost Technology Options for Sanitation in Developing Countries** (see review in Water Supply and Sanitation chapter). Amid a small but growing literature of backlash against appropriate technology values and assumptions, this represents the first lengthy critique of A.T. by an "insider". The author serves up some impressive "aces", but most of his shots are long, wide, or into the net.

Rybczynski begins by attacking Schumacher. "**Small is Beautiful**...did not attempt a reasoned argument but appealed directly to the emotions...(it) was first and foremost a diatribe against modernization." He deplores a "California youth culture" concept of technology that he attributes to spinoffs from the **Whole Earth Catalog**, in a lengthy digression from his main theme. He claims that Illich, Ellul, and others who have significantly influenced A.T. thinking are "modern Luddites", dismissing the original Luddites of early 19th century England as "a kind of antitechnological Ku Klux Klan."

Rybczynski is probably correct when he says that at present "A.T. could be described as an inverted pyramid—a great deal of verbiage and speculation resting on few accomplishments." Having laid waste to what he considers "the excessive claims and unsubstantiated promises of paper heroes", the author switches to a more positive note, favorably describing several specific technologies that might be called A.T. Readers are warned about the problems of appropriate technology strategies that attempt to use conventional aid mechanisms and institutions as the vehicles for reaching the poor. The author suggests alternatives to aid, with technology choice left to the people in the Third World. "A more successful approach, which is particularly evident in soft tech, is the provision of information on intermediate technologies **directly to the individual**...it permits the individual to decide what is appropriate, it supports decentralization, and, almost by definition, it ensures that the individual establishes a healthier control of his technology...It could also be argued that the successful A.T. antecedents such as rural medicine in China, the Vietnamese sanitation program, or Gandhi's hand-spinning campaign, have all been primarily **information** strategies. The decentralization of technique has been the result of the much more important strategy of the decentralization of knowledge."

An excellent chapter on China takes a hard look at imaginary and real lessons to be learned from that nation's experience. (This chapter is not without its wild statements: "It is likely that any Chinese state, whatever its political goals, would end up with some kind of small, decentralized, rural industry approach." Any Chinese state, he neglects to add, with a genuine **political** commitment to sharing the benefits of economic development among the whole population. Unfortunately, such a political commitment is very rare indeed in the world today.)

Rybczynski also reminds us of a distinction between "social change" and "social reform", arguing that technological change always brings some social changes, but no technology, in and of itself, brings social reform. "Better technology (of any kind) can certainly not be a substitute for social reform. Landlordism, powerful rural elites, conservative banks, and rapacious money lenders all conspire to maintain the poverty of the landless peasants."

Curiously, the author disregards the oppressive effects of these same and allied forces at the national level. He makes much of the fact that many Third World governments have called ''for an increase in the transfer of advanced technology from the more developed to the less developed nations'', ignoring substantial evidence that imported technology has given landlords and money lenders increased leverage with the poor.

In fact, for this author ''appropriate technology'' is simply supplementary to, not a genuine alternative to, modern industrial technology. Throughout the book, he subscribes to the traditional view that technological development follows a single evolutionary path; in this view, intermediate technologies are simply way stations along the road to conventional industrialization. He is particularly annoyed by claims that more appropriate technologies could come out of conscious, intentional processes: ''The technology that is in use today is the result of scientific, philosophical, and cultural history. What is the 'new' technology to be based on? Astrology, superstition, magic? How can it fail to rely on existing institutions and ways of thinking, and how, as a result, can it claim to be capable of developing in a different direction?'' Thus Rybczynski implies that there can be no alternative to current economic patterns in which industrial technology leads to ever-increasing consumption of energy and resources. The author simply waves away the limits to growth arguments. His failure to deal seriously with these issues significantly flaws many of his other arguments.

Near the end of the book, Rybczynski argues against ''self-reliance''. He equates self-reliance with a kind of isolated self-sufficiency, and thereby fails to make a distinction found in most of the literature. This attack on self-reliance is only one of a number of semantically-confused thickets that detract from the book. His use of the terms ''modernization'' and ''demodernization'' is another case in point. When finished, he has obscured as much as he has clarified.

In the end, Rybczynski does not succeed in what he sets out to accomplish — to discredit the propositions that 1) many technologies contain within them cultural and political biases, and 2) there are other paths to the future than that of the Western industrial model.

Stepping Stones: Appropriate Technology and Beyond, large paperback, 204 pages, Lane de Moll and Gigi Coe, eds., 1978, $7.95 from Schocken Books, Inc., 200 Madison Avenue, New York, New York 10016, USA.

This is a collection of some of ''...the philosophical stepping stones which have helped shape the techniques, values, tools, and politics of appropriate technology'' edited by associates of Rain Magazine and the California Office of Appropriate Technology. Most of the articles are written by well-known thinkers and practitioners from the U.S. Reprinted pieces cover the tools and approaches of appropriate technololgy, and application of these in our lives now. A set of mostly new essays examines this for our individual and collective futures.

The writings offer a wide range of opinion on a broad spectrum of issues: restraint and re-evaluation of our lifestyles; small vs. large scale; permanence vs. economic efficiency; and simplicity vs. fashion and convenience. It is important to note that these are not the central issues in appropriate technology for the Third World. In many of these countries, appropriate tools and techniques are playing a role in an ongoing uphill struggle for better livelihoods by enlarging the realm of what is actually possible. Such technologies do reflect concern for ecological principles and remain at a scale

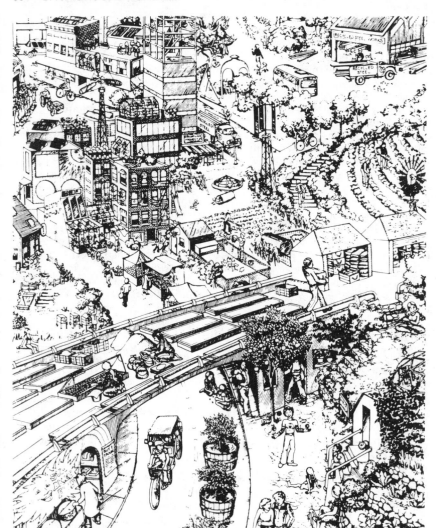

affordable to the poor majority, but they do not usually come as a result of a process of scaling down or cutting back.

For these reasons many of the essays in this book do not address the circumstances faced by appropriate technology groups in poor countries. The essays do offer insights into the North American appropriate technology movement. They show that the philosophical foundations and everyday applications of appropriate technology not only have roots in our past, but also represent a starting point in our work to build a better future.

Repairs, Reuse, Reycling—First Steps Toward a Sustainable Society, Worldwatch Paper 23, booklet, 45 pages, by Denis Hayes, 1978, $2.00 from Worldwatch Institute, 1776 Massachusetts Avenue, N.W., Washington D.C. 20036, USA.

This report critically examines the flow of most materials from their sources (a mine, forest, or crop) to the dump. The imperatives for recycling materials are reviewed: increasing scarcity and energy expense of recovering non-renewable resources; political tensions caused by uneven distribution of resources worldwide; and escalating environmental costs and hazards.

Three basic approaches to sustainable resource use are waste reduction (emphasizing more durable appropriate technologies), waste separation, and waste recovery. Examples of recent recycling programs illustrate the importance of scale for recovery systems. Centralized high technology recovery facilities depend on long term guarantees of a steady flow of waste materials. Any programs which actually reduce the flow of waste then threaten the financial viability of the high cost recovery facilities. "A more sensible approach would be to first see how much of the problem could be solved by comprehensive programs for reducing waste, recycling, and composting. Appropriately-scaled resource recovery facilities could then be constructed to process the remaining waste."

A well-documented paper, pointing to the importance of both "technical fixes" and social reorientation.

Ecology and the Politics of Scarcity, book, 303 pages, by William Ophuls, 1977, $8.50 from W.H. Freeman and Co., 660 Market St., San Francisco, California 94104, USA.

This book offers a good summary of the ecological arguments for a dramatic change in the industrial way of life, and a shift towards "soft" environmentally-sound, appropriate technologies. Most of the book presents the essential arguments on the limits to growth, mineral resource depletion, energy as a key factor, and the limits of sustainable systems for food production. The author concludes with an evaluation of the nature of the political problems this creates, and the elements of a new politics that could handle these problems while allowing both a modest standard of living in a steady-state economy and the protection of civil liberties.

"The essential political message of this book is that we must learn ecological self-restraint before it is forced on us by a potentially monolithic and totalitarian regime or by the brute forces of nature."

In agriculture, the author states that:

"Farming that is both productive and ecologically-sound seems very likely to be small-hold, horticultural, essentially peasant-style agriculture finely adapted to local conditions (especially in the tropics). It should be obvious that many of the developing countries are well placed to make the transition to this modernized version of traditional agriculture. Except for the excessive use of insecticides and chemical fertilizers in some areas, the agriculture of China, Taiwan, Korea, Ceylon, and Egypt, and others is already close to this mode, and has per-acre yields to show for it. By contrast, the United States and some other developed nations (Japan is major exception) appear to face a great deal of 'dedevelopment' in order to change over to this style of agricultural production, so that the transition would be socially painful."

In discussing the politics of a steady-state, sustainable society, the author argues that the kind of democracy that allows the wholesale waste of energy and resources in the pursuit of happiness will have to come to an end. The planet simply will not be able to sustain the ecological destruction brought on by billions of merry resource squanderers, and frugality will have to be compulsory, not voluntary. "The steady-state society will not only be more authoritarian and less democratic than the industrial societies of today—the necessity to cope with the tragedy of the commons would alone ensure that—but it will also in all likelihood be much more oligarchic as well, with only those possessing the ecological and other competencies necessary to make prudent decisions allowed full participation in the political process." These are alarming propositions, differing greatly from the views of an ecologically stable future taken by others such as Schumacher and Lovins. Shall we have a priesthood of environmentalists in place of a nuclear priesthood? The author himself asks the question, "who will watch those in power?"

Ophuls argues that the present agendas for action are basically two: education (consciousness-raising) of the general population, so measures to avoid eco-disasters can be politically acceptable, and 2) intensified experimentation with soft technologies and new social forms, which will need to be available when the political conversion takes place.

An exceptionally clear statement of the environmental crisis facing humanity, with some very troubling political conclusions.

Questioning Development: Notes for Volunteers and Others Concerned with the Theory and Practice of Change, booklet, 48 pages, by Glyn Roberts, 1974, ₤1.65 (British Sterling only accepted), from Returned Volunteer Action, 1c Cambridge Terrace, Regents Park, London NW1 4JL, England.

A discussion of some philosophies of development currently in practice around the world; "ideas which may be useful to anyone who wonders about the changes he is helping to bring about." The author suggests that "we shall clearly have to come to an agreement as to what we mean by development. Paradoxically, this is something which many 'development' personnel have never faced up to. Despite years in the Aid business, they have always been too busy getting on with the job to worry much about the overall picture." People use the word 'development' to describe many different types of activity, many of which do more harm than good to the people affected. "We are all agreed, no doubt, that Development means healthier, happier, fuller and more meaningful lives for everyone. Earlier this was simply rephrased as 'Development = the more equal distribution of power among people.' " As a result of "looking at development in terms of power, we may gain insight into the cause of poverty in our own countries. We may find that the differences traditionally noted between the 'advanced' and the 'less developed' nations are less important than the similarities."

This book asks us to ask questions. It also presents some specific questions worth asking, and some persuasive answers to these questions. "Development is about making such choices, at home or overseas. It is about challenging those who reject those priorities. It is about taking sides; and in this game, nobody is a spectator."

Village Technology in Eastern Africa, book, 60 pages, UNICEF, June, 1976 seminar report, ₤1.50 from ITDG, or UNICEF, Eastern Africa Regional Office, P.O. Box 44145, Nairobi, Kenya.

Here is an excellent introductory book, relevant to most Third World countries. It includes a review of the basic concepts of appropriate technology, and an overview of potential A.T. tools and techniques for agriculture, food preservation, preparation of nutritious infant foods from local sources, and water supply. Criteria for evaluating rural energy needs and affordable alternatives are presented.

Anyone interested in starting a village technology center should read the section on the Karen Village Technology Unit. This demonstration center has a large number of working tools and machines on exhibit, and a workshop outfitted with only simple woodworking and metalworking tools. There is also a simple laboratory section for testing of A.T. devices. Although this center has been criticized as a lifeless "museum" that attracts primarily foreign visitors, it represents a significant effort that should be learned from.

The extension systems discussed in this book differ from the conventional 'top-down' approaches. "Thinking based on 'introduction' of appropriate technology tends to foster an attitude that the technology is something brought in from outside. Whereas, it would probably be more useful to think in terms of the 'generation' of the technology within the society."

Highly recommended.

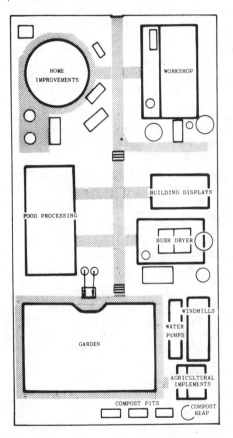

Karen (Nairobi) Village Technology Unit site plan

Appropriate Technology for African Women, report, 101 pages, by Marilyn Carr, 1978, available from the African Training and Research Centre for Women, United Nations Economic Commission for Africa, P.O. Box 3001, Addis Ababa, Ethiopia.

"An increased emphasis on 'intermediate' technologies promises to do much to lessen the inequalities between the urban and rural areas, and between rich and poor families. Its effect, however, will be limited unless increased emphasis is also given to the women who, especially in the rural areas, have the major responsibility for lifting their families out of poverty. Agricultural, rural and national development will be a slow and difficult process if the women, who form half the population and, in some countries, represent up to 80% of the agricultural labor force, continue to be denied access to knowledge, credit, agricultural extension services, consumer and producer cooperatives, labor-saving devices and income-generating activities."

Extension programs that neglect the roles of women often have disappointing results. "Thus, in one West African country, although extension workers had shown the men the correct depth to dig the holes, coffee continued to die due to bent tap-roots because it was the women who were doing the digging."

Many improved village technologies could distinctly help rural women, "who are the drawers of water, the hewers of wood, the food-producers and often the overall providers for the families of Africa." In the main part of this report, the author identifies some of the activities for which intermediate technologies are needed to ease the burdens of rural women, and some of the possible technologies to choose from. Also included are descriptions of a wide variety of village technology-related programs in Africa, and an annotated bibliography on women and technology in Africa.

Rural Women: Their Integration in Development Programs and How Simple Intermediate Technologies Can Help Them, booklet, 84 pages, by Elizabeth O'Kelly, 1978, $4.00 plus postage, from the author at 3 Cumberland Gardens, Lloyd Square, London WC1X 9AF, England.

Ms. O'Kelly discusses the daily tasks of women in Asia and Africa, the concept of intermediate technology, and particular technologies that would tend tend to make life easier for rural women without reducing their role and status. She recommends hand-operated seeders, push carts and wheelbarrows, fencing, threshers, winnowers, improved hand-operated rice mills, corn mills, improved grain storage units, heavy gauge black polyethylene sheeting for sun-drying, fuelwood plantations, biogas plants, solar dryers for vegetables and fruits, improved stoves, haybox cookers, rooftop catchment water tanks, pumps, water filters and latrines.

She notes that "the part that women play in village life in general and in agriculture in particular, is consistently underestimated and many programs are drawn up on the assumption that it will be the man who will be carrying them out when, in fact, it will be the women." And when technologies are directed towards women's work, "care needs to be taken that these do not unintentionally reduce their standing."

Organizations for rural women should be created "beginning in a small way in one or two neighboring villages and continuing by working outwards in ever-widening circles." This is more likely to be successful than top-down initiatives which often lose their thrust before they have filtered down through the bureaucracy to the local level. She describes successful efforts of this kind, in the creation of corn mill societies in the Cameroons, and women's groups in Sarawak. A list of A.T. organizations, equipment manufacturers, and useful books is included.

"Technology for the Masses", January-February 1977 issue of **Invention Intelligence** magazine, Rs. 6 postpaid in India, $3.00 surface mail or $5.00 airmail outside India, from Invention Intelligence, National Research Development Corporation of India, 20, Ring Road, New Delhi 110024, India.

This special issue of the magazine **Invention Intelligence** deals with the prospects for affordable technology for rural India. Strategies for rural industry and rural development are discussed. One proposal includes a national upper consumption limit for individuals. In evaluating the role of science and technology, one author states: ''We have yet to make properly documented studies on our traditional skills and practices and systematically explore the possibilities of both learning from and contributing to them, in order to evolve appropriate technology for the masses.''

Several articles on energy sources describe the progress of the Indian biogas programs, the scope for the use of windpower, possible direct solar devices, the substitution of organic fertilizers for energy-intensive chemical fertilizers, and the increased use of waterways for transport. Tree-and-pasture plantations are proposed, to make maximum use of solar energy for fuel, food and fodder.

Commenting on the importance of the bullock cart, one author notes that the total investment in carts and animals exceeds the total investment in either the railroad system or the road network in India. He proposes a number of design improvements for the bullock cart.

Low-cost housing, dairy farming, aquaculture and increased water use efficiency in irrigation are among the other topics discussed.

Relevant reading for much of the Third World. Recommended.

Appropriate Technology: Technology with a Human Face, book, 220 pages, by P.D. Dunn, 1978, $5.95 from Schocken Books, Inc., 200 Madison Ave., New York, New York 10016, USA; or Ł2.50 from ITDG.

Professor Dunn is chairman of ITDG's Power Panel, and his book is based upon a cornerstone of ITDG's philosophy: the importance of putting people to work. ''The Appropriate Technology method...is essentially rural-based rather than urban-based as is the Western technology. One important feature is the emphasis on the creation of workplaces. The work place in the developed countries will cost, typically, Ł2,000 to Ł3,000 in terms of capital investment. (In fact, capital investment per job in U.S. manufacturing ranges from $20,000 to $60,000—ed.) This is clearly an impossibly high figure in a developing country, which will have little capital and even less foreign exchange with which to import equipment. Such a country will, however, have a plentiful supply of labour even though this will be unskilled and unfamiliar with industrial practice. Appropriate Technology can be regarded as centering around job creation for these people.''

This book provides some background on the A.T. movement, with general coverage of worldwide conditions of poverty and resource depletion, and the failures of conventional development strategies to address these conditions. Also discussed are some of the most immediate problems facing rural communities in agriculture, water supply, energy, and health care, along with examples of tools which have helped solve some of these problems in certain areas.

Professor Dunn does not address himself to how communities in developing countries might participate in and control the path of technological innovation so as to better address their own needs. He proposes no basic changes in organizational, educational or political strategies. "This book is concerned essentially with poverty and methods by which it might be alleviated, and inevitably this has political implications which often in a particular situation cannot be ignored. Social, political, and economic circumstances differ very widely and views on them differ even more; so, whilst recognizing its local importance, no political view has been expressed here. Most of the book is devoted to the actual practice of Appropriate Technology." By adopting this approach, Prof. Dunn may "miss the forest for the trees" by zeroing in on tools and techniques without considering how they might become part of a program for self-reliant progress toward an improved quality of life. Readers seeking an effective introduction to A.T. as a community-based strategy for development should look elsewhere.

ADDITIONAL MATERIALS FOR BACKGROUND READING

The first books reviewed in each chapter are often background reading for that topic area. See especially:

GENERAL
REFERENCE BOOKS

General Reference Books

As most of the publications reviewed throughout this book can be considered references, this chapter was created for special kinds of publications. These include books and periodicals that span many topic areas, several bibliographies, and directories of appropriate technology organizations.

The first five entries in this chapter, from Nepal, Bangladesh, France, Peru, and India, each contain information on a wide variety of technologies relevant to many developing countries. (In Volume I, see especially the excellent **Liklik Buk** from Papua New Guinea, and **Appropriate Technology: Directory of Machines, Tools, Plants, Equipment, Processes and Industries** from India.) Books that contain information on a number of technologies within a single general subject area have been placed in the relevant chapters; the enormous catalog **Tools for Homesteaders, Gardeners and Small Farmers**, for example, can be found in the AGRICULTURAL TOOLS chapter.

Three bibliographies have been included here. **Guide to Convivial Tools**, intended for librarians, identifies the books of a new discipline—the study of the cultural, social, and political conditions necessary to allow democratically determined limits to industrial technology and the industrial mode of production. **Non-Agricultural Choice of Technique** leads the reader to the pre-1975 literature on the economic implications of technology choice (see also **Economically Appropriate Technologies for Developing Countries** reviewed on page 56.) **Appropriate Technology Information for Developing Countries** is an attempt to cull and re-evaluate old U.S. Agency for International Development research reports for possible relevance to appropriate technology efforts, a difficult task. Written for North American audiences and conditions are **Soft Tech, The Book of the New Alchemists**, and the various publications from the National Center for Appropriate Technology. **The Formula Book: 1** contains a number of household product formulas that would also be relevant in the Third World.

Four directories of appropriate technology institutions are included. The only low cost one is **Rural Technology in the Commonwealth**, which lists institutions in 26 British Commonwealth countries only. The **Appropriate Technology Directory** from OECD is probably the most accurate, while UNEP's **Directory of Institutions and Individuals Active in Environmentally Sound and Appropriate Technologies** is the most extensive.

Completing this section are the low cost two-volume **Chambers Dictionary of Science and Technology** with some 50,000 English technical terms from 100 fields of activity, and **Small Technical Libraries**, a short booklet that may help in organizing the small libraries of well-equipped appropriate technology groups.

In Volume I:

A Handbook on Appropriate Technology, by Brace Research Institute and the Canadian Hunger Foundation, 1976, page 32.

Village Technology Handbook, VITA, 1970, page 41.

Other Homes and Garbage, by J. Leckie et al, 1975, page 42.

Appropriate Technology: Directory of Machines, Tools, Plants, Equipment, Processes, and Industries, ATDA, 1977, page 43.

Cloudburst 1 (A Handbook of Rural Skills and Technology), edited by Vic Marks, 1973, page 43.

Cloudburst 2, edited by Vic Marks, 1976, page 44.

First Steps in Village Mechanization, by George MacPherson, 1975, page 44.

Simple Working Models of Historic Machines, by A. Burstall, 1968, page 45.

Making Do, by Arthur Hill, 1972, page 46.

Technical Information Handbook, page 47.

Pictorial Handbook of Technical Devices, by O. Schwarz and P. Grafstein, 1971, page 47.

New Alchemy Institute Journals 2 and 3, 1974 and 1976, page 48.

Indonesian Village Technology Booklets, by BUTSI, page 48.

Village Development Notes and Booklets, by National Development Service, Nepal, page 48.

Appropriate Technology and Research Projects, by M.M. Hoda, page 49.

World Neighbors Newsletter, page 49.

Liklik Buk—A Rural Development Handbook/Catalogue for Papua New Guinea, Melanesian Council of Churches, 1977, page 50.

The Last Whole Earth Catalog, Portola Institute, 1971, page 51.

The Whole Earth Epilog, page 51.

CoEvolution Quarterly, page 52.

The Cumberland General Store Catalog, 1975, page 52.

Dick's Encyclopedia of Practical Receipts and Processes, 1870, page 54.

2,000 Down Home Skills and Secret Formulas for Practically Everything, circa 1900, reprinted 1971, page 55.

Village Technology Handbook, Rural Communications, 1976, page 55.

Economically Appropriate Technologies for Developing Countries: An Annotated Bibliography, by Marilyn Carr, 1976, page 56.

RAINBOOK: Resources for Appropriate Technology, editors of RAIN, 1977, page 56.

Mini Technology, booklet, 76 pages, by B.R. Saubolle, S.J., and A. Bachman, 1978, US$4.00 airmail overseas, $3.00 surface mail overseas, from Sahayogi Prakashan, Tripureshwar, Kathmandu, Nepal.

"This booklet is based very largely on the experience of the author, who was born and bred in India. It gives mostly Indian solutions to problems encountered in an earlier age before the onrush of modernity. It tells how to cool a house without air-conditioning, how to chill beer without a refrigerator, how to produce gas for cooking and lighting where there is no town supply, a way to make crows trap themselves, several ways of getting hot water at no expense, and so on and so forth." There are, in fact, four ways to cool food and yourself, four solar water heaters, an unusual solar dryer, several cookers and ovens, a self-closing water standpipe, some water filters, a one-man desk fan, a hand-held corn sheller, a fly trap, a fluorescent light insect trap, and an African bee hive.

Delightfully written and full of unusual devices that are otherwise largely undocumented.

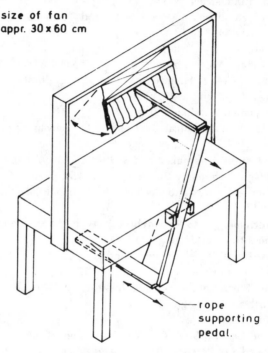

size of fan
appr. 30 x 60 cm

rope
supporting
pedal.

Foot-operated desk fan

Simple Technologies for Rural Women in Bangladesh, book, 70 pages, by Elizabeth O'Kelly, 1978, free from UNICEF, GPO Box 58, Dacca, Bangladesh.

This is a compilation of simple equipment that can be made or purchased in Bangladesh and many other developing countries. Only single drawings or photos are included for most examples; for some items this is sufficient information to make them.

The book begins with a description of the activities of rural women in Bangladesh, and the tools and equipment they use. Some employment-generating activities that could benefit rural women are suggested. Potentially relevant technologies presented have been taken from a variety of sources (FAO, ITDG books and equipment catalogs, the A.T. Sourcebook); these include vegetable coolers, cooking stoves, threshers, winnowers, and water pumps. Some manufacturers' addresses and a bibliography is included.

The author notes that "the division of labor between the sexes...needs careful study especially as in many countries the women enjoy considerable prestige as the growers of food for their families—which they will lose if the pattern of living is changed too drastically.''

Foot-operated grain huller

Fichier Encyclopedique du Developpement Rural, folders of leaflets, available for 60 Francs per year, from Groupe de Recherche sur les Techniques Rurales (GRET), 34, rue Dumont d'Urville, 75116-Paris, France.

These are sets of leaflets, in French, on a wide variety of village technology topics. The information is taken from French sources and international sources such as VITA and Brace Research Institute. References for additional information are given in each case. These leaflets offer an introduction to the concepts and applications of many successful technologies. Some of the topics covered: soil-cement block making, raising grapes and making wine in the tropics, water supply, bamboo construction, and cane crushing for sugar production.

GRET has a large collection of other French language publications on village technologies, and we urge readers in French-speaking countries to write to them for their publications list.

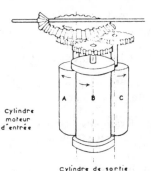

Cylindre
moteur
d'entrée

A   B   C

Cylindre de sortie

Sugar cane crushing mill

MINKA: A favor de una autentica ciencia campesina, Boletin de la Comision Coordinadora de Tecnologia Adecuada, 4 issues per year, $10 per year (international subscription) from MINKA, Apartado 222, Huancayo, Peru.

This Spanish-language journal provides the rural people ("campesinos") of Peru with information on locally successful appropriate technologies. MINKA is committed to a search for local solutions to local problems, through the development of a more scientific approach among the campesinos. It emphasizes technologies that have come out of the people's own experiences, such as the waterpumping windmills built at Miramar. An attractive mix of drawings, cartoons, photos and articles present information on a wide variety of subjects. Each issue concludes with a project for children, with simple plans for a working model of a tool or machine. Past issues have covered topics such as: "Is Mechanization Progress? For Whom?", plans for a locally-designed spade that is easily made and repaired, plans for a chain pump that can be built in a

Operating a chain pump

village, and a description of traditional Inca and pre-Inca water technologies. There are also reviews of illustrated manuals that can be of practical use to campesinos.

To the question "Do the campesinos read?", the editors answer that publishing is always done in cities, on topics that city people want to read about. By publishing a journal in a popular format on topics of direct concern to the campesinos and with their input, the editors hope to encourage a wide readership in the rural areas of Peru.

An outstanding example of a local communications resource for the sharing of local ingenuity and information on appropriate technologies.

Techniques Appropriate for the Villages: Some Examples, booklet, 50 pages, 1977 (revised 1980), Rs. 10 ($1.20) from Centre of Science for Villages, Magan Sangrahalaya, Wardha—442 001, India.

"Herein are collected some examples of techniques appropriate for the villages which could be tried." This booklet presents 33 short descriptions (no drawings) of tools and techniques for housing, motive power, and small

REFERENCES 377

chemical industries currently being developed at several Indian research institutes. Some of the techniques are simple enough to be tried without further information. Many (e.g. mushroom cultivation in paddy straw; chemical treatment of bamboo for handicrafts and bamboo-cement construction; fungi treatment and fire protection for thatch roofs, etc.) are of interest in countries other than India as well.

This short, low-cost booklet is an interesting example of information sharing among people and institutions working in village technology. "This is, no doubt, a very ambitious scheme and requires a tremendous amount of work. But the hope is that once the idea catches the mind of the people concerned, beneficial results...will flow out of it in a short period of time."

Guide to Convivial Tools, Library Journal Special Report #13, 112 pages, by Valentina Borremans, 1979, $5.95 (payment with order), from R.R. Bowker Company, 1180 Avenue of the Americas, New York, New York 10036, USA.

This annotated bibliography was produced by Valentina Borremans, director of the Centro Intercultural de Documentacion (CIDOC) in Cuernavaca, and a close associate of Ivan Illich. (It was Illich who coined the term "convivial tool" — see review of **Tools for Conviviality** on page 33). The bibliography "lists and describes 858 volumes and articles that, in their turn, list books on alternatives to industrial society or people who write on that subject."

"This new discipline deals with the cultural, social, and political conditions under which use-value oriented modern tools can and will be widely used, and with the renewal of ethics, politics, and aesthetics which is made possible by the democratically decided limitation of the industrial mode of production."

There are three kinds of people in the intended audience: 1) the librarian attempting to create a specialized research library away from the large general libraries; 2) the librarian in the industrialized countries who wishes to expand the reference section on this topic; and 3) the individual researcher without access to a library at all.

Non-Agricultural Choice of Technology: An Annotated Bibliography of Empirical Studies, book, 84 pages, by Gareth Jenkins with an introduction by Frances Stewart, 1975, Institute of Commonwealth Studies, Oxford University, £2.75 from ITDG.

Provides access to a fascinating list of studies on technology choice, with implications for many of the debated economic aspects of appropriate technology theory. The annotations make very interesting and valuable reading even without going to the original articles.

Appropriate Technology Information for Developing Countries: Selected Abstracts from the NTIS Data File, 425 pages, second edition November 1979, edited by Paul Bundick, available from Bureau for Latin America and the Caribbean, Agency for International Development, Washington D.C. 20523, USA.

This bibliography contains 2000 annotated entries, chosen from the materials held by the National Technical Information Service (NTIS). The editor hopes that this information can be adapted for "direct benefits which foster self-reliance and a sense of dignity among the poor." This publication came out of a collaborative effort between NTIS, the U.S. Agency for International Development, and VITA.

Unfortunately, few of the reports included were originally written with any sensitivity to the concept of appropriate technology. These are mostly reports on research projects; there are virtually no practical publications with information that can be directly applied. Recommended only for those willing to work their way through a lot of extraneous information in search of a few valuable items.

To label this collection of government (mostly A.I.D.) research papers "appropriate technology" is to ignore the dramatic shift of development strategy represented by the appropriate technology movement. An awkward dilemma facing A.T. supporters is the question of how to extract the valuable technical information from past research efforts which neglected social factors, and which were based on now-discredited assumptions (e.g. about acceptable levels of investment per job created). Because this bibliography does not help the reader to identify the relevant portions of these research papers, we recommend that you handle it with great caution.

Soft Tech, book, 175 pages, edited by Jay Baldwin and Stewart Brand, 1978, $5.00 from Penguin Books, 625 Madison Avenue, New York, New York 10022, USA; or Penguin Books Ltd., Harmondsworth, Middlesex, England.

A compendium of the "soft technology" sections—articles and reviews of products and books—from past issues of the **CoEvolution Quarterly** (itself an extension of the **Whole Earth Catalog**—see reviews on pages 51-52). What is "soft technology" to the editors? " 'Soft' signifies that something is alive, resilient, adaptive, maybe even lovable."

The emphasis here is on technologies that can be used in the U.S. Much of this book provides access to **products**—identifying who is making and selling the best quality and most unusual practical tools—from hand tools to machines and renewable energy measuring devices. In this sense **Soft Tech** is a buyer's guide for a "highly evolved toolbox". You'll also get a look at solar gadgets for U.S. homes, the 1891-1930 California solar water heater boom, energy-efficient cars and mopeds, folding bicycles, wood-burning for space heat, underground buildings, owner-building strategies, passive solar design, and the New Alchemy Ark.

I didn't believe in the alternate-energy future until I saw how dull it was gonna be and how stupid the slogans were gonna be and how much I wasn't gonna like it. Then I knew it would come.

— Steve Baer

The editors have a wide-open attitude towards technology, avoiding absolute criteria for "soft tech" and fascinated with things that work, searching for tools and attitudes that are earth-nurturing rather than earth-destroying. They are looking to "a future in which technology is steered with an eye towards helping to make Experiment Earth a success...we are going to have to learn how to live here in large numbers without trashing the place."

The Book of the New Alchemists, 174 pages, edited by Nancy Jack Todd, 1977, $6.95 from E.P. Dutton, 2 Park Avenue, New York, New York 10016, USA.

A collection of articles on the work of the New Alchemy Institute on Cape Cod, Massachusetts. This group carries out probably the most scientifically sophisticated work of any of the U.S.-based appropriate technology groups. The articles are about gardening and small-scale farming, aquaculture in small ponds, and "bioshelters" (primarily the Ark, a complex unit that combines passive solar heating, greenhouse food production, fish raising and human living quarters). Twenty pages are devoted to strategies for ecological farming in Costa Rica, where some of the Institute members work part of the year. Most of these articles are reprinted from the earlier issues of the annual **Journal of the New Alchemists** (see reviews on page 48).

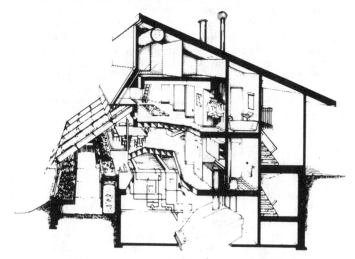

A section through residential greenhouse, heat storage composting toilet, and living areas of the Ark

NCAT Publications, from Publications Section, National Center for Appropriate Technology, Box 3838, Butte, Montana 50701, USA.

The National Center for Appropriate Technology (NCAT) is a U.S. Government-funded non-profit organization supervised by the Community Services Administration. It was established in 1977 to promote appropriate technologies among low-income and poor communities in the United States. Most of NCAT's programs are in response to the problems of rapidly increasing energy costs. They include simple techniques for conservation of energy in existing homes (weatherization) and use of renewable energy sources.

NCAT offers bibliographies on alcohol fuels, wind power, alternative waste/recyling systems, methane production, solar energy, management of solar greenhouses, and organizing community gardens. They also offer some longer publications (several of them reviewed in the A.T. Sourcebook) and a bi-monthly newspaper describing demonstration projects and new developments in technologies relevant to low-income people in the United States.

The Formula Book 1, 2 & 3, three volumes, 209, 193, 185 pages, by Norman Stark, 1975, 1976, 1978, $5.95 from Sheed Andrews and McMeel, Inc., 4400 Johnson Drive, Fairway, Kansas 66205, USA.

Book One is filled with 220 formulas for household products, many of which are relevant to the Third World. All have been chosen to be made in the home, with simple tools. Thus many of these might be appropriate for production in small scale industry efforts in the Third World. Equipment needed is very simple: double boilers (one pot sitting on top of a second pot filled with water), wooden spoons, mixing bowls, measuring cups, thermometers.

The author claims that some of the formulas have been "modified from large scale manufacturing quantities to small batches that are suitable for the do-it-yourselfer", and that "all are tested under actual use conditions."

Most of these formulas are for very simple mixtures. When covering such topics as making soap, candles, and composting, the text is a bit too brief to prepare the reader with all that would be important to know.

There is a listing of the usual sources of supply for the chemicals used (mostly drugstores, hardware stores and grocery stores in the U.S., though sometimes chemical supply houses). All these chemicals are defined in an appendix. Metric equivalents for all formula measures are provided.

Most of these formulas are for safe things, but sometimes hazardous ingredients have been included, which the author has identified. However, it is beyond the scope of the book to explain the exact effects of these hazardous chemicals. (For example, the formula for preventing algae bloom in ponds provides no clues as to what is likely to happen to fish or crops downstream.)

Books Two and Three are disappointing, averaging less than half as many formulas as Book One, with some duplication, and very few products useful in the Third World.

Book One is recommended. It includes the following most relevant formulas: waterproofing mixture for concrete, waterproofing mixture for canvas (using soybean oil and turpentine), waterproofing mixture for leather, mixtures to protect wood from fire and termites, biodegradable laundry detergent, mixture for fireproofing cloth, chimney soot remover, safe cockroach poison, airtight seal for canning, bay leaves used in stored flour and cereals to repell insects, liquid glue, mixtures for the repair of holes in galvanized roofing sheets, heavy duty mechanic's hand soap, deodorant, contact lens fluids, engine cleaning additive, automobile radiator leak sealer, and tire leak sealer.

Appropriate Technology Directory, book, 361 pages, by Nicolas Jequier and Gerard Blanc of the Development Center of the Organization for Economic Cooperation and Development (OECD), 1979, $22.50 in English or French, from OECD Publications Office, 2 rue Andre Pascal, 75775 Paris Cedex 16, France; or OECD Information and Publications Center, Suite 1207, 1750 Pennsylvania Avenue, N.W., Washington D.C. 20006, USA.

"The idea for such a 'Who's doing What' in the field of appropriate technology grew out of hundreds of requests for information addressed to the OECD development Center...In trying to provide these answers, we soon discovered that the number of organizations involved in developing and diffusing 'appropriate', 'intermediate', or 'soft' technologies was considerably larger than anyone had suspected...What we have attempted to do here is to present in a standardized way...all the basic information about organizations involved in the promotion of appropriate technology, both in the industrialized and developing countries."

280 groups and organizations are listed alphabetically by country. Text on each organization includes information about origin, funding, main objectives, examples of technologies worked with, and future plans. Data on scale of activities, budget, and staffing are also given when available.

Directories are an important tool for the building of horizontal information-sharing networks; this directory is one of the most useful, comprehensive, and up-to-date of its type. A lower price would help it achieve worldwide circulation.

Directory of Institutions and Individuals Active in Environmentally-Sound and Appropriate Technologies, hardcover book, 172 pages, 1978, revised edition 1979, by the United Nations Environment Programme, $20.00 from Pergamon Press Ltd., Headington Hill Hall, Oxford OX3 OBW, England; or Pergamon Press Inc., Maxwell House, Fairview Park, Elmsford, New York 10523, USA.

This directory was compiled as a preliminary listing of potential sources for UNEP's International Referral System for environmental information. It contains names and addresses of more than 600 individuals and groups around the world. The revised edition (expected to be available in 1980) is to include some 2000 entries. The quality of the project descriptions varies, some giving complete accounts of the activities, while others give only brief lists of topics of interest. The information was compiled from a variety of secondary sources and much of it is unchecked.

While it is much less well-researched than the **Appropriate Technology Directory** published by OECD, this book represents a more extensive listing.

Rural Technology in the Commonwealth: A Directory of Organizations, 127 pages, 1980, Ł1.50 from Director, Food Production and Rural Development Division, Commonwealth Secretariat, Marlborough House, Pall Mall, London SW1Y 5HX, England.

A listing of 118 institutions in 26 Commonwealth countries. Consists of clear short descriptions, giving address, major functions, programs, and specializations of each organization. This directory concentrates on organizations working in agriculture, forestry, and water resources for rural development. A useful index of equipment and processes is included. "We hope that this directory will fill an immediate need for two groups of people. The first group comprises those who are working on appropriate technologies, who we hope will be assisted by the directory to find out less laboriously who is doing what and where...the second group which we felt would make use of this directory are the 'travellers' of the development business...this directory will, we hope, enable you to make contact with people who share your interests. Equally useful contacts can be made by correspondence."

International Directory of Appropriate Technology Resources, large loose-leaf book, about 340 pages, compiled by Brij Mathur, 1978, $15.95 from VITA.

A compilation of 250 responses to a one-page questionnaire sent by VITA in 1977. "Part One lists the organizations alphabetically by countries. Part Two provides lists of publications, reports, papers, etc. published by these organizations. Part Three is a subject index to facilitate use of the material." Responses to the questionnaires were uneven, and many of the entries consist of one or two sentences—lots of white space here. The publications lists give prices (now somewhat outdated) and acquisition information.

Chambers Dictionary of Science and Technology, Volume 1. (A-K) and 2(L-Z), books, a total of 1300 pages, edited by T.C. Collocott and A.B. Dobson, 1974, $10.95 for the set of two volumes, from The Two Continents Publishing Group, 30 East 42nd St., New York, NY 10017, USA.

This dictionary contains ''50,000 entries from 100 fields of activity covering every aspect of scientific and technological knowledge, all carefully integrated into a single alphabetical list.'' It may not cover every aspect but it sure comes close. Definition and field of use are given for each word. This two volume set would be of great use to anyone trying to read, understand and use books printed in English on science and technology topics. The fields covered include forestry, botany, chemistry, plumbing, printing, veterinary medicine, textiles, architecture, fuels, surveying, and **many** others.

Highly recommended.

Small Technical Libraries, booklet, 40 pages, by D.J. Campbell, 1973, reprinted 1980, UNESCO, $6.00 from UNIPUB.

In this booklet you can find good common-sense ideas that will save some time and money in best organizing a small technical library, to support real village technology adaptation and development work. Most of this booklet describes libraries that are better funded and beyond the scope of an appropriate technology unit.

LOCAL SELF-RELIANCE
AND A.T.:
STRATEGIES FOR THE FUTURE

Local Self-Reliance and Appropriate Technology: Strategies for the Future

Many of the publications reviewed in this book contain evidence that community involvement and increased self-reliance in problem-solving go hand in hand with appropriate technologies. Together they can make a large contribution in solving the problems of poverty, particularly in the Third World. But one should not forget that the community exists within the political and economic confines of the nation-state. Because of the great economic and coercive power of these states, national decisions about development strategy and allocation of resources will deeply affect all the choices open to communities.

This fact of life—that decisions and choices made "at the top" crucially affect what is possible "at the bottom"—is the focus of this section. **Appropriate Technology: Problems and Promises, Part I** (reviewed on page 31) provides an insightful examination of what this means for the development of technologies. Many other publications reviewed here give concrete evidence of the importance of integrating high-level development policy with locally-based decision-making if a project is to be successful (see, for example, **Participation and Education in Community Water Supply and Sanitation Programmes**).

If a genuine effort is to be made to support community-based development of skills, problem-solving capabilities, and institutions, it will be necessary to reorient current structures that channel technical information and assistance, training and education, capital, government revenues, research and development work, and political power. If the point of initiative and problem-solving is to be at the community level, then the community must have access to these supportive systems that make it possible for things to happen. The writings in this chapter address these issues by discussing the most practical sizes of political and economic units and illustrating how the level at which initiatives are taken is a determining aspect of any development strategy.

Leopold Kohr has done some of the pioneering thinking on the differential nature and functioning of social organizations—communities, cities, nations—as they grow larger and larger. In **The Breakdown of Nations** he argues that as social units increase in size, social problems increase faster, until these problems reach a magnitude and complexity at which they can no longer be understood and controlled by human beings. He urges a return to small political states, small cities, and small communities, in which problems can be broken down to a manageable size. Unlike many writers, Kohr acknowledges that there are opportunities for tyranny in small political units; he argues that tyranny at this level is far less costly to overcome than tyranny at the national level, where it means the oppression of an entire people and the launching of large scale wars.

Some of the environmental benefits of smaller states are suggested in **A Landscape For Humans**, which examines the potential for "ecologically guided development" in a region of the southwestern United States. **Local Responses to Global Problems: A Key to Meeting Basic Human Needs** describes many local initiatives that are being undertaken around the world to solve what are often thought of as primarily global problems.

Appropriate Technology in Social Context is an annotated bibliography that reviews the literature on the socio-cultural aspects of technological choice. The author concludes that to ensure that socially appropriate technology is chosen, it is necessary to involve "the community itself in the mechanics of technology choice, even if new procedures and institutions have to be created for this purpose."

Three books reviewed here reveal the importance of rural self-reliance in increasing agricultural production and developing rural small scale industry in China. The decentralized production of cement has been an important factor in enabling the Chinese communes to carry out at reasonable cost a wide range of public works projects such as irrigation canals and building construction. Small cement plants also employ 10 times as many people as modern large scale plants. (See **Small Scale Cement Plants: A Study in Economics**.) **Rural Small Scale Industry in the People's Republic of China** reports on the decision-making process which has supported the remarkable growth of rural industries. These industries have brought with them the development of valuable technical and managerial skills among the rural population, and allow better support of agriculture and more productive use of rural manpower when it is not needed in agricultural activities. **Learning From China: A Report on Agriculture and the Chinese People's Communes** examines the participatory structures (such as research and development teams that include farmers as members) that have been keys to the advances of Chinese rural development.

The possibilities and potential pitfalls that surround intermediate technology when applied in the context of Tanzanian ujamaa villages are explored in **Technology for Ujamaa Village Development in Tanzania**. A look at the many and complicated factors likely to affect the success of appropriate technologies at the local level is provided in **Soft Technologies, Hard Choices**.

The last three entries in this chapter are concerned with the identification of policy measures that can foster the development of appropriate technologies and support alternative, people-centered development strategies. Conventional development strategies are criticized and the assumptions underlying a different approach are described in **Alternative Development Strategies and Appropriate Technology: Science Policy for an Equitable World Order**. Need-oriented, culturally-linked development that aims at liberation is the topic of **Another Development: Approaches and Strategies**. The ILO volume **Technologies for Basic Needs** notes that decentralization has particular advantages in carrying out basic needs strategies, and points to lack of contact with the real problems and experiences of farmers as a major reason for the disappointing contribution of R and D institutions in the Third World.

There are a number of particularly relevant entries that have been placed in other chapters. **Health in the Third World: Studies from Vietnam** (see review on page 715) offers an intriguing look at a system that attempts to meet basic health care needs through the high-level allocation of scarce professional resources to all rural areas, combined with local research, development, and application of preventive health and sanitation measures. Many of the entries in the NONFORMAL EDUCATION AND TRAINING chapter provide insights into effective roles and strategies for outside groups that wish to support the growth of community-based concientizacion and problem-solving.

In Volume I:

A Strategy for the Development of Village Industry, by Berg, Nimpuno, and Van Zwanenberg, 1978, page 55.

First Steps in Village Mechanization, by George MacPherson, 1975, page 44.

China: Science Walks on Two Legs, by Science for the People, page 37.

Churches in Rural Development: Guidelines for Action, by Peter Sartorius, 1975, page 36.

The Breakdown of Nations, book, 250 pages, by Leopold Kohr, 1957, reprinted 1978, $4.95 from E.P. Dutton, 2 Park Avenue, New York, New York 10016, USA.

It is in this volume that Kohr first develops, and in delightful fashion, his theories of scale. Despite his unconvincing attempt to explain away **all** social problems as due to bigness, he ably defends his thesis that scale is an important variable and that systems and institutions that are too large inevitably fail to function properly. The author's humorous tone at times distracts the reader from the seriousness of his points, but makes for a book that is hard to put down.

"If the great powers had at least produced superior leadership in their process of growing so that they could have matched the magnitude of the problems which they produced! But here, too, they failed because, as Gulliver observed, 'Reason did not extend itself with the Bulk of the Body.' "

"Neither the problems of war nor those relating to the purely internal criminality of societies disappear in a small-state world; they are merely reduced to bearable proportions. Instead of hopelessly trying to blow up man's limited talents to a magnitude that could cope with hugeness, hugeness is cut down to a size where it can be managed even with man's limited talents."

A Landscape for Humans, book, 149 pages, by Peter van Dresser, 1972, $5.95 postpaid from The Lightning Tree, P.O. Box 1837, Santa Fe, New Mexico 87501, USA.

Here is "a case study of the potentials for ecologically guided development in an uplands region." Chosen was an area in the northern part of the state of New Mexico, USA, long a secluded zone of Spanish culture. "It is no longer possible to 'solve the problems' of such a regional community by expediting wholesale outmigration and assimilation of its population into the urban, metropolitan, or industrial areas of the nation...Neither is it possible to rehabilitate provincial regions such as the uplands by importing big industry and its works. The dominant characteristic of modern primary and extractive industry (including "agribusiness"), geared to the national market, is labor-conservative, machine-intensive, and moving toward maximum automation. Very large investments are required per job created (e.g., $175,000 for a modern pulp mill). Regions dominated by such industries tend to depopulate except for company towns of varying degrees of cultural and social impoverishment."

Van Dresser suggests a variety of environmentally sound, community supportive economic activities that could be carried on or expanded within the region, to fill the needs for goods, services, and employment. He notes, for example, that a decentralized timber industry would be well suited to the existing distribution of timber resources and population. It might be possible to "vertically-integrate" such a timber industry, so that more employment and value-added would remain in the region, when processed timber products are sold outside the region.

"...The bulk of the livelihood needs of such a region must be met within the region itself by skilled, scientific, intensive, and conservative use of the lands, waters, and renewable biotic and environmental resources of the region. The long-term strategy for economic development should be gradual de-involve-ment from the mass logistic machinery of the continental economy, with its enormous and ever-increasing consumption of energy and irreplaceable na-tural resources.''

The author's recommendation of gradual de-involvement from the national economy is contrary to the thinking of most economic development institutions. The conventional wisdom of the latter is that increased trade is primarily beneficial. Van Dresser makes a persuasive case that the recent effects of such economic ties have been, on balance, quite negative.

"Such an evolution calls for a new technological, agricultural, and industrial orientation, stressing small-scaled and diversified primary production, adapted to the land and natural resource patterns of the region, to the ecologic balance and health of the total biotic community, and to the needs of a decentralized and dispersed population of effective and vital small communities. This type of productive economy will be manpower-, skill-, and science-intensive, rather than capital-, energy-, and machine-intensive.''

Van Dresser argues that an important part of the foundation for building such an economy is the high level of non-commercial "primary production" already taking place in the region. This is particularly strong in the growing and processing of foodstuffs and in the construction of homes and farm buildings. The author's observations about road building, education, and other aspects of a practical plan for ecologically guided development would be relevant in many rich and poor countries. This short book offers a remarkably broad and stimulating introduction to these issues that affect appropriate technology efforts. Highly recommended.

Local Responses to Global Problems: A Key to Meeting Basic Human Needs, Worldwatch Paper 17, booklet, 64 pages, by Bruce Stokes, 1978, $2.00 from Worldwatch Institute, 1776 Massachusetts Avenue N.W., Washington D.C. 20036, USA.

Self-help and self-reliance are the important keys to better living conditions and broader opportunity for people the world over. This paper shows how individuals and communities are meeting their basic needs with little or no help from outside institutions.

In housing, carpentry, plumbing, and bricklaying skills shared at the neighborhood level provide the technical basis for self-help homebuilding in urban slums around the world. Successful projects have often involved the government providing land, credit, and basic services to poor families who can then build their own dwellings.

For food production: "Whether judged by yield per acre or by the cost of production, small farms compare favorably with large farms on all continents. Most of the economies of scale associated with size can be achieved on units small enough to be farmed by a family...A 1970 survey for the United States Agency for International Development (USAID) showed that small farms in India, Japan, Taiwan, the Philippines, Mexico, Brazil, Colombia, and Guatamala had higher productivity per acre than large farms.''

Energy conservation and examples of self-reliance in small-scale energy production and consumption are also discussed.

The author concludes that while foreign aid and other forms of international cooperation can be constructive, problems of basic needs must be addressed at the local level. "In 1975, public and private official development assistance... totaled $18.4 billion, not even enough to meet yearly basic housing needs according to the (World Bank's) estimate. The political will does not exist to solve problems through a large transfer of resources. Any development strategy based on the assumption that the rich will more than double their foreign aid is doomed to failure. This does not mean that foreign aid should be abandoned. But if the resources to fully meet basic needs are not forthcoming from national and international sources, then they must come from communities and individuals. While ready capital is scarce at this level, there is a reserve of labor and ingenuity that money cannot buy''.

Appropriate Technology in Social Context: An Annotated Bibliography, 33 pages, by David French, 1977, $2.95 from VITA.

Lists 180 books, articles and papers, with short paragraph annotations. Many case studies are identified, and a number of important issues are raised. This is not a review of the technical literature.

"Harmony between technology and social context is important. Abundant evidence shows that implanting a socially 'inappropriate' technology in a village has the same result as implanting a foreign object in a person: either the technology is rejected or the village may 'die' as a social organism...novelists and anthropologists have long recorded the disruption of traditional societies by new technologies."

"To take full account of context implies involving the community itself in the mechanics of technological choice, even if new procedures and institutions have to be created for the purpose."

The materials in this bibliography are "abstracted from four separate literatures, those of development agencies, the applied social sciences, village-oriented programs, and sources of technical information...There is a need to break down the walls (between these groups) if appropriate technology is to be kept in social context...Perhaps the first job here should be design of an appropriate institutional technology for 'technology' transfer."

The author uses a relatively full definition of appropriate technology, noting the importance of people's participation, low costs, and use of local resources; thus, his reviews are more interesting and valuable than those in many other bibliographies.

Recommended.

Small Scale Cement Plants: A Study in Economics, booklet, 28 pages, by Jon Sigurdson, 1977, L1.60 from ITDG

"Small scale cement plants have recently been attracting more and more attention from international agencies and industrial economists concerned with rural development. In China there are more than 2800 active small scale plants and more than 200 in Europe (Spain, Yugoslavia, France, Germany and Italy). This booklet examines the criteria which would justify the establishment of mini cement plants in developing countries and specifically compares the situation in India with that in China, where more than 57% of cement is produced in small plants...A short bibliography is provided as well as designs of vertical shaft kilns taken from a Chinese book on small scale cement plants."

In China "the initial smallness of a plant enables the capacity of the plant to grow with the local demand. This may make overall costs lower than if a large capacity plant had been set up from the very beginning."

"When deciding location, size and technology for the cement plants it appears that in China transportation costs are much more important than investments costs per ton of finished product." (Other studies of the Chinese rural development effort indicate that the savings to a public works program can more than offset the investment in a small kiln within a year or two.) Sigurdson notes that the freight policy in India makes cement the same price at all rail depots. (Transport costs are 'pooled' and assigned equally to all cement sold in the country). This eliminates the advantage that small kilns would have in local marketing areas, as the substantial costs of transport are not reflected in the price of cement produced by distant large kilns.

"The viability of small cement plants is at least partly a reflection of demand created through substantial public works programs and other construction activities in rural areas." (Conversely, it must be noted, in many countries the high cost of cement prevents the undertaking of rural public works programs.)

The author raises the question of "appropriateness of product." The Chinese small vertical shaft kilns apparently do not produce cement of Portland Cement quality (in strength and uniformity). However, for most rural area uses, higher quality is not needed.

The small rural cement plants employ at least 250,000 people directly. "This number is at least 10 times higher than employment would be in a small number of modern large scale cement plants producing the same quantity of cement."

A short but important case study that illustrates many of the issues surrounding appropriate technology.

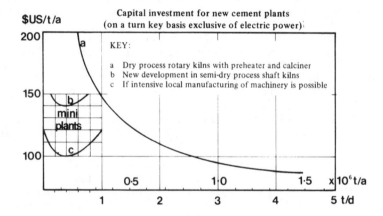

Technology for Ujamaa Village Development in Tanzania, book, 64 pages, by Donald Vail, 1975, $3.50 from Publishing Desk, 211 Maxwell School, Syracuse University, Syracuse, New York 13210, USA.

A thoughtful discussion of the social/political/economic circumstances that affect intermediate technology in one country of Africa attempting to develop a decentralized, self-reliant village socialism.

There is an interesting look at the potential for creative exchange between a testing unit and both 1) international sources of ideas and information, and 2) local people using adapted tools in real farming activities. How can a testing unit learn from both the enormous variety of small tools already in existence worldwide **and** from the farmers themselves? (Learning from the farmers ensures their participation and greatly increases the chances that the equipment developed will be relevant.)

The relationship between intermediate technology and the strengthening— or undermining—of Ujamaa village development is explored. The author argues that without policy backing for Ujamaa as the dynamic mechanism for

A.T. development, new small-scale technology seems likely to strengthen private enterprise at the expense of the cooperative Ujamaa villages. This in turn would have the effect of a concentration of land holdings and stratification into a relatively small group of haves and a much larger group of have-nots.

Soft Technologies, Hard Choices, Worldwatch Paper 21, booklet, 41 pages, by Colin Norman, 1978, $2.00 from Worldwatch Institute, 1776 Massachusetts Avenue, N.W., Washington D.C. 20036, USA.

A good overview of the arguments in favor of the development of appropriate technology. Full of sensible observations such as: "Skewed income distribution leads to the development and adoption of technologies that meet the demands of the privileged...Without social and political changes that redistribute income, overhaul inequitable land ownership patterns, reform credit systems, and provide support for small farmers and manufacturers, appropriate technologies will be difficult to introduce. Powerful vested interests support large-scale manufacturing, mechanized farming and other symbols of modernity...By stimulating local innovation and reinforcing other development efforts, simple technologies can lead to self-sustaining development...No technology— however appropriate—will solve social problems by itself...Nevertheless, the choice of inappropriate technologies can only exacerbate social, economic, and environmental problems...The entire innovation process, from basic research to the introduction of a new technology, is conditioned by such factors as the profit motive, prestige, national defense needs, and social and economic policies. Those forces must be understood in any discussion of appropriate technology...The unfettered workings of the market·system cannot be relied upon to promote the development and adoption of appropriate technologies."

Rural Small-Scale Industry in the People's Republic of China, book, 310 pages, by Dwight Perkins et al., September 1977, $15.00 from University of California Press, 2223 Fulton St., Berkeley, California 94720, USA.

Much can be learned from the remarkable success of the Chinese efforts to develop grass roots skills and innovative capability, and improve the standard of living through promotion of rural small-scale industry. This report of a distinguished group of American visitors in 1975 offers many valuable insights into the successes and problems of these efforts.

The authors discuss the administrative systems, worker incentives, economies associated with small-scale industries, the relationship between these industries and agriculture, and their impact on Chinese society. Special treatment is given to agricultural machinery, chemical fertilizer technology, and small-scale cement plants.

"First and foremost, China is developing a rural small-scale industry because this strategy is believed to be doing a better job of supporting agriculture than did the large-scale strategies of the past."

"The rationale for the use of small-scale factories in rural areas begins with a recognition of the inadequacies of China's rural transport and marketing systems...Reinforcing the effects of high transport costs is the nature of China's rural commercial system. Even when communes are prepared to pay the going price for some desired item, it won't necessarily be available...it may get it faster if it builds one on its own."

"The planning system seems to be in part a nested or hierarchical system of rationing of technically advanced products in such a way that the demand for scarce, high-technology products in the production of products for rural life is minimized." To the extent that rural production units can meet their own equipment and other capital needs, the Chinese can avoid "wasting Shanghai talent producing small threshers for the whole country."

"No research institute in Peking will be able to design machines suitable for all environments and conditions. Local production facilities coupled with design inputs from two directions have largely alleviated this problem. Assistance from above is readily available—e.g., for 12 h.p. diesel engines—or for electric motors and pumps. From below comes a flow of comments and suggestions as to how trial machines perform and what tasks need to be mechanized. The local factories, especially the commune level machine shops, seem ideally suited to wed these two inputs into locally adapted machines."

"Instead of leaving innovations to technicians alone, 'three-in-one' groups consisting of administrators, technicians, and senior workers are organized to attack technical problems and produce innovations in factory technology."

"A reasonably strong argument can be made that the major contribution of the agricultural machinery industry...has been through an indirect process of 'scientification' of the rural masses. A hand tractor imported from Japan would have the same physical productivity as one made in China, but it would certainly not have the same impact as one made in a brigade or commune machine shop where every peasant knows someone who helped build it."

The Chinese have deliberately followed a strategy of starting rural industries small and gradually making them bigger and more modern. The larger, more modern stage could not be reached "without the industrial experience, the chance to mobilize the masses in technical renovation, and the capital funds from profits in the meantime, that are the products of its first period."

The demand for electricity that has accompanied the spread of rural small-scale industries has led to the construction of a very large number of small hydroelectric plants, some 60,000 in south China alone.

"It is not the techniques themselves that the Chinese are adding to the world's storehouse of knowledge, but the fact that these techniques can be adapted to rural conditions on a widespread scale."

Highly recommended.

Learning from China: A Report on Agriculture and the Chinese People's Communes, book, 112 pages, by a U.N. Food and Agriculture Organization Study Mission, 1977, available from FAO, Director, Publications Division, Via delle Terme di Caracalla, 00100 Rome, Italy.

"This is a nation that, within the short span of 27 years, has succeeded in banishing starvation. It is now providing food, clothing, shelter and reasonable security for over 800 million people. It has mobilized the world's largest agricultural labour force, reversed the flood of people into cities and kept people on the land." A multi-disciplinary team of FAO officers compiled this report, which is focused on the participatory structures that have enabled China's dramatic achievements in meeting basic human needs. Present-day organization of production along commune, brigade, and production team lines is presented as part of the history of traditional and revolutionary Chinese collectivism.

Much of the report is devoted to the educational, research, and mechanization strategies employed by the collectives to boost agricultural output. Mobilization of the productive workforce is key to this strategy: "Most other developing countries...are impaled on the horns of a cruel dilemma: there is massive unemployment precisely at a time when so much needs to be done. China has largely solved this dilemma through a development approach that, among other things, designed the commune. In the process, it unleashed a tremendous force for development."

An important conclusion emerging from this report is that technological changes cannot substantially change the position of the small farmer unless they are part of a genuine structural or organizational reform in the countryside. "The Chinese experience suggests that developing countries should consider a temporary and selective moratorium on current plans for comprehensive diffusion or 'transfer' of technology, among these most disadvantaged farmers. These farmers need instead more intensive policies of tenurial improvement and selective, if not widescale, measures of land reform; progressive upgrading of traditional tools and equipment; more intensive use of local resources such as organic manure, and compost and small bio-gas plants; and the mobilization of traditional forms of peasant cooperation and mutual aid for both production and rural capital formation."

Alternative Development Strategies and Appropriate Technology: Science Policy for an Equitable World Order, book, 255 pages, by Romesh K. Diwan and Dennis Livingston, 1979, $25.00, from Pergamon Press, Maxwell House, Fairview Park, Elmsford, New York 10523, USA.

Here is a summary of the criticisms of the conventional industrialization-led, GNP-measured development strategies, and a description of the elements of emerging alternative development strategies. Despite the subtitle, the authors do not really explore science policy except in the broadest sense, arguing for the development of indigenous capabilities to generate environmentally-sound, culture-linked appropriate technologies for the poor. Some science policy issues and other political and economic issues are identified, but specific policies are not proposed. Nicolas Jequier's **Appropriate Technology: Problems and Promises** (see review on page 31) gives the policy issues associated with appropriate technology a deeper look.

The authors draw from a broad relevant literature, touching on so many problems and issues that many sections seem too brief. (An extensive bibliography is included.) Useful distinctions are made between high-income developing countries, high-technology developing countries, and the rest of the developing countries. Insights into the behavior of the international organizations will be of particular interest to many readers.

Because the authors have taken a broad view of the concepts of "development" and "appropriate technology", their conclusions are not crippled by the timid definitions that tend to emerge in international conferences and the publications of UN agencies, where polite fictions must be observed. "In the literature on international science and policy, there is a tendency to confuse the 'interests' of the governments with the 'appropriateness' of technology...However, the concept of 'appropriateness' as discussed in A.T. literature is quite different, and may even be poles apart."

"The conventional development strategy...leaves the bulk of Third World peoples dependent on institutions and forces, within their countries and abroad, that are unreachable and unaccountable." The authors recommend "delinking of developing countries from debilitating global networks dominated by the affluent." They observe that much of the international debate on codes for technology transfer and the New International Economic Order are simply part of the failing conventional development strategy; unless domestic structural change takes place the elite will reap all of the benefits.

A basic assumption of the authors, which lies at the heart of the emerging human-centered concepts of development and appropriate technology, is that ''people, even those who are poor, illiterate, and unemployed, are intelligent. They are capable of defining their own needs and given opportunities, they can and will solve their own problems.''

This is an inherently optimistic book. Diwan and Livingston identify areas for cooperation where many might see inevitable, tragic, sources of conflict. ''It is...in the self-interests of the governments and elites of both developed and developing countries to work cooperatively towards the formulation of an alternative international order which reduces and eventually eliminates inequalities, armaments, biases in the price system, and technological inappropriateness, both nationally and internationally.''

Another Development: Approaches and Strategies, book, 265 pages, edited by Marc Nerfin, 1977, $19.00 surface mail or $21.00 airmail, from the Dag Hammarskjold Foundation, Ovre Slottsgatan 2, S-752 20 Uppsala, Sweden.

Here can be found much of the emerging thinking on new development goals and strategies for the rest of this century, in a collection of 10 articles by well-known writers.

''Another Development would be need-oriented (geared to meeting human needs, both material and non-material)...endogenous (stemming from the heart of each society, its values)...self-reliant...ecologically sound...based on structural transformations (in social relations, in economic activities, and in the power structure, so as to realize the conditions of self-management and participation in decision-making by all those affected by it, from the rural or urban community to the world as a whole)...These five points are organically linked...For development is seen as a whole, as an integral cultural process... Another Development means liberation.''

Part one begins with the concept of Another Development, and examines: the positions of peasants and women, alienation in industrial societies, and emerging simpler alternative life styles in those societies. In part two, national case studies and proposed strategies include: a look at growth and poverty in Brazil, the history of achievements and backsliding following the Mexican Revolution, an alternative framework for rural development in India, a strategy for Another Development in Chile (requiring major political change and based on the lessons of the early 70's), and a discussion of structural transformation in Tunisia.

''The New International Economic Order...makes full sense only if it supports another development...If it lacks a development content, it is bound to result simply in strengthening the regional or national subcenters of power and exploitation.''

''Resources to meet human needs are available. The question is that of their distribution and utilization...The organization of those who are the principal victims of the current state of affairs is the key to any improvement. Whether governments are enlightened or not, there is no substitute for the people's own, truly democratic organization if there is to be a need-oriented, endogenous, self-reliant, ecologically-minded development.''

Highly recommended.

Technologies for Basic Needs, book, 158 pages, by Hans Singer, ILO, 1977, $11.40 from International Labor Office, 1750 New York Ave., N.W., Washington D.C. 20006, USA.

This book was inspired by discussions held at the World Employment Conference in Geneva in 1976. It is largely concerned with top-down national planning for 'appropriate technology,' covering national policy, programs and institutions that might be able to contribute. The suggestions made are mostly aimed at the large-scale and small-scale industrial sector. While mention is made of the informal and rural small farm sectors, the assumption seems to be that technology will be created **for** the people in these sectors without their real participation. The author sees no potential conflict of interest between government-determined priorities and those that might most benefit the largest number of poor people. He is relatively uncritical of the possible role of multinational corporations in developing A.T., impressed by the strong research and development capabilities of these institutions. He claims that beneficial effects such as reduction in transfer payments for technology and the spread of the results of MNC-financed technology research among indigenous producers would be forthcoming—without seeing that such actions are simply not in the interests of the multinational corporations.

There are, however, a few points made that would have relevance in rural grass-roots development strategies. Some of them are:

"The experience of countries which have tried to implement a basic needs strategy (e.g. China, Cuba, Tanzania) suggests that the improvement of simple village technologies is the only feasible approach to the gradual modernization of the rural economy."

There is "growing evidence...that formal technical training plays a smaller part than was previously assumed and that experience and on-the-job training are the main vehicles for implanting new skills."

"A major reason for the disappointing contribution of R and D institutions to the creation of appropriate new technology is the lack of contact" with real problems and actual experience. The R and D institution itself is a modern import from the rich countries, "and disregards past experience...(when) the bulk of technological innovation arose directly from within the production plant" or workshop in response to needs and opportunities perceived there.

"Decentralization has obvious advantages, to help to eliminate the communications gap...This is particularly so in the context of a basic needs strategy, when local needs and the nature of local poverty problems may differ greatly in different regions of a country, when the use of appropriate technology involves the participation of innumerable small production units, and in the context of rural development generally (when the obvious need for community involvement and grassroots identification of problems has led to many variations of decentralized administration)."

ADDITIONAL REFERENCES ON STRATEGIES FOR LOCAL SELF RELIANCE AND APPROPRIATE TECHNOLOGY

THE WORKSHOP

The Workshop

Tools and techniques for small workshops are the subjects of this chapter. **Equipment for Rural Workshops** *is intended to help in the choice of tools when equipping a workshop. Several books offer an illustrated inventory of a great variety of hand woodworking tools, and techniques that would be useful in small industry furniture production, for example. Other books describe the proper uses of a wide range of other hand tools and machine tools for both wood and metal work (see especially* **Amateur's Workshop**). **Basic Machines and How They Work** *introduces basic mechanical principles that would be valuable in the design of simple machines.*

Many of these entries concentrate on particular workshop activities (such as working sheet metal), and discuss techniques and owner-built tools to aid in this work. In particular, **Bearing Design and Fitting** *should prove of use to appropriate technology groups improvising simple bearings to make their devices more efficient.*

Another group of publications are plans for easily-made workshop equipment. Many of these require electric motors, but could be adapted for use with other power sources.

It is hoped that these materials can serve two functions: first, as a set of reference books, containing thousands of ideas, for use by an appropriate technology group in its own workshop; and second, as a source of learning materials for improving skills, increasing versatility, and expanding available tools and equipment among local craftsmen. It should be remembered that many workshop crafts, such as blacksmithing, cannot be easily learned from a book; these reference materials can only supplement skilled instructors. However, even the most experienced blacksmith will find many unusual and valuable ideas in a reference such as **Practical Blacksmithing.**

In Volume I:

Fabricating Simple Structures in Agricultural Engineering, CoSIRA, 1955, page 64.

LeJay Manual, 1945, page 65.

3 Welding Jigs—Complete Technical Drawings, by R. Mann, ITDG, page 66.

Metal Bending Machine, ITDG, page 66.,

Treadle-Powered Wood-Turning Lathe, by W.C. Lecky, page 66.

Hand-Operated Drill Press, by A. Howe, 1948, page 67.

A Foot-Powered Saw, 1929, page 67.

10-inch Table Saw, Popular Mechanics, page 68.

4-Wheel Band Saw, Popular Mechanics, page 68.

Heavy Duty Drill Press, Popular Mechanics, page 68.

Gilliom Build-It-Yourself Power Tool Plans, 10-inch arbor floor saw, 9-inch tilt/table bench saw, 6-inch belt sander, 12-inch band saw, 18-inch band saw, drill press/lathe, page 69.

Woodwork Joints, by Charles Hayward, 1974, page 70.

Diary of an Early American Boy, by Eric Sloane, 1965, page 70.

A Museum of Early American Tools, by Eric Sloane, 1964, page 71.

China at Work, by R. Hommel, 1937, reprinted 1969, page 71.

Traditional Crafts of Persia, by Hans Wulff, 1966, page 72.

Country Craft Tools, by P. Blandford, 1974, page 72.

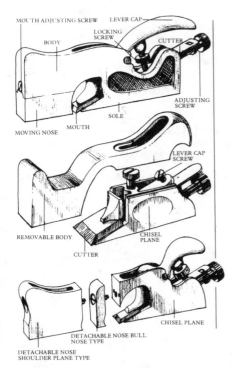

Three kinds of small planes

Tools and How to Use Them: An Illustrated Encyclopedia, book, 352 pages, by Albert Jackson and David Day, 1978, $8.95 from Alfred A. Knopf, Inc., 201 E. 50th Street, New York, New York 10022, USA.

This beautifully illustrated book is the best one available for descriptions of the wide range of useful hand wood-working tools. It also covers a few gardening tools, power tools, plumbing tools, and includes more than 1500 excellent drawings. For each tool the authors list other commonly used names, size, the material it is made from, and its purpose; this is followed by a short but valuable description of how it is used. Most of these tools originated in Europe or North America, but many are in common use all over the world. A good source of ideas for local adaptation.

Highly recommended.

Handtool Handbook for Woodworking, book, 184 pages, by R.J. DeCristoforo, 1977, $5.95 from H.P. Books, P.O. Box 5367, Tucson, Arizona, USA.

This book shows how to use woodworking tools commonly found around the world. These include measuring devices, saws, hammers, drills, screwdrivers, chisels and planes. The author also discusses safety, sharpening, shop math and how to choose good tools.

"You won't find this (a crown) on all saws, but many experts look for it as an indication of careful designing and superior quality. A crowned saw is one where the silhouette of the toothed edge shows a gentle arc rather than a straight line from the heel to the toe. The reason for the shape is to obtain maximum cutting effect with minimum drag. The arc brings fewer teeth into contact with the wood fibers. While you don't have as many teeth in full contact, those that are cut deeper, faster, and easier."

This book is full of tricks and tips for woodworkers, and the 400 illustrations make it easy to understand. Safety measures are very well covered.

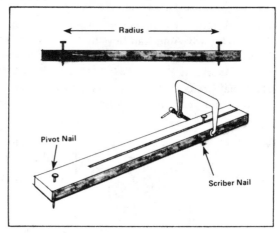

Trammels you can make: Fixed trammels can be made by driving two nails through a strip of wood. A variable type shown at bottom has a fixed pivot nail and an adjustable scriber nail held by a small C-clamp in a saw-cut groove.

The Use of Hand Woodworking Tools, book, 273 pages, by Leo McDonnel and Alson Kaumeheiwa, 1978, $6.80 ($5.10 to schools), from Delmar Publishers, 50 Wolf Road, Albany, New York 12205, USA.

This introductory book presents only the basic hand tools used in carpentry, and most of the book is devoted to explaining how to properly use them. No previous knowledge is assumed. Designed for use in teaching, the book contains questions at the end of each section.

The author begins with measuring tools (from the T-square to the builder's transit level—the only rather complicated tool presented), and continues with saws, planes, edge cutting tools, and boring tools. Sharpening is discussed in detail for each of the cutting tools. Also covered are nails, screws, and dowels. Well-illustrated.

Chamfering an edge with the work held in a clamp

DeCristoforo's Book of Power Tools, Both Stationary and Portable, book, 434 pages, by R. DeCristoforo, 1972, $14.95 from Book Division, Times Mirror Magazines, 380 Madison Ave., New York, New York 10017, USA.

A clearly written, extensively illustrated guide to the use of power woodworking tools, both stationary and hand-held. Includes table saws, drill presses, lathes, band saws, belt sanders and more. Each chapter describes the safe operation of a tool, and standard techniques, as well as many innovative applications of that particular machine. Most useful are photos and plans for simple jigs and accessories which increase the versatility of the power tools and allow the production of many identical pieces. For example, there are plans for adjustable wooden frame for cutting large panels easily with a power hand saw.

This book only provides instruction in the proper use of commercially available machines. No design and construction details for such machines are provided. The greatest weakness of this book is that it includes almost no information about repair or even routine maintenance of the power tools. May be useful to those wishing to teach themselves wood shop techniques, especially for small industry furniture production.

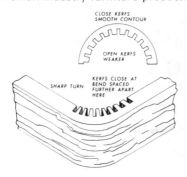

Bending wood

Tools and Their Uses, book, 186 pages, by the U.S. Navy Bureau of Naval Personnel, 1973, $2.75 from Dover Publications, Inc., 180 Varick St., New York, New York 10014, USA.

This book covers a very wide selection of common hand and power tools. The purpose is to "identify tools and fastening devices by their correct names; cite the specific purposes and uses of each tool; describe the correct operation, care and maintenance required to keep the tools in proper operating condition; and finally, perform accurate measurements." This book will not substitute for books covering in detail the techniques and equipment used in woodworking and soldering, for example. It is just a general introduction, but one that would be useful to any workshop education program using common tools. Safety information is included throughout.

Proper procedure for pulling adjustable wrenches

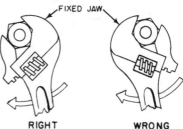

Basic Machines and How They Work, book, 161 pages, prepared by the Bureau of Naval Personnel, 1971, $3.00 from META.

Written as a reference manual for sailors in the US Navy, this book explains basic mechanical principles and their applications in simple and complex machines. Illustrated examples are used to show how these principles work in common devices. For example, oars, wheelbarrows, handpump handles, and the block and tackle are forms of the lever; the brace and bit, wrench, and winch are forms of the wheel and axle. Many of the examples include explanations of how mechnical advantage is obtained, and how to calculate that advantage using simple arithmetic. Later chapters explain the uses and combinations of basic machines elements (like bearings, linkages, cams) in mechanisms such as the typewriter and automobile engine.

Useful as a reference and as a practical way to study the physics of mechanical devices. Slang and sailing terminology may sometimes be difficult for the non-native speaker of English.

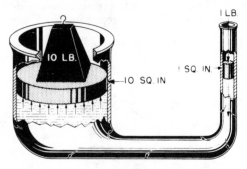

Mechanical advantage in a hydraulic system

Amateur's Workshop, book, 265 pages, by Ian Bradley, 1976, MAP Publications (UK), $11.00 from META.

This book is intended for people who already possess at least the basic skills

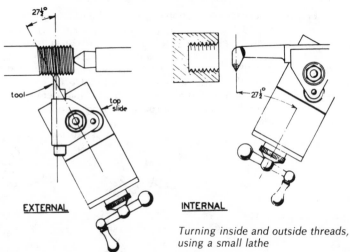

Turning inside and outside threads, using a small lathe

of metal working using hand and power tools. Despite the use of the term 'amateur' in the title, this book contains an enormous amount of useful information. The text is clear and well illustrated. Workshop skills are thoroughly covered. There are many detailed plans for tools and attachments to tools; many of these are complex devices.

The Amateur's Workshop will help someone with basic skills become a skilled craftsperson, after lots of practice. The book would be especially useful in programs for training metal workers. There is no glossary.

This is a step beyond **The Beginner's Workshop**, by the same author. An excellent book.

The Beginner's Workshop, book, 244 pages, by Ian Bradley, 1975, MAP Publications (UK), $7.00 from META.

This book will introduce you to the basic tools and machines in a metal-working workshop. The author provides suggestions on buying tools and building some small items. The basic uses of the tools and machines are very well described.

Recommended for the reader wishing to develop basic workshop skills. Readers who already have such skills may find some new information here too.

There are many illustrations, an adequate index, but no glossary.

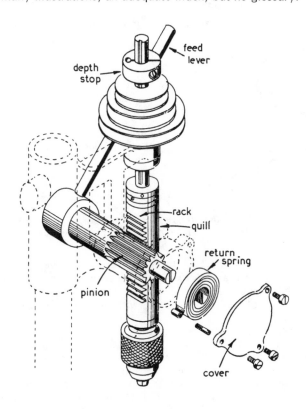

Essential parts in a drill press

TURNING OUTSIDE DIAMETER

CUTTING THREADS

TURNING INSIDE DIAMETER

FACING

DRILLING

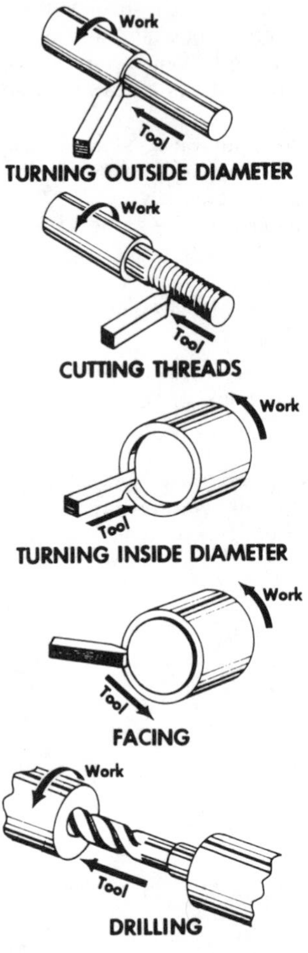

*Principal
lathe operations*

Metalworking Handbook: Principles and Procedures, book, 480 pages, by Jeannette T. Adams, 1976, $12.95 (plus 75 cents postage and handling), from ARCO Publishing Co., 219 Park Avenue South, New York,. New York 10003, USA.

This book is both a manual for the beginner and a reference book for the skilled craftsperson. For example, you'll find an introduction to working with sheet metal, soldering, rivetting and metal spinning. A variety of machine tools are discussed: drilling machines, milling machines, shapers, planers, lathes, and grinding machines. Electricity is required for most of these.

The appendix offers many useful tables. Unfortunately there is no glossary.

This book would be useful in a community metalworking shop.

The Recycling, Use, and Repair of Tools, book, 112 pages, by Alexander Weygers, 1979, $6.95 from Van Nostrand Reinhold, 135 West 50th Street, New York, New York 10020, USA.

''The scrap steel yards across the country are full of every conceivable metal object discarded for reasons of wear, obsolescence, or damage. Much of this material can become useful stock for the beginner, as well as the skilled metal craftsman, who intends to 'make do' with what can be gleaned from this so-called junk.''

In this book the author uses more than 600 drawings to show how to make useful wood and metal-working tools and other implements from steel scrap and discarded machine parts. Punches, chisels and gouges can be shaped and forged from steel tubing, automotive shafts, and spring steel. Files, rasps, fireplace tools, candlesticks, and other decorative implements can be made from mild and high-carbon steel scrap. Detailed perspective drawings show how to make a wood turning lathe from salvaged materials, and adjustable bearings from fruitwood. The final third of the bood discussed rehabiliting and operating metal turning lathes, and how to use and make inserts for a trip-hammer. A short section on how to temper high-carbon steel is included, but in general it is assumed that the reader has basic blacksmithing skills. The author's previous book **The Modern Blacksmith** (see review on page 60) provides a good introduction to the use of the hammer, anvil, and forge.

"It is through actual demonstration, seeing how to manipulate tools to make tools, that I believe the student benefits most. But short of that one can learn from books in which the illustrations come as near as possible to live demonstrations. I have tried to present the information in such a way that the reader can imagine he is watching me making things in the shop."

A practical book illustrating a creative craftman's approach to repair and re-use.

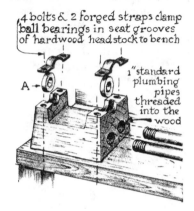

4 bolts & 2 forged straps clamp ball bearings in seat grooves of hardwood headstock to bench

A→

1" standard plumbing pipes threaded into the wood

a wood-turning lathe made from salvaged materials & inexpensive surplus items

The Blacksmith's Craft: An Introduction to Smithing for Apprentices and Craftsmen, book, 116 pages, by Council for Small Industries in Rural Areas, 1955, £4.25 surface mail from Information and Publications Section, CoSIRA, 141 Castle Street, Salisbury, Wiltshire SP1 3TP, England.

This well written book is divided into 37 lessons to teach people the basic skills of blacksmithing. These sections cover the tools and techniques very thoroughly, and the many photos show each step in making chain links, u-bolts, harrow bars, and many other things. There is a list of books recommended for further reading.

This is possibly the best introduction to blacksmithing available.

CHAIN STAPLE and U-BOLT

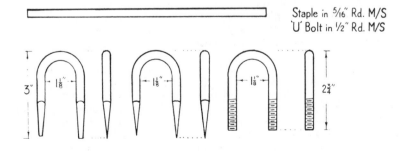

Staple in 5/16" Rd. M/S
'U' Bolt in 1/2" Rd. M/S

Practical Blacksmithing, book, 1089 pages, compiled and edited by M.T. Richardson, originally published 1889, reprinted 1978, $7.98 plus $1.50 postage and handling, from Crown Publishers, 34 Engelhard Avenue, Avenel, New Jersey 07001, USA.

Originally published in 1889, **Practical Blacksmithing** is a compilation of a great variety of articles on different aspects of the craft. These articles, originally printed in the 19th century journal **The Blacksmith and Wheelwright**, were submitted by hundreds of blacksmiths from all over the United States. This book thus represents an extraordinary attempt to collect, preserve, and make available common "hands-on" wisdom about a critically important craft. Hundreds of drawings show tools, layout of blacksmiths' shops, and methods of working steel and iron which had previously been passed from individual smiths to a few apprentices.

This is an outstanding reference book, to go with the teaching books **The Modern Blacksmith** (see review on page 60) and **The Blacksmith's Craft** (see review in this section).

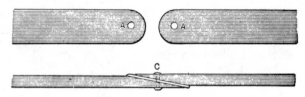

Fig. 97—Welding Springs by the Method of " W. B."

Metallurgy, book, 472 pages, by Carl G. Johnson and William R. Weeks, 1977, $13.75 from META.

Intended for those who require cast metals of high strength and durability, this book is biased towards the high-technology metals industries. There is, however, plenty of background information that could be useful to small-scale operations. A general introduction to the science of metals, the book covers the properties and testing of materials, and treatment and production of a variety of ferrous and non-ferrous metals and alloys. There is a glossary of terms used in the metals sciences.

This book could be useful as a reference in a large blacksmithing, casting, or smelting operation. Some basic knowledge of chemistry is required.

Sharpening Small Tools, book, 114 pages, by Ian Bradley, 1976, MAP Publications (UK), $3.50 from META.

"In this book the sharpening of tools in general use is dealt with, and, whenever possible, simple and well-tried methods have been adopted, bearing in mind that usually the aim when sharpening a tool should be to restore, as accurately and as consistently as possible, the original form of the cutting edge."

Bradley begins with an introduction to the materials and equipment used in sharpening. Then he explains (with illustrations) the proper sharpening techniques to use with metal-working tools—lathe cutting tools, shears, drill bits, and other tools used in boring. This is followed by sharpening techniques for woodworking tools—planes, saws, chisels, and drill bits—and some common household tools such as knives and scissors.

When a sharpening "stone is used dry, it will soon become filled with metal particles and...have little abrasive action...water or oil is applied to the stone to enable the metal dust to be carried away..."

"The four common forms of cold chisels...are generally sharpened on the grinding wheel, [although] when the edge is but little blunted it can readily be restored on a coarse emery bench stone."

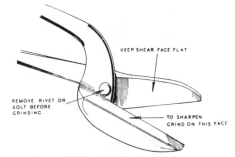

KEEP SHEAR FACE FLAT

REMOVE RIVET OR
BOLT BEFORE
GRINDING

TO SHARPEN
GRIND ON THIS FACE

Sharpening metal shears

A Manual on Sharpening Hand Woodworking Tools, large booklet, 48 pages, by J.K. Coggin, L.O. Armstrong, and G.W. Giles, $1.00 from The Interstate Printers and Publishers, Inc., Danville, Illinois 61832, USA.

Written and published as a shop manual for students in woodworking and industrial arts classes in rural schools. Drawings and simple instructions show how to grind and sharpen chisels, plane irons, saws, augurs, knives, axes, and screwdrivers. Many of the illustrations are clear and complete enough to be used without the text. Includes simple explanations of types of steel used in hand tools, and an illustrated glossary of sharpening terms.

An excellent low-cost teaching tool and reference.

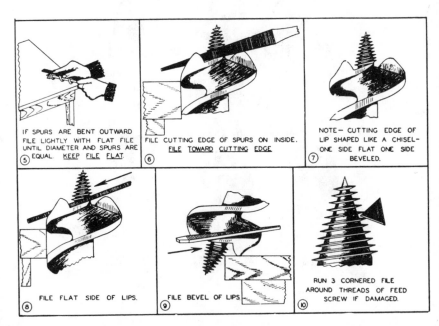

IF SPURS ARE BENT OUTWARD FILE LIGHTLY WITH FLAT FILE UNTIL DIAMETER AND SPURS ARE ⑤ EQUAL. KEEP FILE FLAT.

FILE CUTTING EDGE OF SPURS ON INSIDE. FILE TOWARD CUTTING EDGE ⑥

NOTE— CUTTING EDGE OF LIP SHAPED LIKE A CHISEL— ONE SIDE FLAT ONE SIDE ⑦ BEVELED.

FILE FLAT SIDE OF LIPS. ⑧

FILE BEVEL OF LIPS. ⑨

RUN 3 CORNERED FILE AROUND THREADS OF FEED ⑩ SCREW IF DAMAGED.

How to Work Sheet Metal, book, 142 pages, by Herbert J. Dyer, 1963, $3.50 from META.

This book outlines methods of sheet metal working using techniques and equipment that stress low cost efficiency. Riveting, soldering, brazing, and other metal joining techniques are discussed as well as sheet edging and shaping methods. Includes sections on equipment, materials, and metalworking machines. Also included are dimensional drawings of a few basic sheet metal tools and equipment.

A beginner may at times have difficulty following the text; but, on the whole, an excellent book.

HOLLOWING AND BLOCKING

A typical block — Start about ¼ in. from edge and work inwards — Ease out crinkles and carry on in circles

Raise the angle and hit square — Work the bottom right in to centre — Change ends with hammer and block all over again to final shape

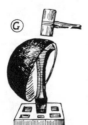

Work all over with mallet, using domed stake — Repeat previous operation, using flat faced hammer rubbed clean on emery cloth — Unless hammer is clean every little dust particle on it will be shown with each blow

The Procedure Handbook of Arc Welding, book, 630 pages, by The Lincoln Electric Company, 1973, $5.00 plus $1.00 postage, from The Lincoln Electric Company, 22801 St. Clair Ave., Cleveland, Ohio 44117, USA.

This remarkably thorough and detailed book tells you probably everything you would ever want to know about small and large scale electric arc welding. It "is directed toward those people who have day-by-day working interest in arc welding—to the supervisory and management personnel of fabrication shops and steel erection firms; to welders and welding operators; to engineers and designers; and to owners of welding shops. The editorial aim has been to be practical—to present information that is usable to those on the job." The authors have also attempted to make the text as understandable as possible to the beginner.

Following an introduction to the fundamental principles of electric arc welding, the topics covered include preheating, relieving stress, welding different types of metals, safety, and welding underwater, in addition to power sources, equipment and supplies for arc welding.

There is an extensive reference section containing data on weights, hardness of different materials, and etching methods. There are many illustrations and a good index. Readers should have some understanding of basic welding methods.

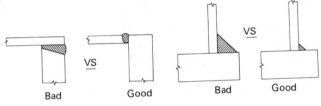

Size of weld should be determined with reference to thinner member

Bearing Design & Fitting, book, 80 pages, by Ian Bradley, 1976, MAP Publications (UK), $3.00 from META.

"Although the subject of this book is complex and covers a very wide field, an attempt has, nevertheless, been made to deal with the main principles involved and, at the same time, to furnish examples of bearing design and application that may be found of use particularly in the small workshop."

Bearings are used whenever something rotates or slides against something

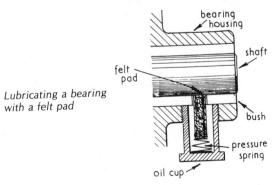

Lubricating a bearing with a felt pad

else and it is desirable to reduce friction and wear. This book is good on the design and production of metal bearings. Many of the ideas could be applied to other materials, such as wood, out of which bearings are often made in developing countries. The book covers casting of bearings from metals and plastics, machining bearings, design, lubrication, different types, and maintenance and repair. There are many useful tips and illustrations. A very useful book for any community workshop. See also **Oil Soaked Wood Bearings: How to Make Them and How They Perform**, on page 111.

Spring Design and Calculation, book, 37 pages, by R.H. Warring, 1973, $2.25 from META.

This book tells you how to make metal springs. Topics covered include spring materials and the following kinds of springs: flat, helical, tapered helical, torsion, clock, constant force, and multiple leaf. Wire sizes and other details of spring construction are also discussed. Simple algebra is needed to design springs using this book.

"Spring design proportions are not something that can be 'guesstimated' with any degree of accuracy—and trial-and-error design can produce a succession of failures. Thus this book on spring design is full of formulas, as the only accurate method of predicting spring performance. However, all are **practical working formulas**; and all are quite straightforward to use."

This is not a book that shows specific spring making procedures—it just covers the **designing** of springs. Once you master the simple math you should have no problem using this book to design springs of good quality. There are numerous illustrations and charts that will help the reader more fully understand the methods and principles described.

TAPERED HELICAL SPRINGS

largest active coil
bottoms first

solid height

Gear Wheels and Gear Cutting, book, 94 pages, by Alfred W. Marshall, 1977, MAP Publications (UK), $3.00 from META.

"An elementary handbook on the principles and methods of production of toothed gearing." The author tells why and how to design and make gears. He discusses gear principles and tooth shapes, a wide variety of gears (including bevel and chain gears), and cutting gears on standard metal-working machine-tools. A knowledge of basic geometry is necessary in order to use this book.

Making gears is a complex task that requires a workshop with at least a metal-working lathe and drill press. This is a good guide to the process.

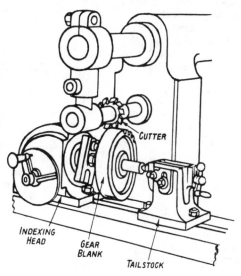

Gear cutting in a milling machine

How to Mill on a Drill Press, Plan No. X422A, 5 pages, 1969, $1.25 from Popular Mechanics.

MATERIALS: bolts, angle iron, strap iron, bearings, steel rod

PRODUCTION: metal turning lathe to make tool holders which support cutting blades; tools to cut, drill, and tap steel

This article includes plans for cutting attachments that can be added to a metal working drill press to enable you to do milling work. It also describes the techniques to use.

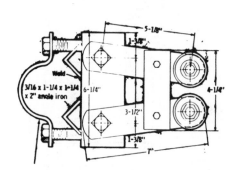

Spindle support attachment

How to Use Metal Tubing, Plan No. X422, 4 pages, 1956, $1.00 from Popular Mechanics.

This article contains lots of valuable hints with drawings, on how to bend, cut, connect, solder, enlarge and generally handle metal tubing. There is a good description of soldering soft copper tubing joints. Relevant for plumbing, solar water heaters, steam engines, and other uses.

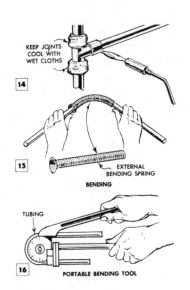

BENDING

PORTABLE BENDING TOOL

How to Work with Copper Piping, Plan No. X198C, 4 pages, 1974, $1.00 from Popular Mechanics.

Good illustrations and text for 'sweat' soldering copper pipe joints. Notes on tools and techniques for cutting copper pipe.

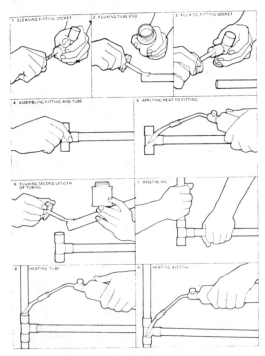

Ten steps for assembling copper piping

Electric Motor Test and Repair, book, 160 pages, by Jack Beater, 1966, $7.00 from META.

This is a guide to rewinding and testing single and poly-phase, plain and split-loop small-horsepower electric motors. Winding diagrams show the sequence and number of coils to be wound into armature and stator slots, and simple schematics show electrical connections to commutator and field windings. Accompanying text gives full instructions: "...any armature with an even number of slots can be wound in the same general manner. The first coil is started in the slot selected as number 1 and comes back in slot 7, then back through slot 1 and so on around until the correct number of turns have been placed. The wire is now cut at the commutator end, leaving ample length to reach the proper commutator bar with an inch or two left over..."

Testing and winding equipment, expensive or unavailable in many areas, can be made by the repairman himself. The author explains how to build a motor test panel, simple hand-operated armature and stator coil winding machines, devices for taping and packing coils, and gear and pulley puller plates. He also discusses the use of small lamps, hand compasses, and home-made induction devices to test armatures and stators. Other useful ideas and information: reversing motor rotation direction; rewinding automobile generator armatures; building a dipping tank and baking hood for application of coil insulating varnish.

This book requires a basic understanding of electric motors. Bound with a durable cover and packed with ideas for improvising equipment, it could be used wherever motors are being repaired.

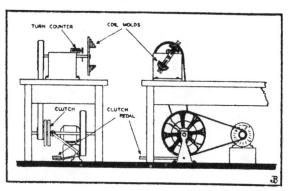

Two views of a coil group winding machine; construction described in the text

Electroplating for the Amateur, book, 106 pages, by L. Warburton, 1963, MAP Publications (UK), $3.50 from META.

Electroplating is a process in which electricity is used to produce a protective coating on metal parts. The author attempts "to provide the amateur engineer with what is hoped will be sufficient data, not only to carry out successful electroplating in the small workshop, but also to provide himself with the essential tools of the trade, i.e., the electrical equipment and plating tanks... reduced to their simplest forms without serious loss of efficiency...The only plating plant obtainable is on a far bigger scale than anything required by even the most enthusiastic amateur." Thus, there is "a detailed discussion of a suitable size of plant, together with details as to how such a plant can be assembled in the small workshop.''

The book covers electrical principles and procedures; the plating tank; chemicals; preparation of surfaces to be plated; electrolytes; chromium, copper, nickel and silver plating; anodizing aluminum; and some other techniques.

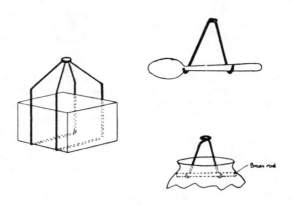

Wire slings for electroplating

Try Your Hand at Metal Spinning, Popular Mechanics Plan X420A, 5 pages, by Sam Brown, 1954, $1.25 from Popular Mechanics.

This is a set of directions for making bowls out of aluminum by bending it into the proper shape on a lathe. No cutting is involved. ''If you begin with soft aluminum and work it over a simple form you can spin a bowl in less than five minutes after the job is set up. Aluminum spins very easily and does not tend to score or buckle under the forming tools.'' Drawings and photos illustrate the techniques and special tools needed (simple to make). Requires 16- to 22-gauge aluminum. The lathe has to operate at about 900 rpm.

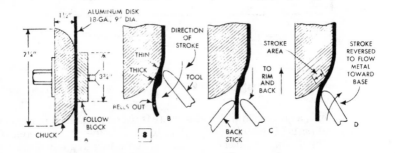

Spinning aluminum bowls on a lathe

A Blacksmith's Bellows, plans, 23 pages, by A.R. Inversin and D. Sanguine, 1977, available in exchange for your publications, from Appropriate Technology Development Unit, P.O. Box 793, Lae, Papua New Guinea; or $2.30 in South Pacific region, $3.50 in Asia, Africa and Latin America, $5.00 in Australia, Europe and North America, from South Pacific Appropriate Technology Foundation, P.O. Box 6937, Boroko, Papua New Guinea.

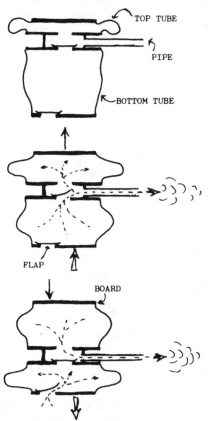

MATERIALS: car or truck inner tubes, wood, sheet metal, nails, 2' length of ¾'' diameter steel pipe.

PRODUCTION: hand tools.

The South Pacific Appropriate Technology Foundation (SPATF) has been organized by the government of Papua New Guinea to develop and promote the use of technologies encouraging individual and village self-help and self-reliance. This booklet, one of SPATF's "how-to-do-it" publications, shows how to construct a hand-operated double-action bellows. Rubber from old inner tubes is used for the flap-valves as well as for the bellows themselves. The simple step-by-step instructions are accompanied by large, clear drawings and an explanation of how the finished mechanism works.

The design in this booklet could be built or adapted at very low cost for any kind of blacksmithing.

Lathe Sanders, Plan No. X388, 2 pages, 1949, $.50 from Popular Mechanics.

This article provides ideas for making simple disc and drum sanding attachments for use with a woodshop lathe. Also shows another drum sander powered by an electric drill.

Sanding on a lathe

Scroll Saw, Plan No. X594, 5 pages, 1945, $1.25 from Popular Mechanics.

This saw is similar to a jig saw, with a narrow, reciprocating blade. The plans have to be studied carefully to be fully understood. This design is made of hardwood and a variety of small metal parts from old automobiles. Some cutting, drilling and tapping steel is required. Uses a ¼ hp electric motor. Appears to be a sturdy machine.

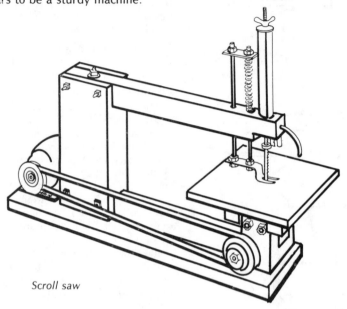

Scroll saw

Motorize Your Hacksaw, Plan No. X334, 2 pages, 1952, $.50 from Popular Mechanics.

''If you have a small metal working shop or use steel bar or shafting to some extent in your home workshop, motorizing a hand hacksaw will save hours of work and can be done at a fraction of the cost of a commercial power hacksaw. The inexpensive drive unit consists of an 8- or 10-inch v-pulley and shaft, a connecting rod and a guide rod, a vise or clamping arrangement to hold the work and a suitable wooden base. When needed for handwork, the saw can be removed from the unit in a few minutes.'' Uses a ¼ hp electric motor.

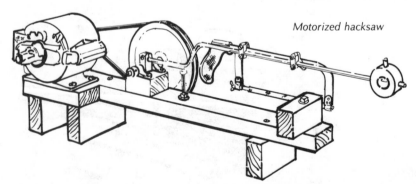

Motorized hacksaw

Two Speed Bandsaw Cuts Wood and Metal, Plan No. X37, 7 pages, 1951, $1.75 from Popular Mechanics.

MATERIALS: wood, water pipe, pipe fittings, angle iron, steel rod, bearings, metal pulleys, wood pulleys, small metal parts, v-belts, bandsaw blades, small electric motor

PRODUCTION: some welding, cutting, drilling steel

This machine can be used to cut wood **or** metal, by shifting v-belts between pulleys to change the speed of the blade. "It has every essential feature of the average dual purpose type machine." The frame is made of water pipe and fittings, while the band wheels are made of hardwood.

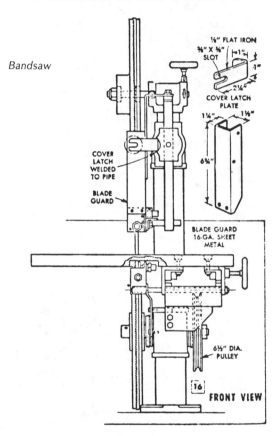

Bandsaw

Wood Planer for $100, Plan No. 802B, 9 pages, 1970, $2.25 from Popular Mechanics.

MATERIALS: bicycle chains and sprockets, steel bar, steel tubing, angle iron, ½'' steel plate, ¾'' steel shaft, water pipe, bearings, small fittings, pulleys, ¼ hp and 1½ hp electric motor

PRODUCTION: metal lathe, drill press, arc welder, tools for cutting steel

This is a workshop machine for planing wood to a specified thickness. Metal working tools are needed to do a lot of precision work to make this machine. Cost $100 for materials in 1970. Useful in converting scrap, low-grade, or recycled lumber into more valuable boards.

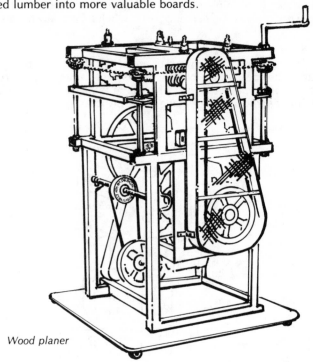

Wood planer

Sheet Metal Former, Plan No. 609, 7 pages, 1966, $1.75 from Popular Mechanics.

MATERIALS: aluminum alloy or steel plate, 1 5/8″ steel rod, ½″ steel rod, gears, channel iron, bearings, small metal fittings.

PRODUCTION: metal working drill press, lathe and milling machine or attachment.

This hand-operated tool allows you to make perfect cylinders in any diameter from 1 5/8″ up. It will handle 20-gauge sheet iron up to 12″ wide, and thicker pieces of softer metals.

Two important pieces are cut from aluminum alloy plate; production will be more difficult if steel is used. This tool will require a lot of precision metal work to make.

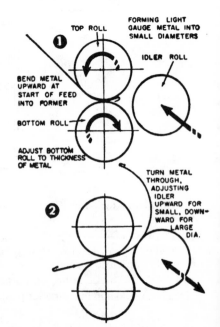

Sheet Metal Brake, Plan No. X606, 8 pages, 1964, $2.00 from Popular Mechanics.

MATERIALS: plywood, concrete, short steel bars and angle iron, steel roller, small fittings, sheet steel

PRODUCTION: metal working tools, including a lathe, are required

This is a valuable, versatile, simple workshop tool for quickly and accurately bending sheet metal. For use in workshops where a lot of sheet metal bending is done, or where precision is important. Hand operated. The tool is 18 inches wide and can bend up to 20-gauge sheet metal the full width, or thicker narrower pieces.

"By using the proper forming block or mold, you can bend sheet metal to any angle, make radius bends, reverse bends, and seams." Part II describes techniques for the effective use of the tool.

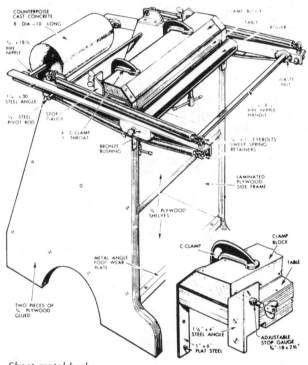

Sheet metal brake

Metal Turning Lathe Built from Stock Parts, Plan No. X387, 4 pages, 1959, $1.00 from Popular Mechanics.

MATERIALS: 1½" pipe and fittings, hardwood, 1/8" steel plate, bearings, steel angle, pulleys, ¼ hp electric motor, steel rod, and small fittings

PRODUCTION: cutting, drilling, and tapping steel, plus some milling and lathe work to make some of the small parts

This metal working lathe is not a precision tool. It can accept work up to 4½" in diameter and 10″ long. Standard pipe and fittings are used to form a frame on which the rest of the lathe is fitted. Precision metal work is required to make this lathe.

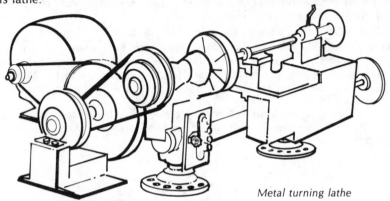

Metal turning lathe

Smelting Furnace, Popular Mechanics Plan X297, 5 pages, by E. R. Haan, 1964, $1.25 from Popular Mechanics.

MATERIALS: fire clay or refractory cement, firebrick, two sheet metal cans, copper tubing, pipe and fittings

PRODUCTION: hand tools for working with tubing and pipes

"With this small furnace you can smelt aluminum, brass and copper; preheat small, thick pieces of iron and steel for brazing or forging; caseharden soft steel; make up alloys...You can use either LP or city gas. The cost is about $25 (1964 prices)." The furnace is about 17 inches high and 12 inches in diameter; it holds a 3-inch diameter crucible. Clear photos and drawings with the text show how to make and operate the smelting furnace. A vacuum cleaner is needed to supply forced air.

This might be of use in a small workshop where casting work is occasionally done.

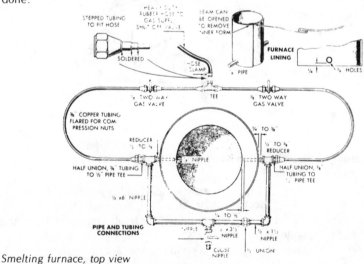

Smelting furnace, top view

Hand-Powered Cement Mixer, Technical Bulletin No. 30, 9 pages, $1.00 from VITA.

Simple drawings and text for the construction of a hand-cranked cement mixer made from a 55-gallon drum with a pipe framework.

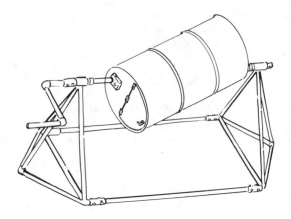

Work Tools, book, 80 pages, by Jehuda Friedmann, 1977, $6.95 from ISBS.

This is a multi-lingual dictionary of hand tools, with more than 300 drawings and the associated tool names in twelve different languages (Arabic, Danish, English, Finnish, French, German, Hebrew, Icelandic, Italian, Russian, Spanish, Swedish).

Stocking Spare Parts for a Small Repair Shop, VITA Technical Bulletin No. 2, 4 pages, by Phil Cady, P.E., $1.00 from VITA.

Basic good advice for systematic stocking and record-keeping for parts.

Equipment for Rural Workshops, book, 94 pages, by John Boyd, 1978, ₤2.95 from ITDG.

This book "is intended to help people choose appropriate tools and equipment. It is **not** an instructional textbook on workshop technology."

Shows workshop building layout and basic sets of tools (primarily hand tools) for 1-6 man workshops, without and with power supply, for woodworking or

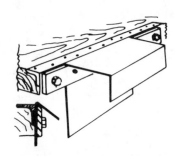

*Simple bench fitting
for folding sheet metal*

metalworking. The simplest level of powered equipment requires an electric drill with attachments to convert it into a circular saw, grinding wheel, jig saw, and power hacksaw. Machine tools are shown for the larger, powered, 4-6 man workshops.

The author notes that "the hand tools in the lists of basic equipment can be used to do the same work as the much more costly power tools...Power tools only speed up the work and are not economic unless there is enough work to keep them in use for a substantial part of each day."

Includes mid-1977 prices of tools. Lists suppliers in Asia/Africa/Latin America. Many illustrations and photos.

Hard-To-Find Tools and Other Fine Things, catalog, 70 pages average length, published quarterly, free from Brookstone Company, 127 Vose Farm Road, Peterborough, New Hampshire 03458, USA.

A commercial catalog, with photos, offering a wide variety of unusual tools. Although some of the listings are expensive gimmicks, most of the tools are of high quality.

ADDITIONAL REFERENCE ON WORKSHOP TOOLS
Construction Manual for a Cretan Windmill contains plans for a pedal-powered wood turning lathe; see review on page 598.

AGRICULTURE

Agriculture

"The effort of the West to make the big landowner the main food producer of the underdeveloped countries has clearly failed."

—E.G. Vallianatos, **Fear in the Countryside: The Control of Agricultural Resources in the Poor Countries by Non-Peasant Elites**

Productive agricultural land is the most fundamental resource of all rural communities and nations. An agriculture which forms a basis for rural and national self-reliance in food production depends upon equitable distribution of this resource. Without secure access to land, the tenant farm family is not in a position to carry out many of the long-term improvements (such as terracing, composting, and tree planting) that may be needed, nor are they in a position to benefit from the multitude of small farm programs sponsored by national agriculture departments and international and bilateral aid agencies. The landless farm laborer is often ignored entirely, though he or she is most vulnerable to unemployment from mechanization. Participation in agricultural production, it has been repeatedly demonstrated, is the only clear guarantee of participation in food consumption.

The concentration of land holdings in a few hands appears to be a major engine of environmental destruction as well, forcing subsistence cultivators onto marginal lands and hillsides. The loss of topsoil that follows is swift and often irreversible. Deforestation becomes a way of life as existence is scratched from land in a final, capital-consuming, desperate process.

The amount of productive land lost to deforestation and desertification is staggering, and the rate of loss is increasing.* A central concern in appropriate agriculture must therefore be a sustainable resource base for agriculture—soil conservation, an assured supply of nutrients, and a buffer from the inflationary costs of inputs that accompany the oil-based agriculture promoted by the rich countries. The elements needed for ecologically responsible agricultural systems exist in most parts of the world. In developing such approaches, indigenous agricultural systems deserve increased attention, for they often reflect important ecological interactions and yield a variety of crops.

In any agricultural system, crop diversity is usually a key to sustainability. There must be a balance between production of cash crops for income, and production of subsistence crops for direct consumption. Cash crops can allow a greater flexibility and access to crucial tools and inputs that would otherwise be unobtainable, and they usually mean a higher value production from a particular piece of land. Yet cash crops often bring with them dependency on global market forces for the sale of produce and for the supply of seeds and

---

The relationship between forestry and agriculture is discussed in the FORESTRY chapter; see especially **China: Forestry in Support of Agriculture.*

fertilizers. They also tend to bring a reduction in crop diversity. All of these factors significantly increase the risks facing farmers. Cash crops can lead to mining of agricultural soils for short-term economic gain, reducing both short-term food supplies and long-term productivity of the land. Cash cropping also contributes to concentration of land holdings, displacement of tenant farmers, and abandonment of traditional social mechanisms of redistribution and collective welfare.

The social aspects of the organization of agriculture are major considerations in the search for appropriate agricultural strategies. What does a new agricultural system do to social relations, the extent to which farm families aid each other, the degree to which extended families continue to take responsibility for all their members? What does a new agricultural system do to the composition and character of rural communities?

When all of these concerns are taken into account, several avenues for appropriate technologies seem evident. There is a need for increased emphasis on intensive food production. Growing fruits and vegetables in home gardens can be done by nearly every family. Relatively high production can be obtained from a small area, and the increased variety in the family diet has clear nutritional advantages.

Farming systems that combine agriculture with forestry bring a varied and higher total production from multiple tiers of plants and trees. Reduced pest problems result, as a more diverse plant environment offers less shelter to pests and more to their predators. More stability of production over time is also assured as differing crops provide protection from weather and market fluctuations. Alternating tree crops with row crops enables a sustainable productive agriculture as protected topsoil and variety of plant life mean that soil fertility can be maintained. The addition of animals, including livestock, fish and bees, into these farming systems can also be important in providing additional food, income and fertilizer.

Renewable energy technologies—including solar greenhouses, water-pumping windmills for small plot irrigation, and solar crop drying systems—also have an important place in appropriate agriculture.

These complementary themes can be found throughout the entries included in this chapter; more synthesis needs to be done in actual programs. This chapter has been divided into general topic areas to better enable the reader to find materials of particular interest. The first three books consider the social, political and economic sides to agriculture, criticizing the conventional narrowly technical approach in rich and poor countries. **Food First** *is a broad-ranging critique that seeks to explode many of the myths about world hunger.* **Fear in the Countryside** *examines the connection between inequitable land distribution in the Third World and the failure of agricultural research institutions to generate agricultural systems and technologies of relevance to rural people.* **As You Sow** *paints a saddening picture of the negative social consequences in small communities that have come with the transition from family farms to huge agribusiness operations in California. This process of decay through growth in land holdings involves a substantial reduction in the number of opportunities for rural people to develop basic business, managerial, and entrepreneurial skills.* **As You Sow** *should be of interest to readers in many countries.*

Small Farm Development: Understanding and Improving Farming Systems in the Humid Tropics *is in many ways the single most valuable volume in this chapter. This highly readable book illuminates the dynamics, characteristics, and constraints of small farms in the tropics; it is required reading for those working on agricultural systems, agricultural tools and equipment, and related* *activities such as farm co-ops.*

The annotated bibliography **Women and Food** *provides access to the literature on women and food production—e.g., research on how changing agricultural technologies affect women's roles and practical information on intensive home gardens that may be used in programs aimed particularly at women.*

Agricultural Extension: The Training and Visit System *documents a low cost extension approach which can help farmers improve their basic practices with almost no cash investment, yet with a high chance of achieving higher production. This approach relies heavily on village-level workers with a low educational background, a strategy somewhat analogous to the use of ''barefoot doctor'' health workers.*

The four following publications (the **IFOAM Bulletin**, *the* **Quarterly Review of the Soil Association, One Straw Revolution,** *and* **Permaculture II**) *all offer insights and practical considerations relevant to the creation of sustainable agricultural systems and agro-forestry combinations. The two journals also provide contact with the worldwide network of enthusiastic and imaginative people working in this field.* **Environmentally Sound Small Scale Agricultural Projects** *helps in thinking about the environmental impact of the various elements of an agricultural project.*

Reference books on soils, seeds, crops and fertilizers are contained within the next sub-section. Soil testing, seed production, recycling of organic wastes as fertilizer, and proper worker protection from pesticides are some of the topics covered here.

Two volumes introduce the technical considerations for small scale irrigation efforts. Irrigation is the biggest single factor in raising farm yields. As its proper planning from a technical and environmental viewpoint can be quite complicated, these are welcome references.

Intensive gardening is the topic of seven entries. The manuals from Bangladesh, Peru and Jamaica are highly recommended references, to go with the **Samaka Guide** *(from the Philippines, reviewed on page 78),and* **How to Grow More Vegetables** *(from the United States).*

Several books on raising small livestock in the tropics (pigs, poultry, rabbits) and a comprehensive volume on the water buffalo have been included. References on the raising, training and use of draft animals can be found in the AGRICULTURAL TOOLS chapter (see **The Draft Horse Primer** *and* **The Employment of Draught Animals in Agriculture**). *The final entry in this chapter is the encyclopedic volume* **Tropical Feeds**, *a unique reference covering nutritional content and uses of 650 tropical feeds, most of them plants.*

In Volume I:

A. Background Reading

Food First: Beyond the Myth of Scarcity, book, 412 pages, by Frances Moore Lappe and Joseph Collins, with Cary Fowler, 1977, $2.95 plus postage from Institute for Food and Development Policy, 2588 Mission Street, San Francisco, California 94110, USA.

"Every country in the world has the capacity to feed itself...food self-reliance is the cornerstone of genuine self-determination." This is the thesis of **Food First**, an extraordinary book that examines the constraints and opportunities facing agriculture in developing countries. Backed by case studies and statistics from all over the world, the authors assert that hunger is not caused by too many people, too little arable land, lack of technology, or overconsumption by greedy Americans. "Inequality in control over productive resources is the primary constraint—on food production and on equitable distribution."

The authors claim that ecological destruction is more closely related to economic exploitation than population pressure. "Soil erosion occures largely because fertile land is monopolized by a few, forcing the majority of farmers to overuse vulnerable soils." Small farmers around the world produce more per unit of land than large farmers; land reform would lead to greater production and equality. And local sources of fertilizers and seeds are essential, because "true food security is further undermined as production is made increasingly dependent on external sources of supply over which there is no local control."

Optimistically concluding that there is plenty of good land and existing technology to enable the world's present population to feed itself, the authors note that 'food self-reliance depends on mass initiative, not on government directives...Self-reliance is not the 'project approach' to hunger." In this context, food aid, controlled by multilateral organizations, "should be used as payment for work that directly contributes to creating the preconditions for food self-reliance."

A story is related that depicts two alternative food production visions. In one, "tens of thousands of entrepreneurial 'farmers' receive corporate credit to raise chickens using hired laborers and all the latest feedgrain and chemical techniques designed to bloat chicks in the shortest time...for a few giant worldwide marketers...In the alternative vision, hundreds of millions of farmers have a few chickens each, in their backyards eating insects and scraps, with, some occasional input such as inexpensive chicken cholera vaccine. In which world...do you think more people are likely to eat chicken? Or agricultural resources less likely to be used raising feed for animals instead of food for people? Or less foreign exchange likely to be lost to the country on imported supplies and through repatriated profits and fees?"

The export of cash crops is identified as one practice that leaves hunger among the producing population. And the plight of the landless is noted: "being excluded from production means being excluded from consumption... People will escape from hunger only when policies are pursued that allow them to grow food and to eat the food they grow."

The authors show an ambivalent attitude towards 'appropriate technology,' claiming that "there can be no separation between technical innovation and social change. Whether promotion of the wealthier class of farmers is deliberate government policy or not, inserting any profitable technology into a society shot through with power inequalities (money, landownership, access to credit, privilege) sets off the disastrous retrogression of the less powerful majority." (A.T. enthusiasts will certainly disagree, claiming that some technologies can improve the relative power of the poor.)

Despite some rhetorical excesses, this book is important reading for those who wish to better understand the causes of hunger around the world.

Fear in the Countryside: The Control of Agricultural Resources in the Poor Countries by Non-Peasant Elites, book, 162 pages, by E.G. Vallianatos, 1976, $16.00 from Ballinger Publishing Company, 17 Dunster St., Cambridge, Massachusetts 02138, USA.

"Central to his book is the thesis that world development and food self-sufficiency can only occur if we succeed in solving the problems of social and economic exploitation of the rural classes by their own elites, frequently professional, technocratic and landowning holdovers of colonial days...He suggests the importance of new strategies geared to encouraging social change, the transfer of 'appropriate' technologies, and the strengthening of the role of the small farmer throughout the world." (From the Foreword)

Taking Colombia as an example, the author notes that the national and international agricultural research institutions are unable to help solve the problems of the small farmer, due to an overly academic orientation, arrogance, lack of understanding of small farmer problems, and concentration on "refining the largely discredited trickle-down technology and knowledge transfer hypothesis." The author also criticizes efforts to alleviate the problems of small farmers through new technology without land reform.

"Agricultural research and institutions in the poor countries are basically irrelevant to local needs," the author claims.

"The philosophical foundation of...research and development centers ought to be based on the active participation and agricultural wisdom of the small farmer...The fact is that the farmer is the best judge of what he does and what he hopes to do better. The farmer will improve his methods of production and accept new technologies only if he has been an active participant in the generation of the new technologies."

An important perspective on technology transfer is given when the author quotes another observer: "...the problem...usually described in the industrial countries as technology transfer...I see more as the problem of a science of one's own versus intellectual colonialism, a social more than a technical question."

Yet in the end the author seems to contradict himself. He recommends massive official U.S. involvement in generating new technology for poor Third World farmers, calling on Congress to "make certain that the Agency for International Development and other U.S. technical assistance institutions transfer to the poor countries only those technologies that do in fact redistribute income at the small farm" level. He overlooks the enormous obstacles that work to prevent U.S. technical assistance agencies from determining what technologies can in fact help to redistribute income in the varied circumstances within and among poor countries. Why should official U.S. agencies be any better at choosing technologies for the poor than the R&D agencies located in the Third World that Vallianatos criticizes? And how could Congress possibly be in a position to oversee this process? The author then suggests that the U.S. use its position as world food bank to "put enough pressure on the food-deficit nations to modernize their agriculture while they reform their social structures." Although the need for human rights, land reform, and major social changes in the poor countries has been evident for many years, U.S. foreign policy makers have always viewed this as a minor concern compared to global questions of political and economic power. There is consequently little reason to believe that such a "food weapon" would be used to the benefit of the poor; more likely, it would be used against the very governments that are genuinely trying to redistribute wealth and power among the poor. The author's recommendations for further U.S. self-righteous intervention make a disappointing ending to a thought-provoking, important book.

As You Sow: Three Studies in the Social Consequences of Agribusiness, book, 560 pages, by Walter Goldschmidt, 1978, $7.95 from Allanheld, Osmun & Co., Publishers, 19 Brunswick Road, Montclair, New Jersey 07042, USA.

For the past several decades, American agriculture has been held up as a model for poor countries. This approach has been criticized for many different reasons. **As You Sow** documents the negative social consequences, within the U.S., of an agriculture that increasingly depends on large-scale farms. Goldschmidt notes, for example, that the number of skilled people in communities with small farms is much higher that in communities with a few large farms. Small farms allow the widespread development of entrepreneurial and management skills that are essential to the development of other rural enterprises. Large farms restrict this process, concentrating management and business learning opportunities in the hands of a few.

Interesting documentation of the relationship between patterns of land ownership and the vitality of rural communities.

Small Farm Development: Understanding and Improving Farming Systems in the Humid Tropics, book, 160 pages, by Richard Harwood, 1979, $16.50 from Westview Press, 5500 Central Avenue, Boulder, Colorado 80301, USA.

The author states, ''In our impatience with 'backward' small farmers and in our haste to rapidly 'commercialize' them, we have overlooked key aspects of their farming systems that could enhance efforts to increase food production and improve rural well-being. To accomplish the development of a greater number of the world's small farms, shifts in emphasis must be made in our thinking, in our technological research, and in our communications with farmers.''

Better understanding and analysis of the bulk of the Third World's small farm production systems is the theme of this important book. The author discusses with great depth and sensitivity the issues and options facing resource-limited small farmers in the tropics. He suggests that a ''purposeful blending of traditional and modern technologies may well prove the key to starting the most disadvantaged farmers along a more rapid development path.''

In the first part of his book, Harwood presents an overview of small farms from subsistence hunting-gathering to primary mechanized operations. He endorses a development approach of scientists, extension workers, and farmers working in close cooperation in farming areas. ''The agricultural development specialist must remain constantly aware of—and on guard against—the natural tendency to superimpose his own values on those of the farmer. The reality that faces the farmer who ekes out his existence from a mere half-hectare of poor land can only be understood—if it is seen as he sees it.''

The second part of the book reviews critical factors in small farm develop-

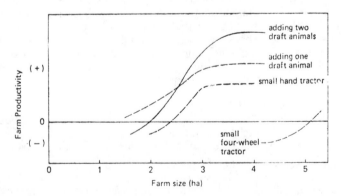

Figure 10.
Change in total farm productivity through addition of various power sources in area where the growing season is six to eight months.

ment which are often overlooked or given little emphasis in development programs. Some examples of these factors include:

- **Animals in Mixed Farming Systems:**
"Despite the almost universal interest of farmers in mixed crop-animal systems, professionals in both crop and animal production commonly pursue research in pure crop systems or pure animal systems, without reference to the interactions between the two that increase the productivity of both. Fortunately, most farmers have no such inhibitions or prejudices. Science should do more for them."

- **Noncommercial Farm Activities:**
"Fencerows are often used for noncommercial plantings as well as for their primary functions as field boundaries, enclosures for containment or exclusion of grazing animals, and erosion controls. There is evidence to indicate that the plant diversity and permanence of the fencerow makes it a refuge for beneficial insects and predators. The relative rarity of pest outbreaks in highly diversified small farm areas where hedgerows and farmyard plantings are extensively used may be due to the net benefits of these traditional features."

Other chapters deal with resource and economic limitations of intensive and multiple cropping systems; economic determinants and resource optimization of micro-enterprises; farm mechanization requirements; and stability in farming systems. An excellent annotated bibliography is also included.

Technical charts and graphs are balanced by photographs of farm families at work. All in all, this book is a fine blending of reasoned arguments for new directions in agricultural development projects. It should receive wide circulation among agriculturalists and development workers concerned with agriculture in the humid tropics of the Third World.

Women and Food: An Annotated Bibliography on Food Production, Preservation, and Improved Nutrition, 47 pages, by Martha W. Lewis, 1979, free to serious groups and individuals concerned with women's development issues, from Office of Women in Development, Room 3243, New State, Agency for International Development, Washington D.C. 20523, USA.

This annotated bibliography "was prepared for program planners working to help women in small scale agriculture and family food production. It describes material that should help raise the level of understanding of the crucial role women play in food producing and gathering. It is directed to encourage rural development workers to appreciate the job that home gardening can do to improve the nutrition of the family and the economy of the community. And it presents information on practical and useful manuals and guides."

"Some of the materials are unpublished papers or publications no longer in print. Sources for publications available and suggestions of resources for additional materials and information are provided."

The author of this directory, Martha Lewis, is a "hands-on" horticulturalist and educator, who has served as a consultant to small-scale gardening projects in Jamaica, Somalia and elsewhere, where she has worked directly with rural women. In this directory, she has thoughtfully compiled practically all the information available specifically on women and food from around the world. It is a document which deserves wide circulation.

Historically in the Third World, "women have been responsible for nearly all stages of food raising and preparation while dominating marketing and processing as well." Yet new agricultural technologies and cash crops have mostly been directed towards, and later controlled by, men. "In countries such as Jamaica, with severe unemployment, low wages for semi- or unskilled labor, a great deal of untilled land, and a high percentage of women-headed households, the answer to survival for these families must lie in subsistence food production, small animals, and the family vegetable garden...Greater attention to garden crops and to marketing of fresh vegetables and fruits should be a priority in any planning for rural development."

Agricultural Extension: The Training and Visit System, booklet, 55 pages, by Daniel Benor and James Harrison, 1977, free from Publications, World Bank, 1818 H Street, N.W., Washington D.C. 20433, USA.

The Training and Visit System "has been put into operation in areas where the need is to improve the level of agricultural production by large numbers of farmers cultivating mostly small farms using low-level technology and usually traditional methods...The cost to farmers is very small...for the initial focus...is on the improvement of basic agricultural practices (such as good seed selection, seedbed preparation, better cultivation, and weeding) which require more work but little cash and brings sure results...The smaller cultivators, who have an abundant supply of labor, may benefit at least as much as the larger farmers."

This low cost extension system "uses village-level workers with comparatively low educational standards supported by subject matter specialists..."

"In the Seyhan project in Turkey, farmers increased cotton yields from 1.7 tons to over 3 tons per hectare in three years. In Chambal, Rajesthan (India), farmers increased paddy yields from about 2.1 tons to over 3 tons per hectare in two years. Combined irrigated and unirrigated wheat yields in Chambal, Madhya Pradesh (India), rose from 1.3 tons to nearly 2 tons per hectare after one season and have since risen higher."

The author describes the common problems with extension programs: multiple roles (not just agricultural) expected of the extension worker, excessively large area of assignment for each worker, and theoretical pre-service training with no in-service training.

For a reformed extension service, the author recommends that extension workers report directly and only to the agricultural department, spend full time on agriculture, and make regular visits to farmers. "Contact farmers must be willing to try out practices recommended by the extension workers and be prepared to have other farmers visit their fields. But they should not be the community's most progressive farmers who are usually regarded as exceptional" and are not often followed by their neighbors.

After the simpler field management practices have led to higher incomes, extension workers should recommend to farmers "the minimum quantity of fertilizer which would noticeably increase their net yields and incomes, and teach the farmers how to make the best use of this amount—for example, when and how to apply it, and how to combine it with organic fertilizers."

"To remain effective, extension must be linked to a vigorous research program, well-tuned to the needs of the farmers. Without a network of field trials upon which new recommendations can be based and without continuous feedback to research from the fields, the extension service will soon have nothing to offer farmers, and the research institutions will lose touch with the problems real farmers face."

There are some significant elements missing from this innovative approach to a low cost extension system. Farmers are not directly involved in the research program, only in final field trials. Participation in defining problems and seeking solutions is missing here. And, what is being recommended is a package of practices which eventually consist of chemical fertilizers and pesticides—to the neglect of organic fertilizers, natural pest control, and practices which protect the long-term fertility of the soil. Taken literally, the author's advice to avoid recommending any activity that does not directly lead to increased farmer income means that long-term strategies to maintain soil fertility and prevent erosion may be sacrificed for short-term profit. Despite the severe need for greater current food production, Third World agriculture departments must not forget that their soils will have to continue to feed their people 50 years from now.

IFOAM Bulletin, journal, 4 times each year, $5.00 per year, from the International Federation of Organic Agriculture Movements, c/o Dr. Hardy Vogtmann, Postfach 4104, Oberwil, Switzerland.

This is the technical bulletin of a very effective organization linking scientists, farmers, students and supporters of what is called "organic, biological, sustainable, permanent, or ecological" agriculture. This group does reputable research on various agriculture topics such as integrated pest management, humus and soil fertility, and lunar influences on plants. They combine this with practical outreach and communication with successful groups and individuals worldwide. Inspiring ideas and individuals abound in IFOAM's ranks.

The Spring 1978 issue contained the following articles: "Ecofarming in Rwanda and Tanzania", "Third World Technology for the United States", and "Organic Farming Research in Europe: Effects of Agricultural Practice on Soil and Plant Quality".

Quarterly Review of The Soil Association, quarterly newsletter, £8/year from The Soil Association, Walnut Tree Manor, Haughley, Stowmarket, Suffolk, IP14 3RS, United Kingdom.

This is the newsletter of a world-wide charitable organization dedicated to the promotion of a fuller understanding of the vital relationship between soil, plants, animals, and humans. From the newsletter's world-wide contributors, a subscriber gains the insights and contact with prominent pioneers and advocates of permanent, biological agriculture. Recommended for agriculturalists and students interested in ecology, organic farming & gardening, and nutrition & health. The reviews of books, abstracts and articles are excellent, and not as technical as the **IFOAM Newsletter.** Some excerpts:

"Known to many as 'The Mother of Mulch,' Ruth Stout is now 93 years old and continues to garden the way she has for more than 30 years—with time left over to write and lecture about her methods. What's her secret? By using a year-round mulch on her garden she avoids the laborious ploughing, harrowing, hoeing, weeding, watering, fertilizing, composting, poisoning and cultivating that other gardeners spend their time and energy on."

"Nowhere is the neglect of poor man's crops greater than in the tropics—the very area where food is most desperately needed. The wealth and variety of tropical plant species is staggering, but most agricultural scientists are unaware of their potential. This neglect...occurs largely because the major scientific research centres are located in temperate zones."

One Straw Revolution, book, 181 pages, by Masanobu Fukuoka, 1978, $7.95 from RODALE.

This thought-provoking book is considered a classic text for advocates of what has been called "natural farming" or "permaculture" (see review of **Permaculture II** in this section). The author was trained in microbiology, specializing in plant disease, in industrializing pre-World War II Japan. His studies stressed high inputs of energy, capital, and chemicals to control and if necessary, combat, natural forces. He began to question the wisdom of these practices, and returned to his village to try an alternative approach. Over the years, Fukuoka, through painstaking observation and experimentation, developed a method of farming which ''closes the gap between agriculture and naturalism.'' This method mimics the natural succession of plant communities and the self-regenerating aspects of ecosystems. He claims that farming units can produce food and fiber in an almost effortless fashion without chemicals or cultivation.

This low-energy system of agriculture contains the following four principles:

- **No cultivation**—do not turn the soil over, and so avoid injuries that divert productive activity;
- **No chemical fertilizer or prepared compost**—let the plants and animals that make the soil go to work on the soil;
- **No weeding by tillage or herbicide**—use the weeds; control them by natural means or occasional cutting;
- **No dependence on chemicals**—insects and disease, weeds and pests, have their own controls—let these operate, and assist them.

One Straw Revolution is a very readable book, with photos of the author practicing his techniques in the fields. While it is inspirational, some caution should be used in considering its relevance to tropical and Third World countries. First, Fukuoka has successfully practiced his ''natural farming'' only in temperate climate Japan. Attempts to make the system work in North America are yet inconclusive. We have heard of no attempts to promote Fukuoka's system in the tropics.

Second, the systems requires a great deal of patience, perseverance, and knowledge—possibly only gained by years of experience. Most traditional Third World farmers do not have the margin of error for experimentation available to nonconformists in developed countries. The immediate problem for most farmers is one of survival, not sustained yields. However, these farmers often do have highly evolved systems of cultivation and extensive traditional knowledge about soils, plants, and local ecology. Quite often, they do practice minimum tillage and marginal use of chemicals. Perhaps a dialogue between concerned scientists, development field workers, small farmers, and natural farming advocates could lead to further refinements and broader applications for farming systems such as this one.

Ideas such as those proposed in this book may be seen by many today as wild and unrealistic. Still, Fukuoka's methods may yet prove to be the last straw if the world's heavily subsidized and centralized food and energy systems were to crumble.

Permaculture II: Practical Design for Town and Country in Permanent Agriculture, book, 150 pages, by Bill Mollison, 1980, $10.95 from International Tree Crops Institute, Box 1272, Winters, California 95694, USA, or Australian $8.00 from Tagari Books, P.O. Box 96, Stanley, Tasmania, 7331, Australia.

Permaculture II is the second, more practical volume, in a series of two fascinating publications that present an approach to permanent agriculture. These books are based on the author's experience in rural Tasmania and the semi-arid areas of Australia. He and his family are part of an intentional community practicing self-reliance in food, energy, and shelter. "Permaculture" is "primarily a consciously designed agricultural system...a system that combines landscape design with perennial plants and animals to make a safe and sustainable resource for town and country. A truly appropriate technology giving high yields for low energy inputs, and using only human skill and intellect to achieve a stable resource of great complexity and stability."

The author argues for species-diversity in combined agricultural-forestry systems, in place of the energy-intensive mechanized monocultures that are standard in developed countries (and increasingly in developing countries). His book is an impassioned appeal, with numerous design sketches, references, and anecdotes to back up his points. "Without permanent agriculture there is no possibility of a stable social order. We can see the departure from productive permanent systems, where the land is held in common, to annual, commercial agricultures where land is regarded as a commodity. This involves a departure from a low to a high-energy society, the use of land in an exploitive way, and a demand for external energy sources, mostly provided by the Third World."

Permaculture II builds upon the philosophy of Fukuoka and his book **One Straw Revolution** (see review in this section): "of working with, rather than against nature; of protracted and thoughtful observation rather than protracted and thoughtless labor; of looking at plants and animals in all their function, rather than treating any area as a single-product system."

This is essentially a design and planning workbook which provides practical details of how plant, animal, and human communities can be organized as a

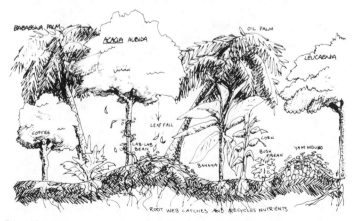

FIG. 5.26: TROPICAL STRATEGIES OF PLANT STACKING AND NUTRIENT RECYCLING. CROPS ARE MULCHED FROM LEAVES AND BRANCHES OF TREES, WHICH HOLD THE LEACHED NUTRIENTS IN AN UNDERGROUND ROOT WEB.

unit. **Permaculture II** claims to be a realistic and optimistic, yet not utopian book.

A strong emphasis is given to water resource management and homesite integration into the overall plan. Mollison's group employs a method of soil and water conservation known as the Keyline System, with which unproductive and sterile soils can be rehabilitated. Soils are reconditioned by the use of chisel plows and no-tillage implements where tractors or animal traction is available (since these may have been the cause of compacted conditions), or with deep rooted plants. These efforts, combined with innovative rainwater catchments, contour irrigation dams, ditches, wells, and fishponds, help to provide adequate irrigation water for the next phase of development. Mixed tree crops and field crops are planted successively, as gardens are laid out and kept nourished by plant litter. Planted and built shelters are devised for humans and livestock, and are incorporated into the perennial-based plant community.

This is, of course, an oversimplified account of the Permaculture system, which becomes increasingly complex and organized over time. The author provides only brief overviews of how a Permaculture system might operate in semi-arid and humid areas of the tropics. Since the species selected are applicable to the southern hemisphere and Tasmania specifically, many adaptations would be required before this system could be attempted in other areas. For its insight and inspiration, however, this book deserves wide circulation.

Recommended.

Environmentally-Sound Small Scale Agricultural Projects: Guidelines for Planning, book, 103 pages, by the Mohonk Trust and VITA, 1979, $3.95 from VITA.

A sustainable agriculture must be ecologically sound. Practices that are not will degrade and consume the natural basis of agriculture. This book explains why this is true, discussing basic ecological principles and the implications of human alterations of naturally stable systems. Much of the book shows the importance of water supply, soil, and pest management in good planning.

"What are the effects of using groundwater for irrigation?" "What is pesticide persistence?" These are examples of questions posed and answered, with clear text and line drawings. Questions aimed at the effects of different alternatives are especially useful. For example, when considering chemical pesticides and/or Integrated Pest Management techniques: "Can a species-specific pesticide be used?...Does the project design recognize the possibility that the target species will develop resistance to the pesticide?...Are similar pesticides being used locally for health purposes, such as malaria control?... Are there plants with pesticidal properties which could be used?"

"Development workers are in a position to pass on awareness of environmental concerns to community groups, government planners, village people, farmers, and students. For example, a development worker may use this manual in class to increase students' awareness of erosion control methods and alternatives."

Also included is an "easy-to-use-in-the-field methodology for planning and benefits/costs analysis of small-scale projects". This chapter emphasizes the importance of intelligent questions, readiness to learn from local experience, and flexibility.

However, this book does not focus on the tropical and semi-arid conditions which are found in most developing countries; and it does not provide specific details on any techniques suggested (such as building terraces for erosion control, or monitoring local conditions).

Recommended for general reference.

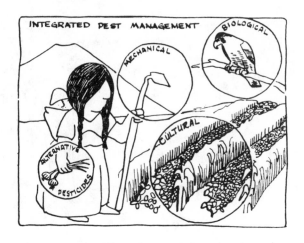

B. Soils, Seeds, Crops and Fertilizers

Better Farming Series, booklets, 29 to 63 pages each, 1977 FAO English edition, $1.50 each from UNIPUB.

Twenty-six titles have been published in this series of handbooks for a two-year agricultural training course. In each case the text is very simple, containing only basic but useful information, and many drawings. The United Nations Food and Agriculture Organization (FAO) has published this English language set. These booklets were originally produced by the Institut africain pour le developpement economique et social (INADES), in French, for use in Africa. (French language editions are available from INADES, B.P. 8008, Abidjan, Ivory Coast.)

1. The plant: the living plant; the root
2. The plant: the stem; the buds; the leaves
3. The plant: the flower
4. The soil: how the soil is made up
5. The soil: how to conserve the soil
6. The soil: how to improve the soil
7. Crop farming
8. Animal husbandry: feeding and care of animals
9. Animal husbandry: animal diseases; how animals reproduce
10. The farm business survey
11. Cattle breeding
12. Sheep and goat breeding
13. Keeping chickens
14. Farming with animal power
15. Cereals
16. Roots and tubers
17. Groundnuts

18. Bananas
19. Market gardening
20. Upland rice
21. Wet paddy or swamp rice
22. Cocoa
23. Coffee
24. The oil palm
25. The rubber tree
26. The modern farm business

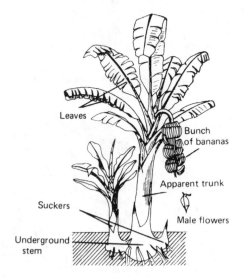

Leaves

Bunch
of bananas

Apparent trunk

Suckers

Male flowers

Banana plant

Underground
stem

Test the Soil First, Popular Mechanics Plan No. X630, 4 pages, 1957, $1.00 from Popular Mechanics.

This article provides a good basic explanation of soil testing, including the preparation of chemical solutions to do the tests and evaluating test results.

No mention is made of local plants which can often be used to measure pH. The author recommends adding chemical fertilizers even when tests for phosphorus, potassium and nitrogen indicate very high levels are already present—a wasteful recommendation. There is also no mention of natural fertilizers or composting.

Guide for Field Crops in the Tropics and Subtropics, book, 321 pages, edited by Samuel Litzenberger, reprinted by Peace Corps in 1976, free to Peace Corps volunteers and development organizations in developing countries, from Office of Information Collection and Exchange, Peace Corps, 806 Connecticut Ave., N.W., Washington D.C. 20525, USA.

"In the tropical and subtropical areas of the world, food grains make up the bulk of the diet for most people. Food grains together with fiber and specialty crops are also principal cash producers. It is with these commodities that this Guide concerns itself...The Guide is designed for use by foreign assistance personnel and cooperators...The text (composed of 40 chapters) is written in layman's language...The first four are general introductory chapters, and treat rather extensively the important subjects of climate, soil, cropping, and farming systems as related to the tropics and subtropics. The other 36 chapters are divided as follows: 6 on cereal crops, 9 on food legumes, 6 on oil crops, 7 on root or tuber crops and bananas, 6 on major fiber crops and 2 on other cash crops. These chapters do not attempt to deal with the factors of providing inputs such as national supplies of fertilizer, insecticides and fungicides."

This manual is quite a balanced textbook for development workers with interests or skills in agriculture. Of special interest are the chapters entitled "The Tropical Environment for Crop Production" and "Farming Systems for the Tropics and Subtropics", which provide useful information on traditional farming models and tropical ecology.

"There is a possibility that the functions of the slow restoration of soil productivity by native vegetation, can be duplicated by man's management of soils without removing them from continued farming. The first step should be to extend the years of continued crop production, by the adoption of technology for individual crops. Such technology is outlined in the 36 chapters on the different crops. An important feature is the addition to soil organic matter by the return of crop residues to the soil, and by the use of manures and compost for producing crops. Adequate fertilization will certainly increase substantially the annual addition of crop roots to the total soil organic matter...A second step when feasible may be to grow green manure crops to restore soil organic matter. These may follow a regular crop, or replace a year of crop production. The green manure crops may be utilized for feeding livestock, but the green manure should be plowed under, so that decaying roots and tops will add to fertility. Small farmers are usually not in a position to grow green manure crops. More appropriate would be for them to produce an economic crop as recent research has shown that with the use of soil amendments most soils can be maintained in food production returning only crop residues to the soil."

This book clearly favors field crops and makes little mention of perennials and agroforestry. We do feel that it can be a helpful supplemental handbook for agricultural students, rural development volunteers and extension agents.

Agro-Forestry Systems for the Humid Tropics East of the Andes, booklet, 25 pages, by John P. Bishop, 1980, available from the author (donations for postage or exchanges suggested), at Estacion Experimental Napo/Centro Amazonico Limoncocha, Instituto Nacional de Investigaciones Agropecuarias, Apartado 2600, Quito, Ecuador.

This is a set of two papers by Dr. John P. Bishop, an agricultural researcher located in Ecuador. Bishop works with traditional farmers, who are called "colonists", "uncontrolled migrants", "shifting cultivators", and other less favorable things. Bishop is convinced that traditional farmers have an understanding of species, soils and ecology that can be put to use in modified "permanent agriculture" models (see review of **Permaculture II** in this section).

The papers are entitled "Integrated Foodcrop, Swine, Chicken and Fuelwood Production", and "Integrated Timber and Cattle Production." The first covers small farmholdings of 1 to 10 hectares. The second describes a supplemental scheme requiring an additional 30-40 hectares. Included are charts of cropping system timelines and systems models.

Since this information comes from monitoring real farms, it could be directly relevant to conditions in the delicate humid American tropics, and of interest to people in other regions of the world.

Growing Garden Seeds: A Manual for Gardeners and Small Farmers, booklet, 30 pages, by Robert Johnson, Jr., 1976, $2.30 from Johnny's Selected Seeds, Organic Seed and Crop Research, Albion, Maine 04910, USA.

The author of this booklet is the founder of a successful small-scale vegetable seed production and distribution company. The booklet is informative and easy to understand and apply. A brief description of the process of selecting, harvesting, and storing seeds is followed by instructions for producing seeds from 33 of the most common vegetables grown in North America and Europe. No special tools, expensive facilities, nor education are necessary to master the techniques described.

"Adaptation, usefulness, and quality characteristics of a vegetable variety can be improved...by selection. The basic type is 'Natural Selection', caused by environmental pressures. For example, in the North in a given year, perhaps only half of the plants of a corn crop will produce mature ears and kernels. Naturally, the ears selected for seed would be chosen from these earlier maturing ears. In this way, Nature forces a crop to either adapt or perish."

"The other type of selection is accomplished by the gardener. For instance, not only would one choose for seed ears of corn which did mature well, but further select the most desirable ear types from what are considered to be the best corn plants." This is of course what traditional farmers have done for centuries in most places.

The main drawback to using this booklet in other parts of the world is that the vegetable varieties are from temperate zones, and many can't be grown in tropical regions except in highland areas. Groups in developing tropical countries could adapt this information to suit their own conditions, by including other crops and consulting with local farmers and extension agents about the best local practices.

Burpee Seeds and Everything for the Garden, catalog, 183 pages, free from Burpee Seed Company, Warminster, Pennsylvania 18991, USA.

The Burpee Seed Company has bred plants and teated seeds for more than 100 years. Their catalog of flower and vegetable seeds, bulbs, and roots, is part of a long American tradition of sales by mail order. The catalog contains short descriptions of the plants (growing season, size, flavor, etc) accompanied by dazzling color photographs. About half the entries are flowers and decoratives; the rest are vegetables: beans, peas, corn, cabbage, greens, herbs, onions, other roots, pumpkins, squash, tomatoes, berries, tree fruits, and others. "How to grow" hints (when to sow, watering and fertilizing, spacing in garden) accompany each of the vegatable sections.

This catalog is clearly intended for the gardeners in temperate zones, particularly in the United States. However, some of the vegetables may be of interest in home gardening programs in other countries.

Tropical Legumes: Resources for the Future, book, 331 pages, by the National Academy of Sciences, 1979, free from Commission on International Relations, JH 215, National Academy of Sciences, 2101 Constitution Avenue, N.W., Washington D.C. 20418, USA.

This book features over 30 members of the **Leguminosae** family of plants, commonly known in English as legumes. These highly valued plants can improve soil conditions and are excellent sources of protein. **Rhizobium** bacteria attached to growths (nodules) on certain legume roots capture nitrogen from the air, which gives the plants the power to grow in areas subject to erosion, low fertility, and other adverse conditions. Root crops, pulses (beans), fruits, forage crops, fast-growing trees, luxury timbers, ornamentals, and miscellaneous species from within this vast plant group are discussed in this well-documented and illustrated text. Brief descriptions of each species—advantages, limitations, and research needs—are provided. There is a very good chapter that illustrates how legumes can be used for green manures, soil reclamation, and erosion control. Also included are charts of comparative nutritional values for the various species; address lists for seed and germplasm sources; and listings of research correspondents around the world.

A National Academy Sciences panel selected each plant on the basis of:
1. Its potential to help improve the quality of life in developing countries;
2. The present lack of recognition of this potential;
3. Its need for greater attention from researchers and farmers, and increased investment by organizations that fund research and development projects.

Some of the more remarkable species include:

African Yam Bean. "This root crop from Africa produces a nutritious seed, as well as edible tubers and leaves. It can be grown in inherently infertile, weathered soils where the rainfall is extremely high. Although highly regarded among people of tropical Africa, the crop is virtually unknown elsewhere. It has received essentially no research attention or recognition from agriculture researchers."

Moth Bean. "An exceptionally hardy South Asian legume that thrives in hot, dry, tropical conditions, the moth bean produces nutritious seeds and green pods, leafy forage for hay or pasture, and a soil-building 'living mulch' to complement orchard crops and to protect and improve fallow land. Nonetheless, the moth bean remains virtually untouched by modern science and unknown outside the Indian subcontinent. It has characteristics that could make it valuable for torrid, semiarid regions throughout the tropics. It is likely to prove very useful in extending agricultural production into marginal regions—especially those bordering tropical arid zones."

Carob. "The sugar-rich, mealy pulp contained in carob pods has for millenia been a favorite of people in hot, dry areas of the Mediterranean basin. The handsome, drought-tolerant carob tree deserves more research and widespread exploitation in semiarid areas, for in addition to pulp it provides a chocolate substitute, high-protein flour, and an industrial gum, as well as shade, beautification, erosion control, and forage."

Sesbania grandiflora. "This Southeast Asian tree grows exceptionally fast and provides an amazing range of products: edible leaves, flowers, and gum, as well as forage, firewood, pulp and paper, and green manure. It is also used as a shade tree, ornamental, nurse crop, and living fence. It has extraordinarily prolific nodulation and could become valuable for village use and for large-scale reforestation throughout much of the tropics."

Root Crops, Crop and Product Digest No. 2, book, 280 pages, by Mrs. D.E. Kay, 1973, ₤1.95 (surface mail), ₤3.00 (airmail), from TPI.

This book contains the same kind of information as in **Oils and Oilseeds** (see review), except that this book covers 40 varieties of root crops.

One underexploited root crop is the Jerusalem artichoke. It is "relatively free from serious attacks of pests and disease in the field, although if grown where the drainage is poor, root rot, Sclerotium rolfsii can be troublesome... The tubers are ready for harvesting when the leaves begin to wither and die and are usually lifted manually with a fork as required, since they can be 'field-stored' without any deterioration in their quality or flavor. When grown for pig feed the animals are often turned loose on the plot and root out the tubers."

Vegetables for the Hot, Humid Tropics, annual newsletter, available from Franklin W. Martin and Ruth Ruberte, Mayaquez Institute of Tropical Agriculture, Mayaquez, Puerto Rico 00708, USA.

"Within the tropics a relatively few major vegetables are emphasized while the majority of minor vegetables are not well investigated nor even well distributed. It is difficult to get reliable information about varieties and culture of many of these. This annual newsletter, distributed free to any research worker or institution that requests it, is designed to fill a special need through emphasizing the lesser known vegetables and making them available throughout the tropics.''

The editors of this newsletter are Dr. Franklin W. Martin and Ruth Ruberte, co-authors of **Edible Leaves of the Tropics** (see review in this section). They invite contributions by field workers from throughout the tropics. The newsletter is also intended as a means of generating exchanges of seeds and propagating material.

Articles from the 1977 newsletter include: ''The Utilization of the Potato in the Tropics'', ''A Preliminary Checklist of Diseases of Some Local Vegetables in Nigeria,'' and ''Traditional Vegetables of Papua New Guinea.''

This newsletter contains moderately technical language. Readers will need to have minimal training in agriculture to appreciate its contents. Articles are published in English, Spanish, and French, and the deadline for contributions is June 1 each year. For those who only wish to get copies of past and present newsletters, we recommend a publications exchange or a letter describing your interests and activities.

Oils and Oilseeds, Crop and Product Digest No. 1, book, 202 pages, by Mrs. V.J. Godin and Dr. P.C. Spensley, 1971, Ł1.15 (surface mail), Ł1.85 (airmail), from TPI.

Here is information on 36 plants which produce oil and oilseeds. Growth requirements, planting and harvesting procedures, products and their uses, yields and trends in world supplies are noted. The book also contains information that can help in ''making a first, tentative selection of possible crops for cultivation in a given set of geographic and economic circumstances.''

There are indexes of botanical names, common names, sources of common oils and fats, and an extensive list of sources for further information.

Jojoba Publications, from the Office of Arid Lands Studies, University of Arizona, 845 North Park Avenue, Tucson, Arizona 85719, USA.

Jojoba, a plant native to the Sonoran desert in North America, produces a liquid wax with a wide variety of potential uses. This liquid wax possesses ''qualities not to be found in any other vegetable oil.'' One major use is to replace sperm whale oil as a lubricant for high-speed machinery. Historically,

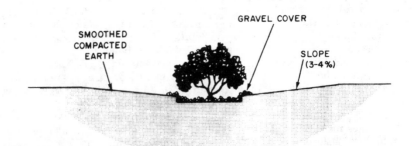

Gravel mulch compacted runoff area

the plant has had a wide range of uses among the native American populations in the area.

Much recent research has focussed on plantation cultivation of jojoba. **Jojoba and Its Uses** (81 pages, $5.00) is a 1972 conference report, including a paper on the potential of using rainstorm runoff farming techniques to increase jojoba yields. A major drawback of plantation cultivation of jojoba is the length of time needed before significant production can be achieved—up to 10 years. Some recent developments indicate that it may be possible to greatly reduce this gap between planting and full production.

Two annotated bibliographies, with a total of 460 entries, were published in 1974 and 1978. These are **Jojoba: A Wax Producing Shrub of the Sonoran Desert** ($10.00), and **Jojoba: An Annotated Bibliographic Update** ($7.50). The Office of Arid Lands Studies acts as a clearinghouse for this and other information on jojoba activity, publishes a quarterly newsletter (**Jojoba Happenings**, $10.00 per year), and arranges for distribution of jojoba seed.

Guayule: an Alternative Source of Natural Rubber, book, 90 pages, by National Academy of Sciences, 1977, free from Commission on International Relations, JH 215, National Academy of Sciences, 2101 Constitution Avenue, N.W., Washington D.C. 20418, USA.

"This report examines the state of knowledge and the future promise of guayule Parthenium argentatum Gray, a little-known shrub native to the desert of southwest Texas and northern Mexico that was a commercial source of natural rubber during the first half of this century."

This perennial shrub thrives in arid conditions and can survive heavy frosts. Guayule, after thorough drying, has been found to contain as much as 26% rubber. This rubber can be used to make vehicle tires or any other item currently made with natural rubber. It is a promising plant for use in reforestation of desert fringe lands and is easy to grow. Extraction of the rubber is not technically difficult; in fact, small-scale household extraction is possible. The plant can be cut down to the ground and will grow again from the roots.

The book covers: background and history, botanical information, rubber extraction, agricultural production, rubber quality, economics, research needs, selected readings and recommendations. There are no lists of sources for seeds.

"When guayule grows actively, it produces little or no rubber. If the plant is stressed, growth slows and the products from photosynthesis are diverted into rubber production. Thus when growth slows during cold weather or because of reduced moisture supply the rubber content begins to increase."

There is no reason why guayule rubber could not be produced and used in many regions of the world. This report is very interesting and highly recommended.

Microbial Processes, book, 198 pages, by National Academy of Sciences, 1979, free from Commission on International Relations (JH 215), National Academy of Sciences—National Research Council, 2101 Constitution Avenue, Washington D.C. 20418, USA.

"Microbes can be marshalled to aid in solving many important global problems including food shortages, resource recovery and reuse, energy shortages, and pollution. Microbiology is particularly suited to make important contributions to human needs in developing countries, yet it has received comparatively little attention. The range of possible applications covers uses by individuals and industries in rural settings, villages, and cities."

This volume contains information on microbial processes within ten subject areas, chosen by a National Academy of Sciences panel. Topic areas include food, animal feed, soil microbes, nitrogen fixation, insect control agents, waste treatment, and antibiotics and vaccines. From each topic area, a small number of examples of the uses of microbes that could be valuable in many developing countries were chosen. These examples include Indonesian tempeh (a high-protein fermented soybean food), and the freshwater fern Azolla pinnata (which harbors a blue-green nitrogen-fixing alga), commonly used to supply nitrogen in the rice paddies of Southeast Asia. For each example, the value, limitations, R&D requirements, suggested readings, and sources of micro-organisms are listed.

Organic Recycling in Asia, book, 397 pages, FAO Soils Bulletin No. 36, 1978, from FAO or $23.00 from UNIPUB.

This collection of papers from a 1976 workshop provides a look at the use of crop residues, animal manures and green manures in agriculture in the region. Brief country papers are included, along with working papers on specific topics (such as the relationship of soil fertility to organic matter, composting of municipal wastes, and economics of sewage sludge composting).

"During the period 1973/74, as a consequence of the energy crisis, mineral fertilizers became very scarce and expensive and, because of this, were out of reach of many farmers particularly in developing countries...For many of the these farmers organic sources of fertilization are the only means available and may remain so for a long time to come."

In the early part of this century, night soil and straw supplied 70% of the nitrogen needs of Japanese farms. Rice production at the time was 3300 kg per hectare, "higher than the present-day production of rice in any country in South and Southeast Asia."

Where artificial fertilizers are now used, several of the authors note that simultaneous application of organic fertilizers tends to increase the percentage of artificial fertilizers available to the plants.

China: Recycling of Organic Wastes in Agriculture, book, 107 pages, FAO Soils Bulletin No. 40, 1977, $5.00 from UNIPUB.

This valuable resource book surveys the use and re-use in present-day China of substances such as night soil (human waste), city garbage, and water weeds —which are often ignored or disposed of in both developed and developing countries alike. Good quality photographs, charts, working drawings, and systems diagrams are used to explain the various methods and installations found in China by an FAO/UNDP study team.

Techniques of special interest include:

• The seeding and innoculation of rice paddies with Azolla Pinnata, a small aquatic plant which harbors nitrogen-fixing blue-green algae. These biological fertilizers are cultivated and stored by simple methods.

• The production of fertilizer directly in the fields in silt-grass manure pits. River silt, rice straw, animal dung, aquatic plants, and small quantities of chemical fertilizers (such as superphosphate) are built up in layers in round or rectangular pits and covered by a sealing layer of soil.

• The composting of night soil and city garbage in concrete tanks and mud-plastered piles. High temperatures, conscientious maintenance, and scientific controls assure that disease-causing organisms are kept under control.

• Extensive use of "green manures", crops which are not harvested for animal or human consumption. These are plowed under to add organic matter, improve soil structure, prevent nutrient leaching, and, in the case of leguminous crops, add nitrogen to the soil.

• The widespread use of bio-gas technology to convert human and animal wastes into fuel and fertilizer. (This topic is covered more fully by other books reviewed in the bio-gas chapter of the A.T. Sourcebook.)

The information presented in this book is easily understandable. It should be remembered that the cost and production figures cited are as reported by the Chinese themselves. It is doubtful that the virtually complete recycling of organic matter as practiced in China can be adopted in many other countries. Incentives may be lacking, and there are often cultural inhibitions against waste handling. Nevertheless, this book identifies effective and proven options which could be attempted throughout the world. Highly recommended.

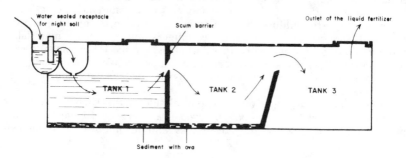

Three-tank fermentation system for treatment of night soil

Composting for the Tropics, booklet, 27 pages, edited by V.L. Leroux, 1963, available from the Henry Doubleday Research Association, 20, Convent Lane, Bocking, Braintree, Essex, England.

Dating from British colonial days in Eastern and Southern Africa, this booklet describes three successful composting methods developed in present-day Kenya, Malawi and Zimbabwe. The simple but effective methods of three former market gardeners and farmers are presented by the Henry Doubleday Research Association in the hope of sharing practical experience with farmers in other tropical countries.

Perhaps the greatest challenge to tropical agriculture is to maintain soil fertility and productivity at the same time. Often the value of both natural and chemical fertilizers is lost, due to rapid processes of decay and leaching. Using sawdust-based composts, these farmers were able to take advantage of the long decay period of sawdust to slow the breakdown and loss of plant nutrients. Thus, these nutrients remained available for food crops.

The information is valuable but may be of limited use in developing countries because sawdust may not be a material readily available to the rural farm population. Also, available sawdust and wood shavings are often used for fuel.

No illustrations are included, but the written descriptions of the processes are easy to understand, if the reader has a basic knowledge of agriculture.

How to Make Fertilizers, Technical Bulletin No. 8, 8 pages, by Harlan Attfield, illustrated by Marina Maspero, $1.00 from VITA.

Drawings and simple text on composting crop residues and manure, adapted from a Bangladesh booklet. Uses bamboo bins.

Soils, Crops & Fertilizer Use, book, 103 pages, by Dave Leonard, 1969, free to PC volunteers and development organizations in developing countries, from Peace Corps (see address below with next review).

Developed for Latin America-based volunteers, this book presents basic information on the physical and chemical characteristics of soils, plant nutrition, and soil fertility. The author is openly skeptical about the practicality of organic fertilizers. He emphasizes the use of chemical fertilizers and soil amendments such as lime, as a means of achieving higher yields in agricultural development projects.

This book should be used along with a training program consisting of actual field analysis of soil structure and texture, chemical soil tests, and pot or plot trials. This will help avoid wasteful use of chemical fertilizers where no net benefits are likely. Although no previous agricultural education is necessary, the reader should have at least a secondary school command of English.

According to the editors, their materials have been designed ''to provide technical support to Peace Corps Volunteers and to share materials on 'intermediate technology' with other participants in the international development community.'' Private voluntary organizations in developing countries are encouraged to exchange their own materials for this book, and send comments and criticisms.

This manual is way over on the chemical side of the chemical fertilizer—organic fertilizer debate. It should be used as a reference if balanced by other publications that describe the advantages and techniques for organic fertilizers.

Pesticides, Resource Packet No. 2, two booklets, 1977 and 1978, available to Peace Corps volunteers and development organizations in developing countries, from Information Collection and Exchange, Peace Corps, 806 Connecticut Ave., N.W., Washington D.C. 20525, USA.

Here are good basic safety rules for handling pesticides. The precise place of pesticides within sustainable, environmentally-sound agriculture is still unclear. Many people feel that pesticides should be avoided completely. Others feel that pesticides should be used sparingly, and only when insect population counts and other indicators suggest that they are needed.

As important debates about pesticides continue, the fact is that pesticides are widely used in the Third World. The lack of local language labels with safety instructions, mis-use, and lack of protection for those who handle pesticides have made these chemicals a major health hazard to farm laborers. The full extent of this health problem is as yet unmeasured, as it hits hardest among migrant farm laborers and others with little access to diagnostic services. Even so, reports indicate a frighteningly high incidence of debilitating disease and death among these people. Compounding this problem is the continuing, cynical sale in the Third World of pesticides banned in the United States; most of this is through the U.S.-based multinational corporations.

Where pesticides are being used, practices for their safe handling should be understood by all agricultural and health field workers. This packet will be helpful in that educational effort.

C. Irrigation

Surface Irrigation, book, 160 pages, by L.J. Booher, FAO, 1974, FAO Agricultural Development Paper No. 95, available from Distribution and Sales Section, United Nations Food and Agriculture Organization (FAO), Via delle Terme di Caracalla, 00100 Rome, Italy.

Relevant to both small and large farming units, this is a good introduction and reference book on surface irrigation. No special technical background is necessary, although general knowledge of agriculture and basic mathematics is required. This volume is more in-depth than **Small Scale Irrigation**, but it does not cover micro-irrigation with catchments, or runoff irrigation techniques.

The sections on soils, land preparation, ditches, and pipeline distribution systems offer good background material for the later chapters on basin, border, wild flooding, furrow, corrugation and drip irrigation. There are helpful guidelines for choosing an irrigation system based on crop, slope, soil, and available water. Charts and tables show how to plan irrigation systems to suit varying conditions (for example, recommended length and spacing of furrows based on soil type and land slope). Photographs and drawings show both mechanized and low-technology tools and equipment for land preparation and water control.

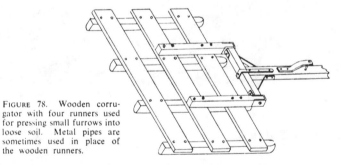

FIGURE 78. Wooden corrugator with four runners used for pressing small furrows into loose soil. Metal pipes are sometimes used in place of the wooden runners.

Small Scale Irrigation, book, 152 pages, by Peter Stern, 1979, L3.95 from ITDG.

A valuable introduction to the technical requirements of irrigation on farms from .1 hectare vegetable plots to 100 hectare units. "The strongest argument in favor of small scale irrigation is that...the human problems are reduced to a manageable scale."

Often people underestimate the quantity of water needed for irrigation. "If all the water consumed in a month by a rural community of 1000 people with 250 cattle and 500 sheep and goats were used for irrigation, this would provide two irrigations a month to an area of about a quarter of a hectare."

The author begins with a discussion of moisture conservation techniques, and maximum use of runoff water. He introduces seven principal surface irrigation methods: basin, border, furrow, corrugation, wild flooding, spate and trickle irrigation. Also mentioned are sprinkler systems (too expensive for most uses in developing countries). To calculate water quantities needed, he discusses crop water requirements and soil infiltration rates. The slopes required for different systems and soils are noted. Other topics include design of drainage systems, channels and pipelines, hand and animal-powered water lifting systems, and measurement of rainfall and water flow in streams. This book gives a basic background, but the reader is expected to get more detailed information either from local agricultural officers or by trial and error.

In areas with very little annual rainfall, micro-irrigation systems can be used. The author gives an example of a farm with annual rainfall of 500 mm, insufficient to produce vegetables. A farmer "could set aside 1000 square meters of his land for catchment irrigation. Of this 1000 square meters, 700 square meters would be prepared as a catchment apron, from which runoff would be fed into a catchment tank, and 300 square meters would be used as a vegetable garden, irrigated by watering can from the tank. In a dry year, with 300 mm of rain, the catchment tank would receive 210 cubic meters of water... (and allowing for losses) the garden would then receive 300 mm of direct rainfall plus 330 mm from the tank."

Recommended.

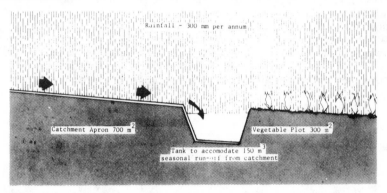

Micro-irrigation system

D. Gardening

Gardening for Better Nutrition, booklet, 64 pages, by Arnold Pacey, 1978, special edition for developing countries is Ł0.50, price to others is Ł1.60, from ITDG.

"The subject of this particular manual is the basic technology of horticulture and vegetable growing, as it applies mainly to family gardens."

This is a thought-provoking overview of the practice of nutrition-oriented agriculture for tropical and developing countries. It summarizes the lessons learned in various projects ranging from Bangladesh to Brazil, and provides a detailed reference bibliography with emphasis on specific regions.

"Although it may include economic activity (such as selling produce at local markets), nutrition-oriented agriculture differs from commercial agriculture in a number of ways:

    a. In growing crops because of their nutritional value rather than because of their market value.

    b. In concentrating on gardens of a size which most families can cultivate.

    c. In appealing primarily to those who produce the family's food—in many communities, the women.

    d. In linking agricultural extension work to health education, social education, and community development."

All aspects of gardening vital to the successful implementation of local programs are touched upon, including crop selection, vegetable agronomy, and problems and techniques. The photos and drawings are excellent, the text clear. Highly recommended as a basic resource book, to be complemented by local technical manuals such as Papua New Guinea's **Liklik Buk** (see review on page 50), the **Samaka Guide** from the Philippines (see review on page 78), **Gardening for All Seasons** from Bangladesh (see review in this section), and **Cultivo de Hortalizas en la Huerta Familiar** from Peru (see review in this section).

Small-Scale Intensive Food Production, book, 135 pages, 1976, persons working on problems of hunger and malnutrition in less developed countries may obtain free copies by writing to: League for International Food Education, 1126 Sixteenth St. N.W., Washington D.C. 20036, USA; others may purchase copies at $5.50 each from VITA.

These are the proceedings of a workshop entitled "Improving the Nutrition of the Most Economically Disadvantaged Families" in developing nations, held in California in 1976, sponsored by L.I.F.E. Participants were representatives of private voluntary organizations, agricultural scientists, and practitioners of a method of intensive gardening known as the Biodynamic/French Intensive Method (see review of **How to Grow More Vegetables**).

The intent of this meeting was to discuss the income and food needs of small peasant farmers, to present the intensive gardening methods (the Biodynamic/French Intensive Method in particular), and to consider the training programs necessary to introduce such methods around the world. As a result of this conference, an ambitious program was initiated to send trained North American volunteers to Latin American countries to create demonstration garden projects. Due to organizational problems and the lack of connection between the volunteers and existing development efforts, this project ended two years later.

This book has some valuable information (paricularly by Dr. Y.H. Yang's well-documented article "Home Gardens as a Nutrition Intervention") and raises some important issues. Anyone interested in home gardening projects, nutrition education, or alternative methods to the Green Revolution in agriculture will find this book useful. L.I.F.E. also publishes a monthly newsletter providing valuable information on food and nutrition.

Cultivo de Hortalizas en la Huerta Familiar, book, 69 pages, by Hans Carlier, 1978, available from Instituto de Estudios Andinos, Apartado 289, Huancayo, Peru.

Beautifully illustrated, this practical guide to intensive home vegetable gardening was written for the Peruvian altiplano (the high plateau of the Andes mountains). All the basics of improved local food production are covered, in simple and straightforward Spanish. Nutritional value of vegetables, varieties, plant propagation, crop rotations, fertilizers, pest management, and simple hand tools are all presented in a systematic and enjoyable format.

This book was written for the direct use of rural people. The safe use of chemical pesticides and fertilizers is stressed, while natural or less dangerous methods of pest control and fertilizer production are preferred. Technically, this is one of the most complete and compact publications on the subject anywhere.

The Institute of Andean Studies is a private voluntary organization working to support rural development efforts involving the indigenous people of the altiplano. We suggest a book exchange for groups in other parts of the world, as this book would be of interest and inspirational even to those who don't understand Spanish.

Highly recommended.

Terracing
(from Cultivo de
Hortalizas en la
Huerta Familiar)

- **Primero separar la tierra buena.**
- **Hacer los muros con piedras grandes.**
- **Nivelar el terreno.**
- **Rellenar con tierra buena.**

Gardening with the Seasons, Technical Bulletin No. 46, 72 pages, by Harlan Attfield, 1979, $3.50 from VITA.

Similar in intent to the **Samaka Guide** (see review on page 78) and **Cultivo de Hortalizas en el Huerto Familiar** (see review in this section), this practical booklet describes gardening techniques and vegetable varieties for Bangladesh. **Gardening with the Seasons** is briefer than the others, though it also is well illustrated and based on extensive field experience. The author has worked on grass roots rural development projects for 8 years in West Africa, South America and Bangladesh. The Bangladesh gardening project has been a key component of that country's Integrated Rural Development Program.

This booklet contains general guidelines for soil preparation using raised beds, seed germination, transplanting, and companion plants. Brief specific information—when to plant, the best soil conditions, spacing, and care—is provided for 36 vegetables grown in Bangladesh.

"Generally people plant the vegetables they like to eat. But good gardeners should also consider food value because some vegetables are richer in value than others. Vegetables should be selected that are easy to grow under local soil conditions, add richness to the soil, and are resistant to insects and disease. Fresh vegetables are an excellent source of minerals and vitamins. They contain many of the minerals, such as calcium and iron, which the body utilizes to make bone, teeth and blood. They also provide important vitamins, mainly Vitamin A, the B vitamins, and Vitamin C."

Highly recommended.

Small Vegetable Gardens, Packet No. 4, 1978, available to PC volunteers and development organizations, from Office of Info. Collection and Exchange, Peace Corps, 806 Connecticut Ave., N.W., Washington D.C. 20525, USA.

(Review by Martha Lewis.) This packet is a collection of materials on vegetable gardening and nutrition for use by volunteers and other development workers. The most relevant items include:

a) Local Agricultural and Nutritional Assessment Tool, 42 pages.

This is "a guideline for examining local food preferences and cultivation practices, for identifying agricultural resources, and for inquiring into the local nutritional situation."

b) **Intensive Vegetable Gardening for Profit and Self-Sufficiency**, 159 pages, written and illustrated by Deborah and James Vickery, 1977.

This gardening manual was prepared for use in Jamaican projects but is useful in any area. It starts with simple botany, soils analysis, components of fertility and methods for soil management and improvement. Instruction concentrates on intensive gardening systems, and describes simple tools, composting, irrigation, rotation and companion planting. Useful charts and illustrations.

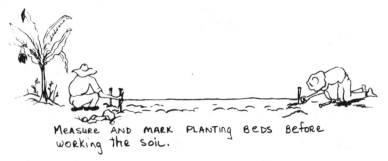

MEASURE AND MARK PLANTING BEDS BEFORE working the soil.

c) **Journal Articles on Vegetable Gardens and Nutrition**, 104 pages.

A collection of articles and unpublished papers by people from A.I.D., the World Bank, and various college and university faculties. Topics include design of vegetable gardening projects, vitamin and mineral deficiencies, seed selection, and 'multimixes' for complementary proteins.

d) **Small Vegetable Gardening/Nutrition Reports from Peace Corps Volunteers**, 48 pages.

This section broadens the scope of the packet by reporting experiences of volunteers in the field. Eight reports give interesting descriptions of cultural diversity and anecdotes on the trials of being a 'change agent'.

One article, "Vegetable Gardening in Zaire", by Stephanie Hannapel, presents solid instructions on how to make a garden. She covers buying seeds, garden design, mulching, seed beds, shade requirements, and vegetable selection, giving cultural requirements for each vegetable.

e) **Indigenous, United States, and International Based Resource Organizations**, 24 pages.

A listing of organizations that operate, fund and/or advise small scale gardening, agricultural, and nutrition projects in developing countries.

How To Grow More Vegetables (Than You Ever Thought Possible on Less Land Than You Can Imagine), large paperback, 115 pages, by John Jeavons, 1979, $6.50 from Ecology Action, 2225 El Camino Real, Palo Alto, California 94306, USA (add $3.00 for airmail worldwide). A Spanish translation of the first edition, 88 pages, is available for $5.50 (add $3.00 for airmail); a French translation of the new edition, 115 pages, is expected by mid-1981 ($6.50).

Ecology Action is devoted to education and research on bio-dynamic/French intensive horticulture. Their gardening classes for the public began on small plots of donated land in 1972. "The series of classes led to the development of information sheets on topics such as vegetable spacings and composting techniques. Many people asked for a book which contains all the information we have gathered...This book is the result. '' In the five years since the publication of the first editon of **How to Grow More Vegetables** (see review on page 80), Ecology Action researchers have revised downward their estimates of the water requirements of the method, to about 1/8 of that used in commercial agriculture per pound of vegetable grown. This new edition has also been expanded to include sections on garden planning and fertilization as well as chapters on history and philosophy, preparation of the double-dug raised beds, compost, seed propagation, and companion planting/backyard ecosystems.

An attractive, easy-to-read book with many good illustrations and a great deal of tabular information on seeds, yields, spacings, time to maturity, fertilizing, and insect pests and their plant controls. While successful gardening relies on experience, this book is probably the most useful single reference for getting started in temperate climates. In tropical and subtropical developing countries, the **Samaka Guide** (see review on page 78) remains the most directly useful manual on intensive gardening. Simple English and clear drawings make **How to Grow More Vegetables** a useful secondary reference book in the tropics, but the important plant species combinations and soil conditions will be different.

Highly recommended.

The Self-Sufficient Gardener, book, 256 pages, by John Seymour, 1979, Faber & Faber (London), U.S. edition $7.95 from Doubleday & Co., 501 Franklin Avenue, Garden City, New York 11530, USA.

This large, beautifully-illustrated book was intended as a companion to the author's **The Complete Book of Self-Sufficiency**, which it surpasses. As a practical manual of planting, growing, storing, and preserving home-grown produce in temperate or sub-tropical regions it ranks as one of the clearest and most concise available. Especially useful are diagrams showing how to convert a conventional row-crop garden into an intensive deep digging bed garden. This book covers practically everything under the sun except pest management, and in a most entertaining and informative way.

In developing countries an indigenous gardening resource manual like the **Samaka Guide** (see review on page 78) will be much more useful than this book. However, we do recommend it as a supplementary reference in developing countries, and a primary resource in industrialized countries.

The AVRDC Vegetable Preparation Manual, book, 103 pages, by Mrs. T.H. Menegay, 1977, free to Third World countries (limited copies available), from Office of Information Services, AVRDC, P.O. Box 42, Shanhua, Tainan 741, Taiwan.

''Our purpose is to produce a manual for use by field workers such as home economics agents, rural teachers, and public health workers throughout the tropics. Thus, the recipes included here should be simple, tasty, and nutritious. We envision that this project, if given your encouragement, may offer one practical means of uniting food production and nutrition specialists in a concerted attack on malnutrition in the tropics.''

This cookbook is an attempt to introduce the soybean, mung bean, tomato, chinese cabbage, sweet potato, and white potato to peoples who have not been exposed to them before, or have not known how to use them in enjoyable meals suited to their culture. The idea of introducing a food crop **along with** suitable recipes for the local culture has been too often ignored.

"Besides being an important source of protein meal and vegetable oil, the immature soybeans can be eaten as a green vegetable, and the dried beans can be consumed in a wide variety of forms...Fresh green soybeans are a good source of protein, calcium, phosphorus, iron, vitamin A, thiamin, and riboflavin. The dry, mature beans are also rich in all these nutrients, except vitamin A, and contain oil as well."

The recipes are inexpensive and are intended to fit into the cooking habits of many different cultures. There is a list of further sources of information at the back of the book.

"White potato has been referred to as a well balanced, well packaged food. Nutritionally, it is close to sweet potato in calorie production per hectare per day and second only to soybean in protein production per hectare. In addition, white potato is an excellent source of vitamin C and vitamin B."

This sort of cookbook is a very important effort to spread nutritional information and plant varieties to parts of the world where they can be adapted to local climate and cultural conditions. We've found the recipes we've tried easy to prepare and enjoyable to eat.

Cowpeas: Home Preparation and Use in West Africa, book, 96 pages, by Florence Dovlo, Caroline Williams and Laraba Zoaka, 1976, free to local groups in developing countries, $6.00 to others, from IDRC.

The cowpea (vigna unguiculata) is an example of an easily grown, protein-rich legume that can be cultivated in temperate or tropical zones. "Cowpeas are extremely versatile and can be used in plain cooking (soups, stews, boiled or roasted) or in processed dishes (boiled, steamed, fried, or baked)...Prostrate (growing flat on the ground), spreading, climbing and dwarf varieties are cultivated and are used as a pulse (bean) or a vegetable, or for fodder and green manure."

Recipes make up more than half of this book. The rest covers such topics as nutritional content of cowpeas, cooking equipment and techniques, and household storage. The authors describe how to combine legumes with grains and other staple foods to increase the usable protein of the food.

The recipes are well explained and have been enjoyed by people of many cultures. Sources for further information are included.

Edible Leaves of the Tropics, book, 240 pages, by Franklin W. Martin and Ruth M. Ruberte, 1975, free to serious groups, from Mayaguez Institute of Tropical Agriculture, P.O. Box 70, Mayaguez 00708, Puerto Rico.

"Green leaves are the most physiologically active parts of the living plant, and as such are usually rich in vitamins and minerals...some contain sufficient protein to supplement an otherwise inadequate starchy diet." The authors discuss green leaves for direct consumption, use of leaves as spices and teas, poisonous tropical leaves, and culture of green leafy vegetables. A lengthy list of tropical plants with edible leaves, giving present geographic distribution, is included.

E. Livestock

Pigs and Poultry in the South Pacific, book, 93 pages, by Ian Watt and Frank Michell, 1975, Australian $2.95, from Sorrett Publishing Pty Ltd., P.O. Box 94, Malvern, Victoria 3144, Australia.

"This book sets out in simple language the information required by extension workers and others responsible for helping the farmer. It deals with all levels, from simple improvisation in a village to semi-intensive and intensive type production."

Two-thirds of the book is on raising pigs. This section covers management systems, pig nutrition, housing, breeds, and diseases. The poultry section covers raising and feeding young chickens, management and feeding of laying hens, timing of replacement of stock, deep litter bedding, ducks, and diseases. Both sections discuss the costs/benefits of home-grown versus commercial feeds. The nutritional needs of the animals are described and some sample home-grown foods are mentioned that meet these needs.

"By delaying maturity, the bird will produce larger eggs when it starts to lay. If a bird is made to lay eggs at too young an age, most of the eggs it will produce during its life will be small eggs...Lowering the protein content of the feed from 21% to 15% for the actual growing period of the bird is probably the easiest way of delaying maturity. So for the first six weeks a layer chicken is fed a 20-21% protein medicated feed, but at the end of six weeks it is changed over to a grower feed which is also medicated (against Coccidiosis), until the bird is about 24 weeks of age when it begins to lay."

The clear, illustrated presentation should make this book valuable to anyone considering pig or poultry raising in the tropics.

Raising Rabbits, book, 82 pages, by Harlan Attfield, 1977, $2.95 from VITA.

"A rabbit raiser can start with two females and one male and produce fifty, or more, rabbits in one year." This rapid reproduction rate and the rapid growth rate of these animals has made rabbit-raising schemes popular in small development projects. Here is a manual that offers good basic advice for most aspects of rabbit raising.

The author stresses the use of locally available plants and grains for food. Because rabbits reproduce and grow quickly, they also consume a lot of

*How to hold
an adult rabbit*

*How to hold
a small rabbit*

food—it takes about 4 pounds of grain to produce 1 pound of rabbit meat (which tastes much like chicken). Record-keeping to aid in breeding, symptoms and treatment of common diseases, skinning and tanning are all discussed. There is no mention of special problems affecting rabbit raising in the tropics. Cages of bamboo, wood, and wire are shown. In all, a well-illustrated, easy to understand manual.

The Rabbit as a Producer of Meat and Skins in Developing Countries, G108, book, 36 pages, by J.E. Owen, D.J. Morgan and J. Barlow, 1977, Ł0.90 from TPI.

This is a brief discussion of rabbit raising in the tropics, not a how-to manual. "Rabbit production on a relatively small scale, involving minimal inputs, could make a substantial contribution to the supply of animal protein for human consumption in tropical developing countries."

The authors discuss the effects of heat and humidity on rabbits, housing, diseases, feeding, breeds and breeding, slaughter and processing, rabbit skins, and problems with escaped rabbits.

"Heat is one of the most important environmental factors which may affect rabbits in tropical developing countries. At ambient temperatures above approximately 30°C rabbits suffer increasing discomfort and physiological stress...(these effects) can be greatly reduced by the construction of suitably designed housing...using locally available materials."

A nice introduction to rabbit raising, with a lot of facts and illustrations.

The Water Buffalo, book, 238 pages, FAO, 1977, $16.25 from UNIPUB.

This book seems to cover everything one might want to know about water buffaloes, including the types, reproduction, nutrition, diseases, parasites, management, training, and milk and meat production.

Tropical Feeds: Feeds Information Summaries and Nutritive Values, book, 661 pages, by Bo Gohl, FAO Feeds Information Centre, 1975, $32.00 from UNIPUB.

"Published information on the nutritive value of feeds in general is scanty and when it comes to tropical feeds, it is almost non-existent. Correct data on the nutritive value of local feedstuffs are essential for the expansion of the livestock industry in the developing countries." This enormous reference book covers 650 tropical feeds, most of them tropical plants. "The summaries include short descriptions of the feeds and the more important points in their use." Many references for additional information on specific feeds are provided.

General considerations for use are given at the beginning of each feed group (e.g., grasses, legumes, root crops, oil cakes). In the miscellaneous categories, feeds such as grain distiller's byproducts (left over when alcohol fuels are produced from grain) are discussed. At the end of the book, charts offer such information as crude protein content, metabolizable energy per kilogram, and mineral & vitamin content of the feeds. The index allows the reader to look up plants under either their botanical or English names.

ADDITIONAL REFERENCES ON AGRICULTURE

Food or Fuel: New Competition for the World's Cropland examines the potential effects of large scale alcohol fuel production on the world food supply; see review on page 569.

Learning from China (page 391) and **China: Forestry Support for Agriculture** (page 498).

AGRICULTURAL TOOLS

Agricultural Tools

Appropriate agricultural tools and equipment should contribute to the broad objective of increasing the viability of the small farm. Where small farmers are currently employing traditional technologies that are inefficient, they often cannot improve this technology because of the leap in scale and capital cost to commercially available equipment. It is therefore the goal of intermediate technology proponents to help fill this gap with good quality tools and equipment that are affordable and suited to the scale of operations of the small farmers.

It is not enough, however, to consider improved tools for single farms in isolation from other factors. The types of tools available affect the scale of farms, and thus help determine whether or not the poor can capitalize a farming operation. Historically, development of agricultural implements has assumed and reinforced consolidation of lands, and increases in the capital intensity of farming. The situation of the smallest farmers, tenants, and landless laborers is ignored and even made worse by such equipment research and development strategies. Appropriate technology advocates must not repeat these mistakes.

It is also clear that the degree of concentration of land ownership* is a key factor in determining if there are opportunities available for appropriate technology strategies in a community. Agricultural technologies developed with and for the smallest farmers can certainly strengthen the viability of their farms. But if most families have no land at all, land reform and the establishment of rural industries may be far more important steps in a positive community development program than the improvement of agricultural tools and equipment. From the national perspective, support for communities of small farms should bring significant benefits. Whereas it has been widely assumed that only the large farm could effectively increase national food production in the struggle against hunger, mounting evidence from many countries indicates that the small farm has higher yields per acre and plays a crucial role in distribution of food. Small farms also help make the best use of national capital resources:

"To maintain...a rational growth of capital in a low-income economy, small farms are better suited than large ones, for the small farmers do not experience the same pressure to substitute capital for labor; no one wants to mechanize himself out of a job."
— Folke Dovring, in **Agricultural Technology for Developing Nations: Farm Mechanization for 1-10 Acre Farms**

In communities where most people have some access to land, the effective-

---

*See discussion of the effects of unequal land distribution on appropriate agricultural technologies for tenant farmers and landless laborers, on pages 345-346 of the Introduction.

ness of efforts to create relevant new tools can be increased by concentrating on some key areas of agricultural technology. Irrigation is the biggest single factor in raising crop yields. Simple pumps, especially those which use renewable energy sources such as locally built water-pumping windmills, are of particular interest on small plots. (See ENERGY: WIND chapter introduction and the two small scale irrigation books reviewed in the AGRICULTURE chapter.) Animal-drawn plows, cultivators, and seed drills tend to satisfy the equipment needs of small farms at a lower capital investment than tractor-drawn equipment; and the importance of good quality hand tools should not be overlooked. Greenhouses can conserve water, and in temperate climates offer an early start on the growing season (see ENERGY: SOLAR chapter for reviews of greenhouse books). Crop processing equipment, including threshers and mills, can reduce losses caused by traditional techniques. Very small scale equipment of this sort could allow the small farmer to retain full crop production instead of paying the common rate of 10% to the thresher/mill owner. Crop storage is a prime area for improvement, as a major percentage of food produced on small farms is lost due to poor drying and storage. Solar crop dryers and low cost small scale storage bins are particularly promising (see CROP DRYING, PRESERVATION AND STORAGE chapter). In many areas it is difficult to move agricultural inputs to the farm, harvested crops from the fields to storage, and surplusses from the farm to markets. Appropriate transport technologies are thus of great importance to the farmer (see TRANSPORT chapter).

In this chapter, the first two entries offer some recommendations as to the kinds of agricultural tools and equipment most needed by small farmers in developing countries. The third volume, **Agricultural Technology for Developing Nations: Farm Mechanization Alternatives for 1-10 Acre Farms** contains conflicting points of view on the effects of mechanization.

An encyclopedic listing of commercially available equipment is presented in **Tools for Homesteaders, Gardeners, and Small Scale Farmers**. This and **Old Farm Tools and Machinery** contain a wealth of ideas that may stimulate the imagination of readers working to produce designs of agricultural equipment for local production. These people will also be interested in **Rice: Post Harvest Technology**, for a better understanding of the technical characteristics of commercially available rice processing equipment, that could potentially be modified to suit smaller farmers.

The Employment of Draught Animals in Agriculture is probably the best single volume anywhere on the training and use of draft animals, animal-drawn implements, animal power gears, and the economics of animal power. **The Draft Horse Primer** and **The Harness Maker's Illustrated Manual** are also valuable references.

On North American family farms the farmer is often expected to act as mechanic and handyman as part of daily farming activities. The well-equipped farm workshop and multiple skills of the farmer have continued to play a powerful role in generating farm equipment innovations. **Mechanics in Agriculture** is a text for vocational courses teaching the skills commonly required on these farms.

A large number of small engines are used in the Third World for power tillers, irrigation pumps, crop processing and other applications. **How to Repair Briggs and Stratton Engines** and **Small Gas Engines** should be helpful references for maintenance and repair of many of these power units.

Most of the remaining entries are plans for threshers, winnowers, corn shellers and so forth, all of them hand or foot-operated, that can be produced in small workshops by local craftsmen. Three publications reviewed in the

ENERGY: GENERAL chapter also discuss pedal-powered agricultural equipment. To better understand the power and stamina potential of the human body as a power source, see **Bicycling Science** *(reviewed in the TRANSPORTATION chapter); this discusses the performance of stationary pedal-powered equipment.*

In Volume I:

Appropriate Industrial Technology for Agricultural Machinery and Implements, book, 159 pages, UNIDO, 1979, Document No. ID/232/4, available from Editor, UNIDO Newsletter, Industrial Information Section, UNIDO, P.O. Box 300, A-1400 Vienna, Austria.

This publication is for policy makers and planners, offering a systematic look at the kinds of farm equipment needed for different sizes of farms, and the levels at which the different ranges of farm equipment can be produced. Some examples of equipment needed and operating costs are given for fully-equipped production facilities to make both simple and complex agricultural tools and machinery.

There is no mention of the tradeoffs between employment and mechanization, no effort to examine agricultural equipment that would especially support organic agriculture (e.g. manure spreaders and bug light traps), and no concern with participation of the rural people in the design of equipment.

The book does support the provision of basic tools and implements for the very small farm. "In farms below 2 ha, where farming is carried out in a traditional way, using hand tools and animal-drawn equipment with little or no purchase of inputs...the mechanization policy should be based on: improved supplies of high-yield seeds and fertilizers and single or double cropping; high-quality hand tools such as spades, spading forks, digging hooks and hoes, shovels, ploughs, and singlewheel hoes; animal-drawn ridgers, cultivator ploughs and seed drills; low-cost simple power tillers; effective irrigation and water supply by means of windmills with up to 5 ft (1.5 m) lift or small electric or diesel pumps of up to 15 ft (4.5 m) lift; hand-drills, sickles, scythes, forks, and rakes; hand-operated threshers, crushers, etc.; storage bins of up to 3 ton capacity."

Also supported is the concept of local production of basic tools by rural artisans (leaving the more difficult equipment to be produced by urban or rural industrial units). "Government policies must be reoriented to assist artisans in the rural areas. Major efforts are needed to encourage and revive production of hand tools by village artisans through provision of loans at concessional rates, technical assistance, provision of simple design and marketing assistance."

Agricultural Technology for Developing Nations: Farm Mechanization Alternatives for 1-10 Acre Farms, Proceedings of a Conference, May 1978, single copy free from Roy Harrington, Product Planning Department, Deere & Company, John Deere Road, Moline, Illinois 61265, USA.

This collection of papers and panel discussions presents the perspectives of a range of people: World Bankers, multinational agricultural machinery manufacturers, agricultural economists, agricultural engineers and others. Useful as background reading on some of the most promising types of mechanization (broadly interpreted to include animal-drawn equipment) and some of the problems that either prevent or follow mechanization.

For mechanization: "When we began to look at agriculture in other parts of the world, we began to realize that the classic notion that labor is displaced when you increase the number of tractors does not show up in the statistics in a number of countries."

Against mechanization: "...although mechanization raises the productivity of labor, in the conditions prevailing in most Latin American countries its benefits have gone mainly to swell the profits and rents of the large landlords and the wages of the few tractor drivers and other machinery operators...it may be roughly estimated that about three workers are displaced by each tractor in Chile, and about four in Colombia and Guatemala."

What to mechanize: "Mechanization seldom contributes much to the level of crop yields, except in the form of pumps for irrigation." "In Japan...the thresher was more beneficial to farmers than the power tiller."

Ensuring socially useful mechanization: "To maintain...a rational growth of capital in a low-income economy, small farms are better suited than large ones, for the small farmers do not experience the same pressure to substitute capital for labor; no one wants to mechanize himself out of a job."

Tools for Homesteaders, Gardeners, and Small-Scale Farmers, (A Catalogue of Hard-to-Find Implements and Equipment), book, 512 pages, edited by Diana S. Branch, $12.95 from Rodale Press, 33 E. Minor St., Emmaus, Pennsylvania 18049, USA.

"Finding the right tools can be the most critical need for a small-scale farmer or a large-scale gardener. It can mean the difference between staying on or leaving the land, between a sense of drudgery or a sense of fulfillment, between a successful harvest or a meager crop, between profit or loss."

"This catalogue will help you to find and use the tools you need to produce food. The tools and equipment described in its pages were selected primarily for their value to the homesteader, truck farmer, and the small-scale organic farmer, but backyard gardeners should also find things of interest."

This very welcome book is the result of a cooperative effort between the London-based Intermediate Technology Development Group and Rodale Press, an American group which researches and publishes many other titles in the fields of alternative energy sources, organic gardening, and waste recycling. "The idea for this book grew out of the ITDG book, **Tools for Agriculture: A Buyer's Guide to Low Cost Agriculture Implements**" (see review on page 112). The new Rodale book is a better investment.

Thoroughly illustrated and referenced, this catalog of over 700 implements from around the world is an impressive accomplishment. Included are tools for cultivation and plowing; implements for draft animals; tractors and accessories; seeders; planters; harvesting implements; threshing and cleaning tools; processing equipment; tools for composting, mulching and handling sludge; woodlot and orchard equipment; livestock and fish-farming equipment.

The sources for these tools are primarily in industrial countries, although this reflects current manufacturing realities more than any bias on the part of the authors. Most of the best hand tools and animal-drawn equipment for developing countries are included: the Grelinette/U-bar digger, IRRI's push-type paddy weeder, Jean Nolle's various tropical cultivators, the Mochudi toolbar, hand corn shellers, CeCoCo pedal threshers and winnowers, etc. Also featured are interesting articles on tropics such as renovating old equipment and experimental stationary winch systems for pulling farm implements. A detailed source list of North American and international manufacturers which can be built by the user.

A minor shortcoming of this book is the lack of price information. Even though inflation would make such prices quickly out-of-date, this would be valuable for comparative purposes.

''There is a strong heritage, especially in the United States but elsewhere too, of the farmer as inventor. A large percentage of our inventors came from rural communities, and virtually all the industries which grew up in the United States in the 1800s started on a very small scale, often as one-man operations. Cyrus McCormick, Oliver Evans, Eli Whitney, even Henry Ford—each grew up on a farm. The inventors of tools we still need will most likely come from the ranks of today's small farmers—and their children.''

A very valuable book. Highly recommended.

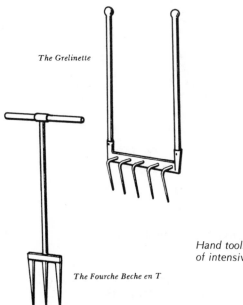

The Grelinette

The Fourche Beche en T

Hand tools for deep digging
of intensive vegetable beds

Low Cost Rural Equipment Suitable for Manufacture in East Africa, booklet, 82 pages, prepared by S. Minto and S. Westley, sponsored by the East African Agriculture and Forestry Research Organization, 1975, small charge for postage only, from Institute for Development Studies, P.O. Box 30197, University of Nairobi, Nairobi, Kenya.

"The purpose of this booklet is to share information about some of the equipment which has been designed and tested for small-scale farmers with the wider audience of those engaged in rural development...The booklet describes, with simple working drawings, illustrations and details of required materials, 23 items of equipment which have been thoroughly tested for their suitability in East Africa." Many of these pieces of equipment have been presented in detail by other organizations, such as the Intermediate Technology Development Group (see reviews of their Agricultural Green Leaflets on pages 104-112).

Equipment presented: Kabanyolo ox toolframe with plow and cultivator, adjustable inter-row cultivator, Kabanyolo seeder/planter unit, hand operated seed planter, oxcart frames/wheels/axles/bearings, bicycle trailer, wheelbarrow, water cart, groundnut sheller, hand maize sheller, rotating maize sheller, flap valve pump, IRRI bellows pump, evaporative charcoal cooler, solar dryer, grain storage crib, feed mixer drum, seed dressing drum, ram for building blocks, post hole auger, oil drum bellows forge, anvil, and hydraulic bender.

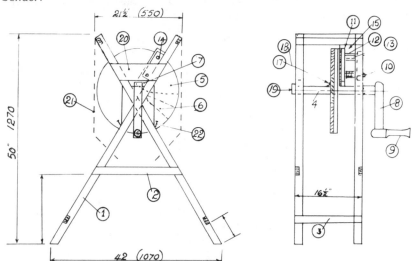

Hand-operated corn sheller

Old Farm Tools and Machinery: An Illustrated History, book, 188 pages, by Percy Blandford, 1976, $22.00 from Gale Research Co., Book Tower, Detroit, Michigan 48226, USA.

This book is by the author of **Country Craft Tools** (see review on page 72). It covers tools and machinery from small farms in Great Britain, Europe, and the United States from the past hundred years. The author briefly looks at animal power gears, carts, steam engines, and the early tractors. Of greater interest, and complete with illustrations, are the chapters on agricultural equipment, most of it capable of being animal-drawn. These include a variety of plows, a cable plow pulled by a stationary steam engine, an excellent collection of seeding devices, manure and fertilizer spreaders, spades, forks, rakes, hoes, harrows, cultivators, reaping machines, hand harvesting tools, mowing machines, and tools related to dairy production. There are 30 photographs and more than 150 simple line drawings. The brief text and drawings are usually enough to communicate the basic ideas and principles used, but you wouldn't be able to make any of this equipment from this information alone. Nevertheless, the book is a great source of ideas.

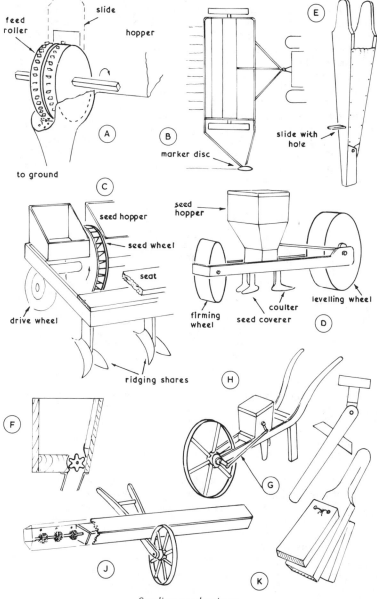

Seeding mechanisms

The Draft Horse Primer: A Guide to the Care and Use of Work Horses and Mules, book, 400 pages, by Maurice Telleen, 1977, $10.95 from RODALE.

The work horse "is a source of power that reproduces itself, with good care is self-repairing, consumes home-grown fuel, and contributes to the fertility of the soil. Horse farming and organic farming are very comfortable with one another."

This book is interesting for several reasons. First, it shows that some North Americans have either stayed with horse-drawn farming equipment (e.g. the Amish) or are now going back to it. (In the United States, "the demand for draft horses has risen significantly since 1960.") Secondly, it captures some of this practical wisdom which normally passed from farmer to farmer. The author draws from his own experience and brings together "material from booklets published by our land grant schools during the twenties and thirties when they had an active interest in heavy horses as a major source of agricultural power."

Telleen discusses the breeds of draft horses used in the United States, what to look for when buying, and basic care of these animals. He presents 70 pages on animal-drawn machinery, 50 pages on harnesses and hitches, and 22 pages on logging with horses. Because horse-drawn equipment has historically been far cheaper than mechanized equipment, a smaller farm can finance it.

A thoughtful book that illuminates the potential role of the draft horse in a small-scale, ecologically-sound agriculture.

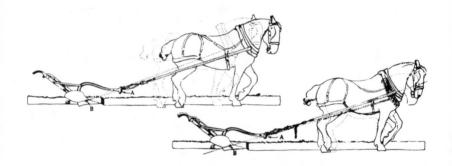

The Employment of Draught Animals in Agriculture, book, 249 pages, by the Centre d'Etudes et d'Experimentation du Machinisme Agricole Tropical, 1968, English translation 1972 by FAO, $16.75 from UNIPUB.

"This manual is mainly concerned with the application of animal draught equipment, a form of agricultural mechanization predominant in the tropical regions of Africa."

The difficulties and disadvantages of introducing engine-driven equipment have become evident in many parts of the world, most notably in Africa, where draught animals historically have been rarely used. Animal-drawn equipment for mechanization appears to represent the more appropriate technology for many of these areas.

This book begins with draught animals (power, training, housing, feeding, harnessing methods). There is an extensive and very good section on animal-drawn implements, and valuable notes on animal power gears. Following this is a discussion of the rural skills and equipment available for implement and harness production and repair. The final section presents economic considerations, and includes a simple method for calculating the costs of animal power. An excellent book.

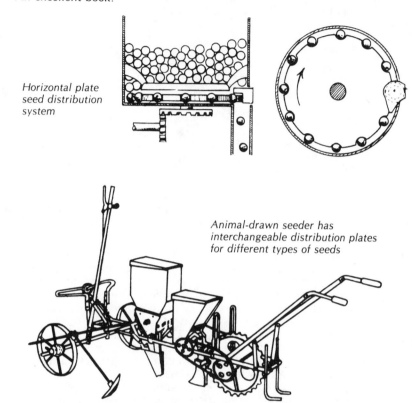

Horizontal plate seed distribution system

Animal-drawn seeder has interchangeable distribution plates for different types of seeds

The Harness Maker's Illustrated Manual, book, 333 pages, by North River Press, 1974, $13.00 from META.

This is a reprint of a book first published in 1875. In those days animal-powered transport was normal in the United States. This book describes how to make and use the harnesses of that time.

"This book originated from a desire to furnish harness makers with a condensed practical guide suited to the workshop, office, salesroom and stable. It treats of leather as furnished to the harness maker by the currier, its texture, strength, adaptability for specific uses; how to cut, fit, and finish; measuring for a harness; complete tables for lengths and widths for cutting the various classes in use, whether for carriage, farm, or road; bridles, halters, horse-boots, mountings, bits, etc."

The language used is slightly out-of-date and might at times present a bit of trouble to the reader. The instructions on harness construction and design are excellent and this book should be considered a detailed manual for this craft.

This should be of interest in the Third World where often better harnesses would mean a great contribution to more efficient use of animal-power. Certain modern materials of course are not included, though some of them will be of interest in particular circumstances.

"Fig. I. represents the United States Government regulation halter, the principal merit of which is its strength and simplicity; it is easy to adjust, and cannot be slipped off by the horse rubbing its head against posts or other objects."

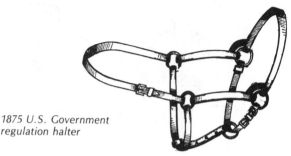

1875 U.S. Government regulation halter

Mechanics in Agriculture, book, 844 pages, by Lloyd J. Phipps, 1967, second edition 1977, $18.00 (educational discounts may be available), from The Interstate Printers and Publishers, Inc., Danville, Illinois 61832, USA.

This is a comprehensive text for courses in vocational agriculture, divided into five parts: equipping and using a farm workshop, engines and implements, buildings, electrification, and soil and water management. Illustrations and excellent instructions on the use of all kinds of hand and power tools make this an encyclopedia of modern American farm mechanics. The text is particularly strong on explanation of principles of operation, maintenance, repair, and safety, for tools and implements. Though it was compiled for use by American secondary school students entering a capital- and energy-intensive agriculture, the book covers many topics of interest to agriculturalists everywhere. Examples include repairing and sharpening hand tools; making sketches and reading blueprints; understanding concrete; soldering and oxy-acetylene welding; blacksmithing and working sheet metal; using rope and leather; and fundamentals of engines and electric motors.

A good reference, possibly useful at centers or field stations for teaching and development of appropriate agricultural technologies.

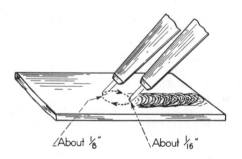

∠About ⅛"   \About 1/16"

Welding: practicing circular motion with a torch—flame is lifted at the front edge of the circle

How to Repair Briggs and Stratton Engines, paperback book, 182 pages, by Paul Dempsey, 1978, $4.95 from TAB Books, Blue Ridge Summit, Pennsylvania 17214 USA.

Briggs and Stratton engines power small pumps and agricultural implements all over the Third World. Though it is written for North Americans, this book could be valuable wherever these small engines are being used.

The author "describes repair and maintenance procedures for all current and many older Briggs and Stratton engines. These procedures extend to all phases of the work, from simple tune-up and carburetor repairs to the serious business of replacing main bearings and resizing cylinder bores. The material is organized by subject and by engine model, and as much as possible, divided into steps that are easy to follow." Clear line drawings and text explain the basics of four-cycle internal combustion engines. 75 pages are devoted to adjustment and repair of ignition systems, carburetors, and pull-starters. The section on engine disassembly and overhaul includes standard machining clearances and dimensions as well as replacement part identification numbers.

A bent connecting rod causes the piston to tilt in the cyclinder and produces a wear pattern indicated by the shaded areas

Small Gas Engines, book, 256 pages, by James Gray and Richard Barrow, 1976, $12.95 from Prentice-Hall, Inc., Englewood Cliffs, New Jersey, USA.

Small gas engines are very common in most parts of the world. They are used in motorcycles, electric generators, water pumps, rototillers, winnowers, boats and many other devices.

This book introduces the theory of small gas engine operation. It does a detailed and thorough job of presenting the basics of repair and maintenance. Understandable to the beginner, and also valuable to those who already know something about the subject.

Recommended.

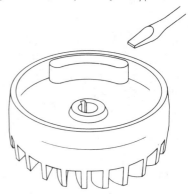

Check condition of flywheel magnet with an unmagnetized screwdriver

Rice: Postharvest Technology, book, 394 pages, edited by E. Arguello, D. De Padua, and M. Graham, 1976, publication no. IDRC-053e, free to local groups in developing countries, $20.00 to others, from IDRC.

This large volume covers all the technical aspects of rice postharvest technology: harvesting, threshing, drying, storage, parboiling, milling, and handling. It also describes ''some of the anatomical and biochemical properties of the rice grain in relation to postharvest processing problems.''

This book was compiled from material used in a training course on postharvest technology in the Philippines. It is not an appropriate technology manual, but rather a reference book on the current state of the art equipment for rice processing operations. It could be useful to small independent groups and university-based organizations that need to know as much as possible about the standard commercial designs, while working to develop lower cost alternatives and an ecologically-sound agriculture that will benefit even the smallest farmers.

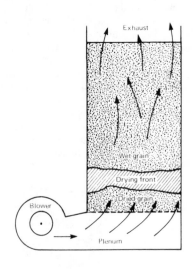

Deep-bed grain drier

Winnowing Fan, Technical Bulletin No. 39, 4 pages, $1.00 from VITA.

MATERIALS: wood, sheet metal
PRODUCTION: hand tools

A portable machine from the Philippines for winnowing rice. This design is hand-operated, but it could be adapted to use pedal power or a small engine. The drawings and text are easy to understand.

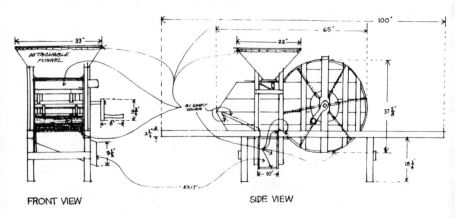

FRONT VIEW   SIDE VIEW

Hand-operated winnowing fan

Rice Thresher, VITA Technical Bulletin No. 22, 5 pages, by Bertrand Saubolle, S.J. (Nepal), $1.00 from VITA.

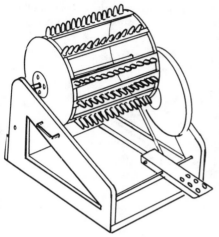

Simple drawings and text for the construction of a treadle-operated thresher. This design ''aims at simplicity of construction, ease of operation and low cost, combined with rapid and thorough threshing. It was built from scrap wood, junk rods, bits of pipe, an old bicycle pedal, discarded motor bearings, and so forth.'' The drawings are so brief that the reader will be challenged to take the final steps needed to make a durable, locally adapted version; for example, nothing is said about the function and materials needed for what appears in a sketch to be a flywheel.

A Wooden Hand-Held Maize Sheller, Rural Technology Guide No. 1, booklet, 8 pages, by G.S. Pinson, 1977, £1.05 (surface mail) from TPI.

This booklet shows how to make a low cost, simple wooden hand-held tool to remove the grains from maize (corn) cobs when they are hard and dry. The tool can be easily made out of hard wood, using hand tools (including a hand drill). The drawings are clear and easy to understand.

Potentially a very useful tool in many parts of the world where corn kernels are still removed by hand, often without any tool at all, using only the thumbs or another corn cob to pry the kernels loose.

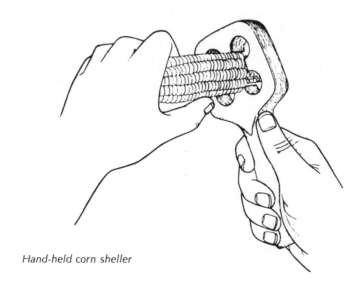

Hand-held corn sheller

Small Corn Sheller, Technical Bulletin No. 21, 3 pages, by Bertrand Saubolle, S.J., (Nepal), $1.00 from VITA.

Simple drawings for the construction of a wooden rotating-drum corn sheller that can be made with simple hand tools. The sheller is small and lightweight, about the size of a hand grain grinder.

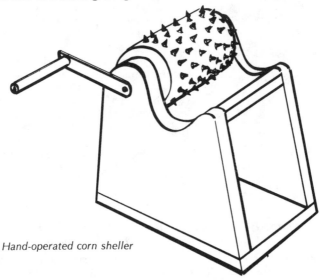

Hand-operated corn sheller

A Feeder to Improve the Performance of a Hand-Operated Groundnut Sheller, Rural Technology Guide No. 4, booklet, 17 pages, by G.A. Collins, L.D.G. Coward and G. Pinson, 1977, £1.00 (surface mail) from TPI.

MATERIALS: ¼'' steel plate, sheet metal, two small pieces of ¾'' steel, steel bar, bolts

PRODUCTION: tools for drilling, cutting and threading steel; welding equipment; sheet metal bending equipment

This is a construction manual for a device that controls the number of groundnuts (peanuts) dropped (fed) into a hand-operated groundnut sheller. The result is less effort in use and fewer broken kernels. Drawings and instructions are clear, and the feeder should be easy to make if the metalworking tools are available. Wood could be substituted for many of the steel parts, but the authors do not discuss this. The feeder is designed to be attached to existing models of groundnut shellers.

Four Person Pedal Powered Grain Thresher/Mill, booklet, 77 pages, by Alex Weir, 1979, available from VITA.

This report begins with a review of the literature on threshing requirements and the energy used in traditional and pedal-threshing systems. The author concludes that for rice, "the use of a treadle thresher multiplies labor productivity by a factor of 2.3 over hand threshing." Tests of a prototype thresher for sorghum indicated that when those holding the grain and those pedalling are counted, labor productivity is increased by a factor of 1.37 over hand methods.

Plans are presented for the combined thresher/mill unit. To make it, you will need a welding machine, a metal-working lathe, and a milling machine. Materials used include bicycle drive mechanisms, steel water pipe, flat and round bar, and ball bearings. The mill unit drawings are not very clear.

If other power sources are available, pedal power is not very attractive for extended use. The total energy produced by six people (4 pedaling and 2 holding the grain) working five hours is less than 2.5 kwh. The advantage of pedal power is in low cost equipment for intermittent use, especially where electricity is unavailable, engine maintenance is difficult, and only small amounts of power are needed. In this case, the author figures the thresher can process 750 kg of sorghum in a 5-hour period, the total crop from approximately 1 to 1.5 hectares in Tanzania.

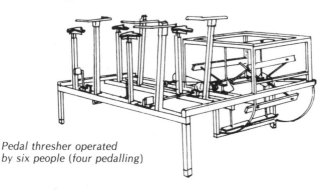

Pedal thresher operated
by six people (four pedalling)

A Pedal-Operated Grain Mill, Rural Technology Guide No. 5, booklet, 32 pages, by G.S. Pinson, 1979, Ł0.80 (surface mail) from TPI.

MATERIALS: two ball bearings, wire mesh, 3 mm steel sheet, 1¼'' and 2½'' steel bar (8 inches long), 6 mm flat and angle iron.

PRODUCTION: small metal working lathe, milling machine, welding set, drill press, metal working hand tools.

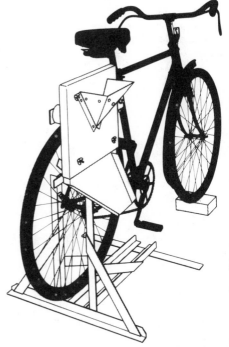

Complete instructions for the production of a grinding mill for grains and legumes. The mill and supporting stand are used with an ordinary bicycle which can be quickly connected and disconnected. The rear bicycle wheel drives a rotor at about 5000 rpm to break up the grain. Wire mesh controls the size of the flour product. "The mill works best on hard, brittle grains such as maize (corn), millet and sorghum and on legumes such as soya beans."

Although the mill is of steel construction, it is intended for use over brief periods to meet the daily needs of individual households. It is not designed for continual, intensive use. No cost estimates are provided. An alternative wooden frame using some bicycle parts is shown. Design modifications could eliminate the more difficult metal working tasks (lathe and milling work), and also reduce some of the other costs.

Bell Alarms and Sack Hoists in Windmills, booklet, 16 pages, H. Clark and R. Wailes, 1973, 45 pence including postage, from The Society for the Protection of Ancient Buildings, 58 Great Ormond St., London WC1, England.

This is a study of the clever ways in which two important functions were accomplished in windmills and watermills. 1) Warning the miller when the grain was low (using bells); and 2) lifting the heavy sacks of grain and flour inside the mill (using hoists that took power off of the windmill or watermill via a drive shaft).

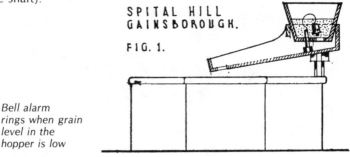

SPITAL HILL
GAINSBOROUGH.

FIG. 1.

Bell alarm rings when grain level in the hopper is low

Cotton Gin, Technical Bulletin No. 41, plans, 5 pages, $1.00 from VITA.

MATERIALS: hardwood

PRODUCTION: wood lathe and hand tools

The cotton gin is a device for separating seeds and other unwanted material from cotton fibers. This design is made entirely out of wood, similar to units in India and China. The machine is operated by hand, by one person. Capacity is 2 kg. of lint cotton per day. The hardwood rollers work well with some varieties of cottons, but other varieties may require use of rollers made from steel rod.

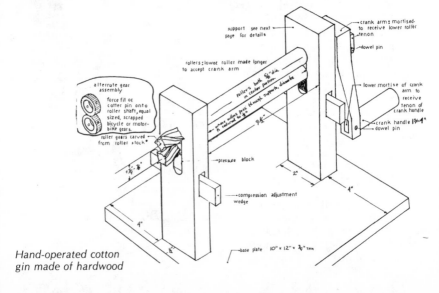

Hand-operated cotton gin made of hardwood

Lightweight Seeder/Spreader, Plan No. 596, 2 pages, $.50 from Popular Mechanics.

These are brief but complete plans for the standard American lawn seeder/fertilizer spreader. It may have some value with modifications for seeding grasses in small farming operations or for other seeding activities. The distribution and rate of seed flow could be modified for other seeding needs.

The seeder is made of 18-gauge aluminum, bent, drilled, and screwed together. Uses two small wheels.

Grass seeder/fertilizer spreader

Making Coir Rope, Technical Bulletin No. 44, leaflet, 8 pages, $1.00 from VITA.

MATERIALS: wood, metal hooks, coconut husks, 3″ nails

PRODUCTION: wood working handtools, wood lathe

A step by step presentation of the process of making coir rope from coconut husks. The necessary equipment can be made out of wood, and most of this is shown: a fiber combing board for separating fibers, hand-cranked single and multiple twisting reels, and a strand block and strand guide for the final steps in rope making. The text is at times confusing and misleading, and the reader will have to be careful.

Some of the basic concepts can also be applied in making wire rope.

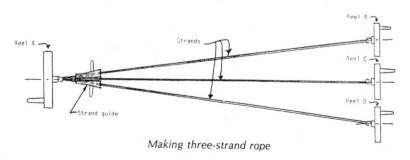

Making three-strand rope

Poultry Feed Grinder, Technical Bulletin No. 31, 6 pages, by Frank Redeker, $1.00 from VITA.

This grinder was originally designed to be powered by a small (⅓) hp motor. It has been adapted also for hand cranking.

The grinder requires some metal working equipment for construction: welding tools, pipe cutting tools, metal working lathe. None of this is precision work, however, so it is quite suitable for the smallest workshop that has such tools.

Materials required include 4½″ and 5¼″ diameter steel pipe, ¼″ steel plate, and 1/8″ steel sheet. Welded ridges on the steel drum and housing (cut from the pipe) create the grinding surfaces.

The plans are clear.

Chain Link Fence Making Machine, VITA Technical Bulletin No. 25, 20 pages, $2.00 from VITA.

These are step-by-step instructions for making and using a hand operated machine to make chain link fencing. There are drawings and photos. "The machine here is designed to produce fencing up to 244 cm (96″) but can be used to produce fencing of any height. The size of the openings (can be varied)...The machine described here requires #12 or #14 wire, but could be modified to take larger wire." A very clever, easily-made device.

"In Botswana, the machine has become the basis of a small fence manufacturing business which serves as a source of employment and produces fencing which is far more affordable locally than is the imported fencing which was the only material previously available."

Of course this unit requires the use of wire (probably imported in most countries). For most fencing needs, traditional alternatives exist and are probably more appropriate.

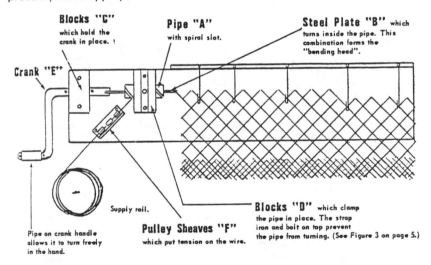

Chain link fence making machine

ADDITIONAL REFERENCES ON AGRICULTURAL TOOLS

Village Technology in Eastern Africa reviews potentially useful tools for agriculture; see page 366.

Fichier Encyclopedique du Developpement Rural contains leaflets in French on agricultural tools, such as a sugar cane crusher; see review on page 375.

The Book of the New Alchemists describes the greenhouse and fish tanks combination used in the Ark; see review on page 379.

Greenhouses used for both food production and home heating are described in **The Solar Greenhouse Book** (reviewed on page 636) and **Proceedings of the Conference on Energy-Conserving, Solar Heated Greenhouses** (reviewed on page 737).

Rural Small Scale Industry in the People's Republic of China discusses the relationship between the decentralized agricultural machinery industry and farming; see review on page 390.

Small Farm Development: Understanding and Improving Farming Systems in the Humid Tropics estimates the effects of various small scale power sources added to small farms; see review and chart on page 428.

Surface Irrigation contains drawings and photographs of low technology and mechanized equipment for use in land preparation for irrigation and water control; see review on page 445.

Grain storage bins and dryers are shown in the chapter CROP DRYING, PRESERVATION AND STORAGE, pages 478-491.

The Management of Animal Energy Resources and the Modernization of the Bullock Cart System includes a discussion of needed cart and harness improvements for these farm vehicles; see review on page 566.

The Use of Pedal Power for Agriculture and Transport in Developing Countries examines the potential applications of pedal power for driving agricultural equipment; see page 572.

Design for a Pedal Driven Power Unit for Transport and Machine Uses in Developing Countries lists crop processing equipment suitable for pedal power; reviewed on page 573.

Water-powered equipment for crop processing is discussed in the ENERGY: WATER chapter, pages 610-625.

Fertilizer is one of the main products of biogas plants; see pages 648-654.

The TRANSPORTATION chapter (pages 680-698) examines small vehicles, many of them important in small farm operations.

CROP DRYING, PRESERVATION AND STORAGE

Crop Drying, Preservation, and Storage

"Experience has taught the small grower in the developing countries that, if produce is stored, it goes bad. This has two effects: a sufficient quantity is grown to feed the family for about three or four months; immediately after harvest (sometimes before it has been dried thoroughly), when there may be a temporary glut of food and prices are low, produce is sold to traders; moreover in many areas the farmers are in debt to the traders and any produce surplus beyond their own food requirements is immediately sold to meet the accumulated debts. Thus one of the major contributory factors responsible for the economic nonviability of farming areas is the farmer's inability to handle and store food efficiently so that he can sell good quality produce when it is scarce and commands a high price. The standard of living in a rural community depends not only upon the range of foods grown, the capacity to grow in quantity, but also upon the facilities for efficient handling, drying, storage and marketing...

In Latin America it has been estimated that there is a loss of 25 to 50 percent of harvested cereals and pulses; in certain African countries about 30 percent of the total subsistence agricultural production is lost annually, and in areas of Southeast Asia some crops suffer losses of up to 50 percent."

—**Handling and Storage of Food Grains in Tropical and Subtropical Areas**, FAO

Low cost technologies for village level preservation and storage of food crops can clearly improve the standard of living for a large number of people around the world. There may be no other problem area in which appropriate village technologies have a more important role to play.

Many observers are now coming to view effective farm level grain storage as the key to both reduced food losses and increased farm family income and security. Landless laborers often stand to benefit as well from good storage, as grain prices flatten out and in-kind wages can be protected from losses in their homes. Centralized government grain storage facilities frequently have proven to be a disappointment, suffering from poor quality control on incoming grain (with resulting high in-storage loss rates) that leads to low prices paid to the farmers. Even with smoothly functioning large-scale grain storage facilities, substantial losses may have already taken place at the farm level before the grain ever reaches the centers.

*Readers who are in a position to help develop or implement appropriate technology solutions in their communities should turn first to two excellent books that detail for different crops the points in the harvest, handling, and storage sequence where losses are most likely to occur. **Post Harvest Food Losses in Developing Countries** takes a look at the potential for reducing losses of a wide variety of foods, while **Handling and Storage of Foodgrains in Tropical and Subtropical Areas** is the better technical reference book, con-*

centrating on grain storage. *Another particularly valuable source of ideas on how to approach storage problems is* **Appropriate Technology for Grain Storage,** *which describes a successful effort to pool community knowledge of grain storage problems and apply it in developing several genuinely workable solutions tailored to local circumstances.*

Proper drying is considered the biggest single factor in determining whether grain will be effectively stored without damage. Simple direct solar drying plays an important role now in preserving a significant portion of Third World production. Usually grain is dried while it stands in the fields, or it is spread out on concrete surfaces, roads, baskets, plastic sheets, or the ground itself. The standard alternative to such direct sun drying methods has been the fossil fuel-consuming artificial drying system. In these units, large quantities of food can be dried with greater speed and greater control over drying rate and product quality. These dryers require a high capital investment and ever-increasing operating expenditures for fuel, but have relatively low labor costs. There are also a number of small artificial dryers that depend on wood, rice straw, or rice hulls for fuel (see **Drying Equipment for Cereal Grains and Other Agricultural Produce,** *reviewed in this section,* **Simple Grain Dryer,** *reviewed on page 121, and* **Small Farm Grain Storage,** *reviewed on page 124).*

Solar agricultural dryers appear to be intermediate in capital cost between direct open air drying and fossil-fueled dryers, combining many of the benefits of both. They are cheaper to operate than fossil-fueled dryers, requiring no fuel. They represent smaller units of capital and capacity, better matched to the small surplusses generated by small farmers. They are more easily made with local materials that may be available free or at low cost. There are basically two kinds of solar dryers. The simplest (for certain crops like corn) are raised bins with roofs, that protect the grain from rain and attack by small animals and rodents while allowing air flow through wire mesh or woven walls to slowly dry the grain. These are "indirect" dryers that depend on air heated by the sun rather than direct exposure to the sun's rays for their drying effects. Much work has been done recently on the design of a different type of dryer that is enclosed with glass or plastic coverings to trap the sun's heat, raising the temperature and lowering the humidity of the air which passes over the crop. With the higher temperature, these enclosed dryers work more rapidly than either ground-spreading or indirect systems. Such dryers usually do bring a higher cost per unit of drying surface than the other simpler systems. They offer some protection against rewetting due to rain, a distinct advantage over ground-spread systems in which the grain must be quickly gathered up whenever rain threatens. An enclosed solar dryer brings the additional advantage of protection from dirt, insects, and animals, which can be very significant in preventing both direct losses and later multiplication of mold and insect populations in stored grain. See **A Survey of Solar Agricultural Dryers,** *reviewed on page 126, for an excellent collection of different designs.*

For fruits and vegetables, the temperatures achieved in an enclosed solar dryer make thorough drying possible when open air drying may not be rapid enough. Drying could extend the low-cost availability of a number of tropical fruits, such as mangoes. In many countries these fruits ripen during a very short period, creating a temporary glut. Much fruit may rot in the absence of expensive canning facilities. (In some places, the absence of a preservation technology actually appears to benefit the poor in the short run, by making fruit either unusually cheap or free for the picking during the fruit seasons.)

The appeal of solar dryers in the Third World will depend very much on the local situation—traditional drying practices, crops produced, food price fluc-

tuations over the year, and weather during the harvest season. (No solar drying system works well under continuously cloudy, humid conditions. When harvests coincide with the beginning of a rainy season a fueled dryer may be necessary.) In some areas, heat absorbing black plastic or canvas sheets that can be quickly gathered up when rain is threatening may be better investments than solar dryers. On the other hand, dryers can be made of primarily local materials (e.g. bamboo, wood, adobe) with the addition of clear plastic sheets, and thus need not be very expensive. One strategy that may make best use of a solar dryer in the Third World is to use traditional direct sun methods but reserve an enclosed solar dryer for rapid final drying in small batches, to be sure that the grain will be dry enough to store well.

In the United States, solar grain drying systems at the farm level are under active testing, and the results to date suggest that they will soon be widely used. These dryers are replacing expensive fossil-fuel burning dryers, and the costs of converting to solar drying can be balanced against reduced fuel bills. Many of these U.S. solar drying systems use large electric fans to circulate air. Two of the publications in this chapter describe U.S. solar drying systems.

There are many good quality storage bins that can be made out of locally available or other low cost materials, that will successfully protect properly dried stored grain from moisture, mold, insects, rodents and birds. One such bin is the Bissa, which appears to be well suited to storage requirements in Sri Lanka (see **Evaluation of the Bissa—An Indigenous Storage Bin**). Lightweight metal bins have proven effective in Guatemala and other countries, and have been promoted in India as part of the Save Grain Campaign (see **Guide to the Manufacture of Metal Bins**). Other traditional and low cost storage bins are described in **Post Harvest Food Losses** and **Handling and Storage of Food Grains**.

Canning, drying and pickling of fruits and vegetables are the topics of the last two publications in this chapter. It appears that the capital costs for containers and the energy requirements of canning make this option out of reach of all but affluent families. Drying may be a lower cost, more universally relevant strategy for fruit and vegetable preservation. Construction details for a solar dryer for fruits and vegetables are presented in **How to Build a Solar Crop Dryer**.

In Volume I:

Small Farm Grain Storage, by Carl Lindblad and Laurel Druben, 1976, page 124.

Home Techniques—Volume 1: Food Preservation, FAO, 1976, page 125.

Smoking Fish in a Cardboard Smokehouse, VITA, page 125.

How to Salt Fish, by D. Casper, page 126.

A Survey of Solar Agricultural Dryers, by BRACE, 1975, page 126.

Plans for Low-Cost Farming Implements, drawings of platform carts with drying pans, page 99.

Post Harvest Food Losses in Developing Countries, book, 183 pages, National Academy of Sciences, 1978, free from Commission on International Relations, JH 215, National Academy of Sciences, 2101 Constitution Avenue, N.W., Washington D.C. 20418, USA.

This valuable, informative book examines the potential of food loss reduction for each of the major food crops, on the small farm or small operator level. Includes cereal grains (e.g. rice, maize, millet, sorghum, wheat), grain legumes (e.g. beans, peanuts, soybeans), perishables (e.g. cassava, yams, bananas, potatoes) and fish. Losses are identified at each step of harvesting, processing and storage, and low cost technology options for reducing these losses are discussed.

For the world as a whole, attention to food losses affecting small farmers holds the greatest potential for benefitting the largest numbers of people. The authors note that improvements must take into account social and cultural factors. Increased losses are often associated with the new high yielding varieties, as they overwhelm traditional processing and storage systems.

Education, training and extension are only briefly discussed. Some important issues are raised, such as the need for improved communication between policy makers and village leadership to insure the development of programs in harmony with village needs.

Highly recommended for the general reader involved in rural development work. Very helpful in understanding the factors affecting food losses, and the opportunities for low-cost technology solutions. The technical vocabulary is not difficult, but the language used may still present problems to the non-native English speaker.

Handling and Storage of Food Grains in Tropical and Subtropical Areas, book, 350 pages, by D.W. Hall, United Nations Food and Agriculture Organization (FAO), 1970, second edition 1975, $6.00 from FAO booksellers, or from Distribution and Sales Section, FAO, Via delle Terme di Caracalla, 00100 Rome, Italy.

This is the best technical reference book on grain storage, providing a good summary of the relevant scientific work up to 1970. It will be valuable for anyone working on grain drying and grain storage problems. The text is in readable (not too technical) English.

"This manual describes the causes of grain loss, deterioration and contamination, methods of drying and storage, the design of small and large storage facilities, and also methods of fungus, insect and rodent control." For all of these topics, the author notes traditional local practices which should be more widely encouraged, in addition to relatively simple improved practices.

"Current knowledge of modern handling and storage techniques is derived from industrial countries, which are mainly in the temperate regions of the world. This knowledge has only limited application under the climatic conditions of tropical countries.''

''Indigenous farmers always have their own methods for assessing the amount of moisture in grain. Some of these provide a fairly reliable estimate of the grain's suitability for safe storage. These methods include pressing the grain with the thumb nail; crushing the grains between the fingers; biting the grain; rattling a number of grains in a tin; obtaining the 'feel' of the grain by smelling a handful and shaking it; or by plunging the hand (fingers extended) into...a sack or heap. With long experience a man can judge whether the grain or kernel is suitable for storage...However...inconsistency can arise due to differences of opinion when the person concerned feels ill.''

''The mixture of wood ash or sand with food grains is carried out in many areas...This method appears to rely for its effectiveness upon the fact that the materials used fill the intergranular spaces and thereby restrict insect movement...Mineral dusts...scratch the thin, waterproofing layer of wax which exists on the outside surface of the insect cutical, allowing loss of water which leads to death.''

''Condensation problems, especially in metal silos, occur in the tropics particularly in areas where the sky is clear during both day and night...Metal silos should be light in color to reflect most of the incoming radiation during the day. The major temperature changes normally required to cause condensation can be avoided by providing adequate shade to prevent large gains of energy in the grain.''

The author does support the use of some insecticides which are now known to be more dangerous and undesirable than previously supposed in 1970.

Highly recommended.

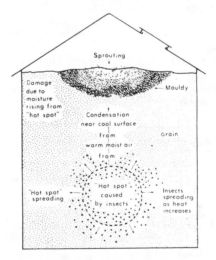

Spoilage of grain in a small
storage bin due to temperature
differences, movement of
moisture, and localized
multiplication of fungi and
insects

Appropriate Technology for Grain Storage, book, 94 pages, by the Community Development Trust Fund of Tanzania, $2.50 from Economic Development Bureau, Inc., P.O. Box 1717, New Haven, Connecticut 06507, USA.

This report documents a very important example of a successful strategy to stimulate villagers to create their own appropriate technologies for grain storage. The wealth of knowledge held by the Tanzanian villagers about their own specific local problems in grain storage emerged from dialogues with a team of outsiders (Tanzanians and foreigners). Potentially relevant experience from external sources was made known to the villagers, who criticized, modified and added to this store of possibilities. The villagers then designed three sets of improvements that matched different needs within the village. The outsiders thus served as resource people and facilitators, yet left the choice of actions to the very people who would best know what constraints they faced and what they could realistically afford to do. The entire process stimulated an awareness among the villagers of the high level of their own collective knowledge and capability for solving their own problems.

Factors affecting grain storage in a Tanzanian village

"The team aimed not to impose an alien analysis of the problem on the villagers but to work from the basis of their perceived and understood reality... The villagers already had 'parts' of solutions to their storage problems. It was the aim of this project to reinforce these existing solutions so that they would be more effective, not to replace them with new solutions."

"Villagers found it hard to understand that the team had not brought a solution to the storage problem, that it did not want simply to convince or force them to do something, and it did not have some gift for them...It was only after having carried a certain line of design (the Nigerian crib) forward in discussions for several weeks only to drop it when the villagers brought up serious criticisms, that the team's credibility was finally established. It was then clear that the team did not have a vested interest in any particular design."

Village discussion groups told the visiting team that home-drying of grain was an essential element of any improved storage system. The grain could not be dried in the fields because the farmers could not prevent the destruction of the crops by wild pigs. Preventing the pigs from entering the fields would require a level of cooperation that the villagers said they realistically did not yet have. "Such an example highlights three important reasons why the dialogue approach places such a problem area as grain storage in the context of the total village reality. 1) The significance of some seemingly technical detail of a development problem can easily be misunderstood. For instance, a well-meaning expert might have argued that farmers should not harvest their maize while moist; they should let it dry in the fields, and then store it in such and such a way. Such an unfortunately common 'outside' approach would be bound to fail because it lays down rules for the farmers and takes no account of the reality of wild pigs. 2) The dialogue approach generates awareness of interrelated development problems that can be taken up in turn. For instance, the planning committee of the project village has already discussed block farming in relation to the problem of protection against pigs...3) By pursuing problems back to their origins, discussion groups confront what are sometimes called 'limit situations', that is, points where they quite genuinely say, 'Tumeshindwa!' ('We have failed!'). By defining and objectifying limit situations and then by focussing human energy on them, they are ultimately overcome. It is the experience of bursting through a previously limiting situation that constitutes the liberating effect of adult education."

At the same time discussions were going on with the adults, grain storage experiments were carried out in the local primary school, in the spirit that 'the school is the village and the village is the school.'

"The final event of the eight-week project was a seminar on village crop storage given by members of the Storage Committee (all villagers, with little formal education) to some fifty third-year crop husbandry and rural economy students at the University of Dar es Salaam, Faculty of Agriculture."!

By the end of the initial project period, the village storage committee "had developed a continuing relationship with such outside resources as the University Agricultural Faculty in Morogoro and knew how to **demand** knowledge and support for future improvements when required."

The team of foreigners and Tanzanians who initiated the project "recommend that none of the design results of this project are immediately applicable anywhere in Tanzania—even the next village—unless such design results are put to the local inhabitants as possibilities in the course of a mobilization programmed somewhat on the lines of the one undertaken in this project."

An excellent example of proper, humble technical assistance in the context of real community participation. Highly recommended.

An improved 2-ton storage bin with rat-guards

Manual of Improved Farm and Village-Level Grain Storage Methods, book, 243 pages, by David Dichter and Associates, 1978, DM 16 ($9.00) plus postage from TZ-Verlagsgesellschaft mbH, Bruchwiesenweg 19, D-6101 Rossdorf 1, Federal Republic of Germany.

This handbook is perhaps most valuable for its good explanation of the important considerations that are keys to better grain storage. The introduction also describes the grain storage problem quite well. Photos and text for the construction of 4 different small storage containers are provided. The descriptions, however, are too wordy and not always well matched with the drawings. Standard designs for sun dryers are also shown, but no cost or output figures are given. As a resource for equipment, this handbook is not as complete as we'd like to see.

The text, which could be shortened considerably, takes the form of lectures with questions and answers. It is intended to be used as a training manual for extension workers in a standard extension effort (in which grain storage designs are chosen by a central agency for dissemination). The book does emphasize the importance of understanding the principles of good grain storage and basing improvements on traditional techniques, rather than the transfer of an alien grain storage technology.

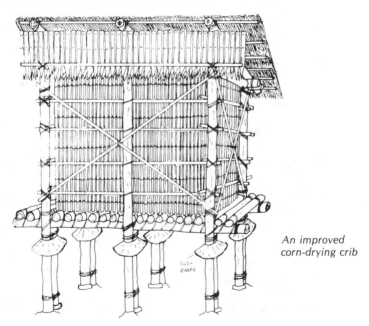

*An improved
corn-drying crib*

Potential of Solar Agricultural Dryers in Developing Areas, paper, 8 pages, T.A. Lawand, 1977, free to serious groups (ask for paper #8549) from UNIDO Industrial Inquiry and Advisory Service, P.O. Box 707, A-1011 Vienna, Austria; also included in **Technology for Solar Energy Utilization**, UNIDO, 1978 (see review on page 630).

This paper, presented to a UNIDO conference in 1977, summarizes the principles of solar dryers, as well as surveying various types of dryers from around the world. Some of the examples: a grape drying rack from Australia, a cabinet dryer from Syria, a glass-roof greenhouse dryer from Brazil, a wind-ventilated dryer from Syria, and a lumber-seasoning kiln from India. There is also a bibliography.

This is a condensation of the information contained in the more complete **Survey of Solar Agricultural Dryers** from Brace Research Institute (see review on page 119). Lawand hopes to stimulate people to adapt these designs and develop their own to fit local conditions.

A good introduction to the subject.

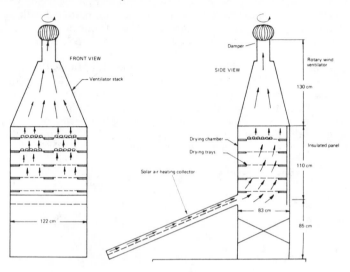

Solar wind-ventilated dryer

How to Build a Solar Crop Dryer, plans, 9 pages, $1.70 from New Mexico Solar Energy Association, P.O. Box 2004, Santa Fe, New Mexico 87501, USA.

MATERIALS: wood, nails, bolts, small hardware, corrugated iron sheet.
PRODUCTION: handtools.

Detailed plans for building a crop dryer. Air is drawn in at the bottom, heated by a collector, and then sent up through the drying chamber. An adjustable vent allows control of temperature (which may reach 120° F). The unit is 4 feet wide and has 36 square feet of drying area, enough for almost 2 bushels of food.

Although the cost is estimated at $60.00, the design can be varied to use cheaper local materials. It is suitable for drying small amounts of fruits and vegetables. Very simple.

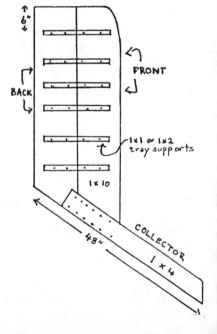

Solar Dryers—Periodical Note #3, 16 pages, 1978, free to individuals and groups involved in development and A.T. work, from Shri A.M.M. Murugappa Chettier Research Center, Photosynthesis and Energy Division, Tharamani, Madras-600 042, India.

Translated from the original Tamil, this report presents 4 very simple solar dryer designs used in rural India for drying paddy (rice), vegetables and fish. All use materials such as bamboo, sand and cement; glazing is provided by a polythene sheet. All were built for Rs. 100-400 (US$12-50) including labor.

The advantages of enclosed solar dryers include faster drying times and protection of food from flies and insects (which helps prevent the spread of disease and reduces crop losses.)

Drawings and descriptions of each dryer are given; these are sometimes hard to follow.

Solar Grain Drying: Progress and Potential, booklet, 14 pages, by G. Foster and R. Peart, 1976, free from Office of Communication, USDA, Washington D.C. 20250, USA.

This describes studies of solar grain dryers, particularly for rice and corn, from the midwestern United States. The dryers were made of inflated polyethylene (soft plastic) shells to heat air as in a greenhouse; the air was then pumped through the grain.

These tests were primarily to determine the feasibility of solar grain drying. Details of the designs are not given. The booklet does offer general descriptions of grain drying systems.

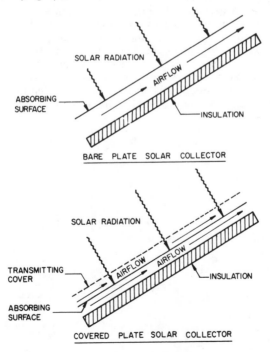

Figure 2.—Schematic of bare-plate and covered-plate solar collector for heating air.

The Performance and Economic Feasibility of Solar Grain Drying Systems, Agricultural Economic Report No. 396, booklet, 33 pages, by Walter G. Heid, Feb. 1978, available from Commodity Economics Division, Economics, Statistics and Cooperatives Service, U.S. Department of Agriculture, Washington D.C. 20250, USA.

This is a summary of the performance of various types of solar grain drying systems from the midwestern U.S. All are for large-scale, temperate climate agriculture. Some of these dryers use electric air blowers and/or auxiliary electric heating. There is a short explanation of the various parts of a crop drying system. Tables compare size, capacity, performance from tests, and costs of 8 different systems now in use.

The emphasis in this paper is on economic evaluation (rapidly becoming more favorable to solar drying since this report was published), rather than on the principles of operation.

Drying Equipment for Cereal Grains and Other Agricultural Produce, plans, 11 pages, by Keith Markwardt, 10 pesos ($1.60) from CARE Philippines, P.O. Box 2052, Manila, Philippines.

MATERIALS: wood, cement, brick, sheet metal, scrap iron, pulley, v-belt, steel shaft, bearings, and 4-8 hp motor or engine for blower
PRODUCTION: welding, hand tools.

The agricultural dryer described in these plans was built by a Peace Corps volunteer in the Philippines. It consists of three components: 1) a concrete and brick furnace 48'' by 24'' by 27'' which uses rice hulls for fuel; 2) an 8' by 16' drying bed; and 3) a gasoline, diesel, or electric powered fan which blows heated air from the furnace through the perforated floor of the drying bed. The dryer has been used for fish, copra, and a variety of grains and vegetables. Drying capacity varies with type of produce. One batch (50 cavan, or 2500 kg.) of paddy dries in approximately 6-8 hours at an operating cost (including engine fuel, maintenance, and depreciation) of about $3.00.

The advantage of this design is the use of rice hulls (plentiful in many rural areas) to cheaply fuel the furnace. The builders estimate that this dryer can be

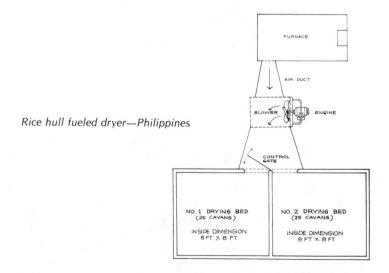

Rice hull fueled dryer—Philippines

constructed for about $500 in the Philippines. Such a dryer might be best used by a cooperative, allowing farmers to collectively meet this initial investment and take advantage of the high capacity and low operating costs.

Guide to the Manufacture of Metal Bins, plans, 17 pages, and **Domestic Storage Bins**, booklet, 25 pages, from Save Grain Campaign, Ministry of Agriculture and Irrigation, Department of Food, Krishi Bhawan, New Delhi, India.

MATERIALS: sheet metal, flat steel bar for clamping rings, rivets

PRODUCTION: rivetting tools, sheet metal rolling and bending tools, soldering tools, metal-bending machine

Complete technical drawings for making four sizes of sheet metal grain storage bins. The booklet gives step by step instructions to be used with the plans. Capacity of the bins ranges from .4 cubic meters (230 kgs of paddy or 300 kgs of wheat) to 1.35 cubic meters (750 kgs of paddy or 1000 kgs of wheat). The lightweight bins are easily transported when empty, and can be lifted by one man. These bins were developed by the Indian Grain Storage Institute as part of the nationwide Save Grain Campaign.

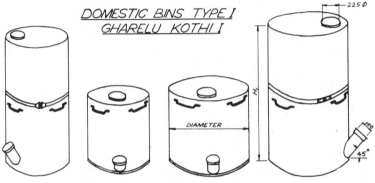

Evaluation of the Bissa—An Indigenous Storage Bin, paper, 38 pages, by K. Pallpane, Rice Processing Centre, 1978, available on request from Food and Agriculture Organization of the United Nations, 202, Bauddhaloka Mawatha, Colombo 7, Sri Lanka.

This paper describes and evaluates the traditional Bissa rice storage bin used by farmers in Sri Lanka. Results of a careful test of the structure are presented. Drawings for construction are included. The bin is made of woven sticks plastered with clay, with a thatched roof.

In Sri Lanka, only 40% of the total rice production is marketed; consequently, there is a high level of on-farm storage for seeds and family consumption. "In improving farm level storage, it is always better to improve and popularize the already existing permanent storage structures, which can be fabricated from material easily available at farm level at a low cost and also whose design and operation is known."

The Bissa is a permanent structure with a capacity from ½ ton to 10 tons. A 5-ton Bissa is estimated to require 164 man-hours for construction, plus the use of local materials, for a total cost in Sri Lanka of $50. The total maintenance, depreciation, and loading/unloading costs for 1 year are about $1.75 per ton. Properly dried paddy (rice) is generally stored for 6 months without any significant loss of quality or quantity.

"The majority of the farmers do not adopt any pest controlling practices because according to them, the damage due to insect attack is negligible if clean dry paddy is stored in the structure...A main defect of this structure is that it has no facilities for aeration to bring down temperature rises." Some minor changes are proposed.

"According to the farmers, a properly maintained Bissa will last for over fifty years."

A good example of a successful, low-cost traditional grain storage bin that could be relevant in many other countries.

Storage of Food Grain: A Guide for Extension Workers, book, 33 pages, by Abdel-Hamid F. Abdel-Aziz, FAO, 1975, $6.00 from UNIPUB.

Based on an FAO farm and community grain storage project and the Save the Grain Campaign in India, this short book is intended to help extension personnel in planning and implementing extension programs for improved grain storage at the farm level. (Some 70% of India's grain is consumed at the farm level, never entering urban markets.) There is little technical information presented; rather, the material covered is the organization rather than the content of an extension effort.

The author takes a conventional information transfer extension approach, but he is sensitive to the value of traditional techniques. He urges creation of a range of storage options for farmers of different incomes levels, including full use of traditional systems with any necessary improvements. The author stresses practical skills training over scientific explanations; he may even be underestimating the importance of understanding principles. A variety of helpful communication aids and strategies are presented.

Home-Scale Processing and Preservation of Fruits and Vegetables, booklet, 68 pages, Central Food Technological Research Institute, 1977 (7th edition), Rs.3.25 (English edition) or Rs. 1.00 (Hindi edition) from Director, Central Food Technological Research Institute, Mysore 13, India.

This Indian publication is a very useful one, both for the material it contains and for the model it presents to other countries. A basic introduction of home-scale food processing technologies (canning, drying, and pickling) is combined with specific fruit and vegetable recipes, a detailed glossary in

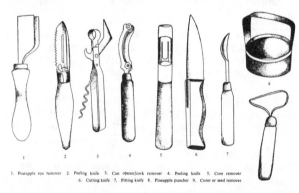

1. Pineapple eye remover 2. Peeling knife 3. Can opener/cork remover 4. Peeling knife 5. Core remover
6. Cutting knife 7. Pitting knife 8. Pineapple puncher 9. Corer or seed remover

Hand tools for home preservation of fruits and vegetables

several important Indian languages, and access information for equipment and supplies.

The wide array of preserved food options is designed to be tasty, reduce produce losses, and improve nutritional levels. Products include: cashew apple extract, mango leather, jackfruit nectar, guava cheese, papaya pickles, and bamboo chutney. Processing time adjustments for higher altitudes are included in the detailed processing charts.

A flaw of this book is its emphasis on a given general technique of preservation: sulphuring for drying, and kerosene burners for canning, for instance. Substituting alternative methods could be considered, such as solar crop drying, and biogas burners in the canning operations. Still, the book presents a low-cost, complete community canning unit.

''An effort has been made to present information in a simple and comprehensive manner, so that an average housewife can use it without any difficulty. It can be used by home science and catering institutions as well as agricultural extension agencies.''

Food Preservation, Series 2, Rural Home Techniques, Volume 5, FAO, $7.00 from UNIPUB, or United Nations Food and Agriculture Organization (FAO) local offices, or Chief, Home Economics and Social Programmes Service, Human Resources, Institutions and Agrarian Reform Division, FAO, Via delle Terme di Caracalla, 00100-Rome, Italy.

This is a folder containing a set of brief 2-page illustrated leaflets on a variety of home food preservation topics. Text is in English, French, and Spanish. The leaflets suffer from being too brief in many cases, leaving the reader without important background information. They are also expensive if obtained from UNIPUB.

This set contains leaflets on blanching of vegetables (cooking prior to drying), drying vegetables and herbs, pickling, making chutney, storage of such processed vegetables, cooling using unglazed pottery or evaporative baskets and cupboards, outside earth silo storage in cold climates, and two kinds of grain storage cribs.

Steps in making chutney

ADDITIONAL REFERENCES ON CROP STORAGE

Village Technology in Eastern Africa reviews some of the simple food preservation and storage technologies affordable at the village level; see page 366.

Rural Women: Their Integration in Development Programs and How Simple Intermediate Technologies can help them suggests the use of enclosed solar dryers, black plastic sheets for direct drying, and improved grain storage units; see review on page 368.

Minitechnology includes a design for a solar dryer; see review on page 373.

The Formula Book 1 suggests the use of bay leaves in stored cereals and flour to repell insects; see review on page 380.

Low Cost Rural Equipment Suitable for Manufacture in East Africa includes designs for a solar dryer and a grain storage crib; see review on page 462.

Rice: Post Harvest Technology describes the technical requirements for rice drying and storage; reviewed on page 468.

FORESTRY

Forestry

"The challenge to forestry of contributing to bettering the condition of the rural poor is...likely to entail a radical reorientation extending from policy all the way through to its technical foundations."
— **Forestry for Local Community Development**, FAO

Wood is a basic resource for meeting human needs. It has always been important as a cooking fuel and building material. But throughout history, expanding human settlements have threatened and eventually destroyed forests. To the individual farmer, the forest is often a nuisance to be cleared away so that the land can be farmed. To the villager, the forest is the provider of plentiful cooking fuel. And to the industrialist, the forest is the source of plywood, paper, cardboard, and lumber to meet the enormous demands of industrial societies. All three of these perspectives have contributed to the rapid consumption of the world's forests, to the point that a major potentially renewable resource has generally been exploited as a one-time boon for the first to arrive.

The consequences of unrestrained deforestation are many. The cultivation of hillsides generally leads to rapid erosion of topsoil and loss of productive potential. The removal of trees reduces the soil's ability to retain water, leading to ever-increasing cycles of flood and drought in the lands below. Inefficient cooking methods and a lack of deliberate replanting of fuelwood trees have forced millions of the poor to spend a large part of each day hunting for fuel and carrying it long distances on their backs. This time-consuming, exhausting work further guarantees their poverty. In search of maximum immediate production from a piece of land, lumbering companies around the world have clear-cut the forests, leaving a devastated landscape vulnerable to erosion, and destroying any potential for sustained production.

"The humid forests of the tropics once occupied at least 1600 million hectares (4000 million acres), and have not only been the main centers for living species on earth, but have held the lands together, moderated and modified world climates, and helped to maintain a desirable balance of atmospheric gasses. Now they are vanishing at an incredible rate. There are reported to be 935 million hectares in actual humid tropical forest, a 40% reduction in total area. They are disappearing at a rate of sixteen million hectares per year..."
— Ray Dasmann, "Planet Earth—1980", 1980

Some observers are convinced that the considerable local, national and international problems associated with deforestation will be accompanied by global climatic shifts if deforestation is not brought under control within the next 10 years. World food production is thus directly threatened by local soil erosion, floods and droughts, and climatic changes that mean shifting rainfall patterns and expanding deserts.

As the problems caused by deforestation are becoming better understood, development planners are scrambling to find temporary and long-term solutions. The skills of the forestry profession are in great demand. Yet on closer examination it becomes clear that more and better funded forestry programs alone will not be enough. Major changes in attitude and strategy will also be required. In particular, foresters and planners cannot continue forest management focussed largely on production for industry. Just as important, forestry programs can no longer be based on the strategy of preventing the community from gaining access to the forest. The FAO book **Forestry for Local Community Development** *marks an historical shift in consciousness, as it describes the strategies and programs that can mean successful sustainable production of forest resources through community involvement. But for the most part,*

"In precious few countries have the energies of the foresters been bent upon helping the peasant to develop the kind of forestry that would serve his material welfare. This is why there are so few village woodlots and fuel plantations. This is why so little work has been done on forage trees, fruit and nut orchards. This is why so few shelterbelts have been created...This is why forestry has been invoked so rarely to reclaim or rehabilitate land. This is why so few of the many possible agro-forestry combinations have been established which are specifically geared to meeting real local needs...

Agriculture-supportive forestry does not by any means exclude forest industries. Small rural industries are an integral part of agriculture-supportive forestry: fuelwood, charcoal, poles, stakes, fencing, hurdles, screens, farm tools and implements, building materials, simple furniture. But these activities, like all other agriculture-supportive activities, are activities that cannot be carried out on the required scale and in the required manner by a conventionally oriented and conventionally organized forest service. They will only be effective, and will only make sense, if they are carried out by the peasants themselves, for themselves. The role of the forester, wherever he may sit in the organizational structure, can only be to stimulate, offer guidance and suggestions, impart techniques and carry out training."

—"Forest Industries for Socio-Economic Development",
by Jack Westoby, 1978, formerly director, Programme Coordination
and Operations, Forestry Department, FAO

The first books in this chapter discuss the extent of the deforestation problem, along with conclusions about sound practices to protect the forests while using them to satisfy human needs. **China: Forestry Support for Agriculture** *offers a fascinating national case study of successful reforestation for maximum agricultural benefit, while* **Reforestation in Arid Lands** *represents a general practical manual.* **Forestry for Local Community Development** *and* **Community Participation in African Fuelwood Production** *shed light on the requirements for successful village woodlots and other fuelwood replanting projects.* **Tree Crops: A Permanent Agriculture** *notes that trees conserve soil far better than row crops on hilly terrain, and argues that conversion to tree crops is the only choice that will maintain the long-term productivity of agriculture in these areas.*

As firewood use continues to be a major factor in deforestation, the promotion of efficient low-cost locally-built cooking stoves appears to be the most cost-effective first step towards conservation in many areas (see the chapter ENERGY: WOOD STOVES). Fast-growing tree species are also getting a great deal of attention. The National Academy of Sciences book **Firewood Crops** *is one new inventory of fast-growing species, and other books reviewed here cover particular species and growing techniques.*

Village forest industries are the final topic in this chapter. Timber drying, through both regular kilns and solar dryers, is an important step in the production of good quality hardwood for tool handles and furniture. The use and repair of chainsaws, and chainsaw attachments for board production, are covered in the last few entries.

In Volume I:

Losing Ground, by Erik Eckholm, 1976, page 76.

Forest Farming, by J. Sholto Douglas and Robert de J. Hart, 1976, page 77.

Selection and Maintenance of Logging Hand Tools, ILO, 1970, page 114.

Alaskan Sawmill, Granberg Industries, 1976, page 114.

Planting for the Future: Forestry for Human Needs, Worldwatch Paper 26, booklet, 64 pages, by Erik Eckholm, 1979, $2.00 from Worldwatch Institute, 1776 Massachusetts Avenue N.W., Washington D.C. 20036, USA.

Eckholm describes how the world's most extensive and productive forests (in the humid tropics) are severely threatened. "There are currently about 75 cubic meters of wood in the world's dense forests for every person. By the end of the century, however, the per capita amount of exploitable timber will be nearly cut in half if the current deforestation rate is maintained..." Furthermore, these patterns of exploitation are of little benefit to those most dependent on wood in the poorer countries, which export most of their commercial timber as unsawn logs. "Though developing countries contain three-fourths of the world's people and more than half of its forest, they account for just 13 percent of global consumption of industrial wood...in fact, each year the average American consumes about as much wood—one cubic meter—in the form of paper as the average resident in many Third World countries burns as cooking fuel."

The author examines the interlocking causes of deforestation, as well as the economic and ecological implications of continued deforestation. "By the turn of the century, at least a further 250 million people will be without wood fuel for their minimum cooking and heating needs and will be forced to burn dried animal dung and agricultural residues, thereby further decreasing crop yields." China and South Korea, however, seem to have reversed serious deforestation trends with aggressive nationwide programs for community tree planting and management. Such strong political commitment to conservation at the top, with broad participation and shared benefits at the bottom, could be keys to successful reforestation in other countries as well.

A timely, provocative view of world forestry problems and possible solutions.

Trees, Food and People: Land Management in the Tropics, book, 52 pages, by J.G. Bene, H.W. Beall, and A. Cote, 1977, $4.00 (free to local groups in developing countries), from IDRC.

"It has been predicted that within the next 25 to 30 years most of the humid tropical forest as we know it will be transformed into unproductive land..."

This book outlines methods that can prevent deforestation, reforest large areas, and constructively utilize forests for production of important raw materials. An extensive list of references is included for people interested in further information.

"Agroforestry is defined here as a sustainable management system for land that increases overall production, combines agricultural crops, tree crops, and forest plants and/or animals simultaneously or sequentially, and applies management practices that are compatible with the cultural patterns of the local population. Trees are the dominant natural vegetation in most of the tropics, and with few exceptions must remain so if the land is to be used for the greatest benefit of man."

The following topics are covered: tropical forest overexploitation and under-utilization; the environmental significance of tropical forests; tree production systems; forest resource utilization; resource development limitations; research needs and priorities; and a proposal for an International Council on agroforestry. There are useful charts and photos. The authors do not discuss the potential use of tropical forests as places for ecologically-sound human habitation, rather than just sources of raw materials. Still, this is an important book and recommended to anyone working on or interested in the concept of agroforestry and the problems of tropical forests.

Man and Tree in Tropical Africa, booklet, 31 pages, by Gunnar Poulsen, 1978, publication no. IDRC-101e, free to local groups in developing countries, others $2.00, from IDRC.

"The first of the three papers in this book examines, in general, the role of the tree in tropical Africa. Highlighted are the variety of products that can be obtained from the forests, the vital role of trees in nutrient cycling and in soil

Table 1. The uses of wood from the plantations and natural forests of Africa.

Forest products	Uses	Approximate consumption (%)
Fuel	firewood, charcoal	90
Poles/posts wood-splits	house building, fencing, transmission (electrical), scaffolding	5
Sawn wood (roughly hewn, pit sawn, machine sawn)	construction, form work (concrete), joinery, flooring blocks, furniture, agricultural implements, exports (mainly fine timber)	4
Carvings	Makonde carvings etc., all kinds of tourist curios, drinking vessels for livestock, saddles	extremely small quantities but partly very valuable wood
Panel products	wood-wool cement, veneer, plywood, particle board (chipboard), fibre board (hard and soft board)	0.3
Pulp, paper, and related products	news print, fine paper, (printing and writing paper), industrial paper (packaging etc.), rayon (regenerated fibres used for clothing, tire cords etc.)	0.7

and water conservation, and their influence on both micro and macro climate. The second essay addresses the ever important question of wood-fuel supplies. In some areas, families are now spending almost 50% of their income on fuel. Wood-fuel shortages have led to a switch to other fuel sources such as manure and crop residues, which has in turn started a vicious circle of decreasing crop yields and environmental degradation. The final essay examines the age-old practice of shifting cultivation. Although it was well-suited to the conditions under which it was originally practiced, recent pressures resulting from increasing population have caused an imbalance in the system with associated problems such as increased erosion and leaching." Some new practices that might be added to this traditional system are suggested, including zero tillage and the use of trees as 'nutrient pumps'.

Highly recommended.

Trees for People, book, 52 pages, by Clyde Sanger, 1977, no price, publication IDRC-094e from IDRC.

Forestry research programs supported by the International Development Research Center are described in this booklet. The history and goals of many of the programs are briefly outlined.

"Each person in towns in the semi-arid areas verging on the Sahara needs one cubic meter of stacked firewood a year for cooking and heating purposes. It takes two hectares of natural forest to supply one townperson's needs and, as cities grow, that means a wider and wider search for fuel. It is estimated that by 1990 people will be hauling firewood into Ouagadougou from a radius of 150 km around the Upper Volta capital. But a single hectare of Australian Eucalyptus in a plantation irrigated from a river can supply the needs of 50 people."

Addresses and names of persons to contact for further information are given for each program described. A useful book for anyone interested in examples of specific reforestation programs.

Tropical Moist Forests Conservation Bulletins 1 and **2**, booklets, 51 and 29 pages, edited by C. Mackie, G. Ledec, and L. Williamson, 1978 and 1979, available from International Project, Natural Resources Defense Council, 1725 I Street N.W., Washington D.C. 20006, USA.

These bulletins "describe briefly some of the major international and national institutions and programs which contribute to the protection and wise use of the tropical moist forests," including efforts to stop deforestation, and programs of reforestation. Full addresses are provided.

China: Forestry Support for Agriculture, FAO Forestry Paper No. 12, book, 103 pages, FAO, 1978, $6.50 from UNIPUB.

This is the report of an FAO/UNDP-sponsored study tour in 1977, "to observe and analyse the Chinese approach to forestry development whereby it is integrated into and supports agriculture."

"Stricken by a series of natural calamities throughout history, China appears determined to tame rivers, regulate water systems, reverse soil erosion, establish a favorable climatological balance and thus banish the feeling of helplessness against natural disasters. Forestry has played a major role in achieving these objectives."

The participation of the people has been a central concept in China's forestry efforts. Research activities concentrate on practical problems, and include commune members; much is learned from the practical experiences of field workers. Education of the people is seen as a requirement for successful tree planting programs. As a result the average Chinese is "much more know-ledgeable about forestry than the average person in any other country," and protection of reforested areas is not a problem. Forested lands have doubled since 1949.

Tree planting has had direct economic benefits in the form of timber, fuelwood, livestock fodder, fruit and other products. In some areas shelterbelt forestry is considered the primary factor in dramatic agricultural gains, ahead of irrigation, fertilization, and improved seeds.

"The team left with impression that the achievement of the PRC in environmental treeplanting would lie beyond the capacity of any organized forest service if such work were to be carried out in a system where labor had to be compensated directly by money wages. In every commune, 'four around' tree planting is an integral part of the commune's economic activity and tree planting is everybody's business."

The authors give particular attention to the organizational framework within which forestry work is conducted, the major kinds of forestry programs (shelterbeds, 'four around' tree plantations, afforestation of bare land, inter-cropping), and fast-growing tree species commonly planted. The report concludes with notes of possibilities for adaptation of Chinese practices in other countries.

Forestry for Local Community Development, book, 114 pages, FAO Forestry Paper No. 7, prepared by an FAO panel, 1978, available in English, Spanish and French, from Distribution and Sales Section, FAO, Via delle Terme di Caracalla, 00100 Rome, Italy.

Forestry for Local Community Development offers a summary of what is known about constraints facing the rural poor that affect forestry, programs that address these constraints, and policy measures that have succeeded in different places. The study concentrates on programs in which rural com-munities process and use the forest products themselves; it excludes large scale industrial forestry.

"Forestry for community development must...be forestry for the people and involving the people. It must be forestry which starts at the 'grass roots' "

"The timescale of forestry is bound to conflict with the priorities of the rural poor, which are logically focussed on meeting basic present needs...Forestry can continue to exist or be introduced at the community level only if it allows for these real present needs."

"The core of the problem for forest communities is...usually that they derive insufficient benefit from the forest...This...is often attributable to conventional forest management objectives and administrative practices, an orientation towards conservation, wood production, revenue collection, and regulation through punitive legislation...The task of forestry for the development of such communities is consequently to engage them more fully, positively, and beneficially in its utilization, management and protection. This may take the form of...logging or sawmilling cooperatives...production of honey...the con-current production of forestry and agricultural crops, or...grazing of animals... This can require quite radical reorientation of traditional forestry concepts and practices."

"A feature of most successful recent community forestry endeavours has been a strong, sustained technical support system, capable of providing advice and essential inputs such as planting stock, and of maintaining such support through the period necessary to generate forestry as a self-sustaining activity in a particular area."

Key factors affecting success or failure of forestry programs are identified and summarized in seventeen brief case studies. An appendix describes forest products other than fuelwood and timber. A list of annotated references is included.

Required reading for anyone who is involved in community forestry programs.

Community Participation in African Fuelwood Production, Transformation, and Utilization, paper, 72 pages, by Marilyn W. Hoskins, November 1979, available on request from Office of Development Information and Utilization, Development Support Bureau, U.S. Agency for International Development, Washington D.C. 20523, USA.

Drawing from actual case studies, this discussion paper reviews the role of the community development approach in fulfilling fuelwood energy needs. Ms. Hoskins acknowledges that there have been many failures in fuelwood projects, and that the time has come to learn from these efforts.

Two very interesting concepts described in this paper are the "Project Package Approach" and the "Management Plan Agreement". The first means that a given fuelwood project will be integrated into larger ongoing community projects. The second concept is very unique. It requires that all parties to a project actively participate in design, negotiation and agreement to terms of a particular fuelwood scheme. This should involve donors or project initiators (local, private voluntary, national, or international groups), national government administrators, and representatives of the local people. The application of this concept may not assure good projects or prevent bad ones, but it at least makes clear what responsibilities each actor is to assume.

Highly recommended.

Reforestation in Arid Lands, book, 258 pages, by Fred R. Weber, 1977, $6.50 from VITA.

Though designed for use by people working on reforestation programs in sub-Saharan West Africa, this book would be useful in other arid areas. It can be considered both as a model for development of reforestation manuals in other regions and as a guide to West African trees potentially useful elsewhere. There are special sections on windbreaks, fire protection and sand dune stabilization.

"Reforestation programs are part of larger conservation efforts. Increasingly they are being conducted with the realization that it is very difficult to separate reforestation from other revegetation efforts—range management, sand sta-

Intercropping trees and row crops

bilization and similar activities. So while reforestation deals mainly with planting trees in locations able to support at least some species, it is important to think broadly of revegetation—planting trees, shrubs, bushes, grasses, and other ground cover in areas which do not have sufficient vegetation."

Fully half of the book is made up of Appendices A and B. The first is a directory of 165 tree varieties found in West Africa, many of which can be found elsewhere in the world too. Appendix B is an expanded look at 30 of the trees in Appendix A, covering details on such topics as their seeds, germination techniques, transplanting, protection and uses.

"This manual assumes basic familiarity with reforestation terms and methods: for example, it takes for granted that the reader will be familiar with laterite soils and with the use of such forestry tools as climate maps and vegetation charts."

Revegetation is a concept that needs wider circulation. The planting of many different suitable types and sizes of vegetation makes the widest possible use of the capacity of a landscape to support plant life. A wide range of plant life also further expands the amount and diversity of animal life that the area can support, and this includes the human animal. A community of plants composed almost solely of trees ignores the potential that shrubs and ground covers can contribute to the productivity of a landscape: animal life, both wild (deer and fowl for instance) and domestic (such as pigs, cattle and geese) will not do as well when there are only trees. Re**forestation** usually creates a place for humans to come and get lumber and firewood and little else. Re**vegetation** creates a place for a greater variety of plants and a larger number of animals, including humans to live...and provides lumber and firewood too.

"A conservation project must be supported by the people living in an area, or it will not work. Local people are the ones who may be asked to give land for a project, or to work on it. And often a reforestation effort will have to be supported by people for years before results can be seen. Therefore, a project should not be started before communities are ready to sustain the effort. And to make this commitment, residents must believe that (1) the project will affect their environment and their lives positively, and (2) the results will be worth the effort."

The book is a very important one. Highly recommended for anyone working on reforestation or revegetation anywhere in the world.

Firewood Crops: Shrub and Tree Species for Energy Production, book, by National Academy of Sciences, 1980, free from Commission on International Relations, JH 215, National Academy of Sciences, 2101 Constitution Avenue, N.W., Washington D.C. 20418, USA.

"To alleviate the growing shortage of wood fuel is one of mankind's major challenges. In this connection, firewood research is vital and deserves concentrated financial support. It will take the combined efforts of government, industry, landowners, villagers, researchers, philanthropic institutions and development assistance agencies. Some activities that must be undertaken include:

• Preventing the extinction of existing forests;
• Instituting policies to relieve the often wasteful use of the firewood now available;
• Testing and developing fuel-efficient stoves;
• Instituting policies and programs to encourage the use of alternative energy sources such as biogas and solar heat;
• Testing the cultivation of native tree species for firewood; and

• Testing appropriate new species such as those identified in this report."

Firewood Crops will be available by August 1980, according to NAS staff. It describes woody species suited for use as fuelwood or charcoal in rural developing areas where firewood shortages are reaching a crisis point.

An introduction by Erik Eckholm points out the urgent need for fuelwood programs throughout the world—and the considerable difficulties in actually implementing them. A chapter entitled "Wood as Fuel" presents an overview of wood energy uses, firewood plantations, fuelwood management, harvesting techniques, species selection, and appropriate research methodologies. The intent throughout is to provide options, not specific recommended solutions.

Approximately 60 species for use in the wet-dry lowland tropics, savanna regions, arid areas, and tropical highlands are presented in the main body of the report. At the time we went to press, these chapters were being completed and organized. The format is similar to previous NAS publications, with extensive photographs, references, seed and germplasm sources, and research contacts. For each species listed, other uses besides fuel will also be cited.

"Woody plants...can also be sources of: vegetable oil and fruits and nuts for food; edible leaves and shoots for sauces, curries, salads, and beverages; forage for livestock and silkworms; green manure for fertilizing soil; medicines and pharmaceuticals; extractives such as resins, rubber, gums, and dyes... In times of hardship, (the tree owner) may sacrifice some tree growth to feed his family or animals with the foliage. In some cases, dense forests can produce a great deal of burnable materials without a living tree being felled. In others, the owner may sell the best-farmed trees for timber or pulp and use the remainder as fuel. Having such options is important to a rural farmer, and in this report we note the main alternative uses for the species selected, even if they conflict with firewood use."

The technical appendices promise to be very valuable: a) a compendium of fuel-efficient stoves; b) a master list of firewood species; and c) firewood success stories from Ethiopia and South Korea.

This book should become essential reading for development workers, extension agents, foresters, and researchers throughout the world.

Tree Crops: A Permanent Agriculture, book, 408 pages, by J. Russell Smith, 1953, reprinted 1978, $5.95 from Harper and Row Publishers, Inc., Keystone Industrial Park, Scranton, Pennsylvania 18512, USA.

"Forest—field—plow—desert—that is the cycle of the hills under most plow agriculture...Field wash, in the United States, Latin America, Africa and many other parts of the world, is the greatest and most menacing of all resource wastes...We are today destroying our soil...faster and in greater quantity than has ever been done by any group of people at any time in the history of the world."

Written 25 years ago, this is still considered one of the most important texts on the agricultural potential of tree crops. "Agriculturalists have completely overlooked the abundant food produced by such trees as the oaks, honey locust, persimmon, and walnut, which...can outproduce, acre for acre, the best efforts of the grass family (corn, wheat, oats) on most lands in formerly forested areas. Moreover, tree crops require less care, bind and improve instead of depleting the soil, and provide a permanent source of income which increases annually."

"If much of the tropic forest is to be preserved, we must make use of tree crops. Tree crops will safeguard fertility while producing food for man. In most cases there can be an undergrowth of leguminous nurse crops of small tree and bush to catch nitrogen, hold the soil, make humus and feed the crop trees—nuts, oils, fruits, gums, fibers, even choice weeds."

"The crop-yielding tree offers the best medium for extending agriculture to hills, to steep places, to rocky places, and to the lands where rainfall is deficient."

"Experiments with trees can be on almost any scale. Two trees, for example, might produce great (hybrid) results. There are thousands of individuals who can experiment and perhaps do something of great value."

Most of this extraordinary book consists of descriptions of the characteristics and uses of what Smith felt were the 35 most promising tree types for temperate and tropical climates. His photos and personal observations from years of traveling throughout the world add considerably to the impact of the book.

Natural Durability and Preservation of One Hundred Tropical African Woods, book, 131 pages, by Yves Fortin and Jean Poliquin, 1974, free to local groups in developing countries, $10.00 to others, from IDRC.

This is a report on the preservation requirements of 100 different tropical African woods. 'Natural durability' refers to the ability of the wood to resist attack by biological agents—fungi, insects, and marine borers were chosen as specific cases. Many woods have certain uses which require little or no preservative treatment, due to this natural durability.

"The protection obtained from a preservative treatment is determined by the effectiveness of the preservative as well as the method of its application. The choice of a suitable preservative is mainly based on the conditions to which the wood is to be exposed. For example, before the wood is utilized, preservatives made of chemicals dissolved in organic solvents, and non-leachable salt preservatives usually give satisfactory protection."

The authors mention the hazards of use of some of the chemical preservatives, plus some safety instructions. Non-commercial preservation techniques are aimed at medium- to large-scale operations, but small-scale operations will also find the book very useful.

Anyone using common African woods will find material of interest in this book. There is an extensive list of sources for further information.

Leucaena: Promising Forage and Tree Crop for the Tropics, book, 121 pages, by the National Academy of Sciences, 1977, free from Commission on International Relations, JH 215, National Academy of Sciences, 2101 Constitution Avenue, N.W., Washington D.C. 20418, USA.

"Of all tropical legumes, leucaena probably offers the widest assortment of uses. Through its many varieties, leucaena can produce nutritious forage, firewood, timber, and rich organic fertilizer. Its diverse uses include revegetating tropical hillslopes and providing windbreaks, firebreaks, shade, and ornamentation. Although individual leucaena trees have yielded extraordinary amounts of wood—indeed, among the highest annual totals ever recorded— and although the plant is responsible for some of the highest weight gains measured in cattle feeding on forage, it remains a neglected crop, its full potential largely unrealized."

Leucaena varieties can grow in arid areas (though they do best in moist conditions) and can tolerate periods of frost and high winds.

"There is a rising belief among agronomists and foresters that tree growing, crop production and/or animal raising should be combined to best preserve structure and fertility of fragile tropical soils. Trees provide the ecosystem, and an agricultural crop, livestock rearing, or fish culture can provide income while the trees are maturing. Combinations of many different plant and animal species seem possible, but versatile leucaena appears to be an outstanding candidate.''

Leucaena is especially useful in reforestation efforts where it's important to get quick results for ecological or economic reasons. The plant grows very rapidly and because it is a legume it also enriches the soil for the benefit of other plants.

''Leucaena helps to enrich soil and aid neighboring plants because its foliage rivals manure in nitrogen content, and natural leaf-drop returns this to the soil beneath the shrubs. Recent experiments in Hawaii have shown that if the foliage is harvested and placed around nearby crop plants they can respond with yield increases approaching those affected by commercial fertilizer.''

The book covers the following topic areas: leucaena botany and cultivation;

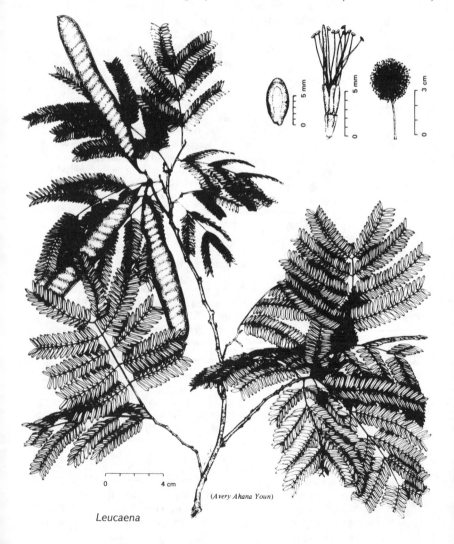

(Avery Ahana Youn)

Leucaena

animal feed; wood products; fuelwood; soil improvement and reforestation; recommendations and research needs; sources of additional information; a list of leucaena researchers; and sources of leucaena seeds, nitrogen-fixing innoculant bacteria, and wood samples. An excellent general survey of the potential of leucaena in tropical, sub-tropical and mild temperate climates.

A Forest Tree Seed Directory, book, 283 pages, by FAO, 1975, $20.25 from UNIPUB.

This is a directory of sources of tree seeds of many varieties. It includes an enormous amount of information on tree seeds, including the number of seeds per kg, germination percentage, and seed treatment applied.

A special remarks section covers such things as germination techniques, ordering delays to be expected, quantities available, and rarity of the seed. This is a very important book for anyone undertaking revegetation or reforestation programs or other nursery projects that require supplies of tree seeds.

All of the text is in English, French and Spanish.

Short-Rotation Forestry, report, 36 pages, by Geoffrey Stanford, 1976, $5.00 from Greenhills Foundation, Route 1, Box 861, Cedar Hill, Texas 75104, USA.

"Coppicing" has a long history in Europe. It consists of growing young trees very close together, and harvesting the growth after 3-5 years during the winter season. New growth comes up from the stump and the cycle is repeated.

This report contains an overview of coppicing history, principles, and yields. Coppicing "was not just a way of increasing the yield of fuelwood from stumps near to the village, it was a means for securing construction timber of the right size other than by selection from a natural mixed forest. These coppices also furnished the wood for the enormous quantity of baskets, barrels, tubs, and pails."

"Coppicing has two important advantages over mature timber: firstly, the yield/hectare/year can be many times greater; and secondly, repeated harvestings at intervals of 3-7 years provide a much shorter-term return on invested capital."

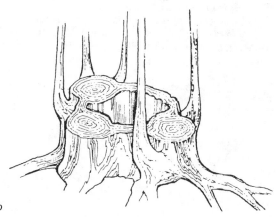

New growth on a stump

Agroforestry Review, quarterly journal, 25 pages average length, subscriptions $6.00 per year in the U.S., $8.00 in other countries, from International Tree Crops Institute USA, P.O. Box 1272, Winters, California 95694, USA.

''**Agroforestry Review** contains articles related to the multiple utilization of trees and shrubs for human food, livestock feed, fuel, conservation purposes, etc.'' Articles examine specific tree species, relate original research findings, and report on applications of tree cropping in the USA and other countries. Also includes reviews and abstracts of books.

Good material on trees for community self-reliance. Probably most useful to researchers and experimenters in the temperate zones.

Timber Drying Manual, book, 159 pages, by G.H. Pratt, 1974, Ł4.60 plus postage, Publication 0.11.670521.3, from Her Majesty's Stationery Office Bookshop, P.O. Box 569, London SE1 9NH, England.

This book is a ''...complete guide to all methods of drying timber. With a total of more than sixty illustrations, this book represents the culmination of nearly fifty years of research on timber drying at the Princes Risborough Laboratory.''

It covers such topics as timber moisture, kiln operation, drying damage, air drying timber, kiln types, drying various types and loads of timber, and other drying methods (including vacuum, steam, vapour and press methods, plus solvent and salt seasoning) for both small- and large-scale drying operations.

''Experience has shown that satisfactory kiln drying can usually be best accomplished by gradually raising the temperature and lowering the humidity of the circulating air as drying proceeds...It has already been indicated that the rates at which different timber species can safely be dried, and the air conditions to which they can be subjected without suffering damage, vary very considerably, and the treatment should, therefore, depend to a large extent on the species that is being dried.'' The appendices cover most of the woods of the world and describe in detail the proper drying procedures for each type of wood. Other technical sections give specific details on testing humidity of wood and air, and redrying timber treated with chemical preservatives. There is an excellent section on troubleshooting timber drying problems, complete with tables of symptoms and cures.

The basic principles of kiln drying described here apply to small- and large-scale operations, solar and traditional heating systems, and community or commercial undertakings. Highly recommended.

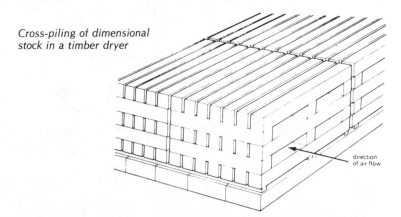

Cross-piling of dimensional stock in a timber dryer

direction of air flow

Constructing and Operating a Small Solar Heated Lumber Dryer, report, 12 pages, Paul Bois, 1977, Forest Products Utilization Report 77, free in limited quantities from U.S. Forest Service, Division of Cooperative Forestry, Box 5130, Madison, Wisconsin 53705, USA.

This report briefly describes the construction and operation of a small solar lumber dryer, designed for cold and temperate latitudes (modifications may be required for tropical operation). The advantage of this solar dryer is that hardwood lumber can be dried in it to a significantly lower moisture content than by air drying alone. Very dry wood is important for uses such as furniture.

Three photos and three small sketches of construction details are provided. The dryer uses an air collector and a fan to circulate the air.

Not intended for large, high-speed drying operations. The dryer has a capacity of 750-850 board feet of 8-foot lengths of hardwoods, requiring about 80 days to dry.

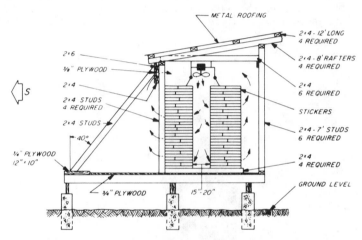

Solar heated lumber dryer: arrows indicate movement of air driven by overhead electric fans

Chain Saw Service Manual, book, 312 pages, by Technical Publications Division, Intertec Publishing Corp., $8.00 from META.

If you're interested in learning how to repair and maintain chainsaws, this is the book for you. The first section (37 pages) covers chainsaw engine principles, troubleshooting, maintenance and repair of all the parts of a

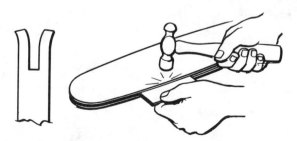

Closing spread bar rails: use steel shim .004" thicker than drive link tangs

chainsaw.

"Small kinks or bends in guide bars can be removed by laying the bar on a large true (flat) anvil or other similar work surface and using light hammer blows to bring the bar back into shape. The technique is very similar to straightening other flat metal pieces."

The second section covers, in detail, how to service and repair specific chainsaws available from these firms: Advances, Allis-Chalmers, Clinton, Danarm, John Deere, Dolmar, Echo, Ford, Frontier, Homelite, Husqvarna, Jonsereds, Lancaster, Lombard, McCulloch, Mono, Partner, Pioneer, Poulan, Remington, Roper, Skil, Solo, Stihl, Tecumseh and Wright.

This book covers repair but not use of chainsaws.

Barnacle Parp's Chain Saw Guide, book, 281 pages, by Walter Hall, 1977, $7.95 from RODALE.

The subtitle accurately states that this book helps you in "buying, using, and maintaining gas and electric chain saws." These are very useful tools where

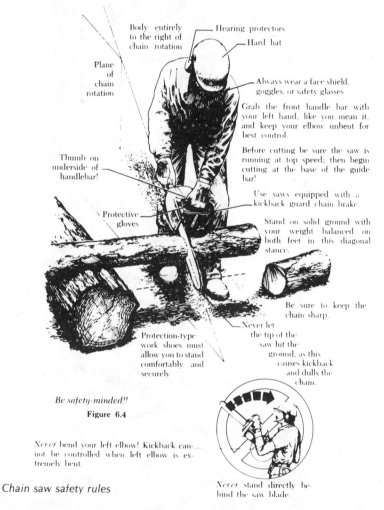

Body entirely to the right of chain rotation

Hearing protectors

Hard hat

Plane of chain rotation

Always wear a face shield, goggles, or safety glasses

Grab the front handle bar with your left hand, like you mean it, and keep your elbow unbent for best control.

Before cutting be sure the saw is running at top speed; then begin cutting at the base of the guide bar!

Thumb on underside of handlebar!

Use saws equipped with a kickback guard chain brake.

Protective gloves

Stand on solid ground with your weight balanced on both feet in this diagonal stance.

Be sure to keep the chain sharp.

Never let the tip of the saw hit the ground, as this causes kickback and dulls the chain.

Protection-type work shoes must allow you to stand comfortably and securely

Be safety-minded!!

Figure 6.4

Never bend your left elbow! Kickback cannot be controlled when left elbow is extremely bent.

Chain saw safety rules

Never stand directly behind the saw blade.

large amounts of timber need to be felled, rapid cutting and processing is important, and a shortage of labor exists.

Basic parts, accessories, safety, and sharpening are presented. Manufacturers addresses, specifications of currently available chainsaws, and periodicals are listed. A very good section on the use of chainsaws is matched with clear descriptions of repair procedures. These make this book a good companion to the **Chain Saw Service Manual** (which presents repair details for specific models).

Highly recommended.

The Chainsaw and the Lumbermaker, booklet, 28 pages, $1.00 from Haddon Tool, 4719 W. Route 120, McHenry, Illinois 60050, USA.

In the few rural parts of the temperate zones where forests are extensive, trees can be cut or bought and used as a low-cost construction material. While trees are usually sawn at a mill, a tool like the ''Lumbermaker'' described in this booklet can be used with a chainsaw to produce rough-cut lumber. The ''Lumbermaker'' allows one person to make straight cuts for boards by guiding the chainsaw along a piece of milled lumber (standard two-by-four) which is nailed to the log. The booklet shows how to use the ''Lumbermaker''

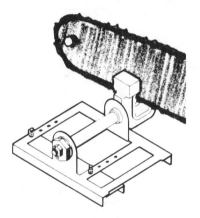

The Lumbermaker attached to a chainsaw

to saw boards of various sizes using different methods of attaching the guide board to the log. Also included are suggestions on sawing angles, braces, making a jig for cross-cutting, and simple log cabin construction. The Lumbermaker is a simpler attachment than that used for the Alaskan sawmill (see review on page 114).

The price of the ''Lumbermaker'' in January of 1980 was $45.00. Matched with a chainsaw, it may have a place in the Third World, between the two-person hand pit saw and the small motorized sawmill. Easily transported, it would seem most applicable in areas where transport of logs to a small mill is impractical. Chainsaws, however, make a wider cut and thus waste more wood than either of the other alternatives. The ''Lumbermaker'' will certainly be most widely used in parts of the U.S. and Canada where wood is still abundant and chainsaws have become widely-owned tools in an affluent consumer society.

ADDITIONAL REFERENCES ON FORESTRY

Fuel-efficient cooking stoves and improved charcoal kilns can both reduce the pressure on remaining trees; they are described in materials reviewed on pages 580-589.

Agro-Forestry Systems for the Tropics East of the Andes, page 437.

The Draft Horse Primer contains 22 pages on the techniques and equipment used in logging with horses; see review on page 464.

AQUACULTURE

Aquaculture

Many people are now pointing to the farming of fish, shellfish and aquatic plants as the solution to the world's food problems. These people note that there are several advantages of water as a growth medium. For instance, aquatic animals can convert more of their food into growth since most of them do not need to support their weight. Most fish don't spend energy to regulate their body temperature. In addition, water is a three-dimensional growing space, so yields per unit area can be quite high when compared to land-based farming.

These advantages have given aquaculture the label of an "appropriate technology". But just as solar energy technology can include solar-panelled satellites beaming microwave energy to the earth, so aquaculture can be approached from a number of technological starting points. Many appropriate technology aquaculture groups in the United States are working with extremely high densities of organisms in recirculating water systems. Some examples of these "intensive" designs are basement fish tanks, backyard fish farms, dome ponds and greenhouse ponds. Most of these require careful biological monitoring and management (because a small problem in the system can kill all the fish) and the economics are not yet acceptable.

While this work is certainly important to our urban areas where little space is available, the high capital and material requirements of such aquaculture strategies make them much less relevant to the developing countries. In fact, many people are convinced that the key to large scale aquaculture development in the United States as well lies in the enormous potential of farm ponds and reservoirs used for irrigation, fire protection, recreation, livestock watering, etc. Throughout the world these unused or poorly managed lakes, ponds, streams and rivers represent a vast resource of harvestable waters. At the same time they are subject to a wide variety of other potential uses. As we manage these water resources, our goal must be expanded from short-term production to long-term stewardship which integrates all potential needs. There are many examples of aquaculture which include sewage treatment, mosquito control, and aquatic weed control. And aquaculture can play a major role in the maximization of traditional fisheries through spawning and ranching techniques—many coastal and inland fisheries are the best producers of cheap protein because the fish raiser does not have to supply the feed.

The following selections partially reflect these views. Much work remains to be done. As an overall source of information to have at your side for constant study during the planning of a fish raising project, start with **Aquaculture: The Farming and Husbandry of Freshwater and Marine Organisms**, and then use Hickling's **Fish Culture** for practical techniques. If you are trying to decide what growing fish might be like, choose between **Fish Culture in Central East Africa** (the most extensive), **Freshwater Fish Pond Culture and Management**, *and* **Elementary Guide to Fish Culture in Nepal**. *The remainder of the selections cover specific topics in aquaculture.* (Mike Connor)

In Volume I:

Aquaculture: The Farming and Husbandry of Freshwater and Marine Organisms, book, 868 pages, by J. Bardach, J. Ryther, and W. McLarney, 1972, $19.00 from Wiley-Interscience, John Wiley and Sons, Inc., 605 Third Ave., New York, New York 10016, USA.

Despite its expense, semi-technical approach and encyclopedic nature, this book is the best investment for any group seriously considering starting an aquaculture project. It is the **only** place where information about aquaculture in every part of the world, including a multitude of species and methods, can be found. The book thoroughly discusses the energy-intensive culture methods used in American catfish, trout and lobster farming, in addition to providing a state-of-the-art treatment of fish culture in Africa, South America and Asia. In many cases this comprehensive coverage, illustrating various approaches to solving particular problems, allows the field worker to combine several different solutions in developing a response to the local situation. The value of this book as a reference guide will be well worth any difficulties encountered with the biological terms.

The book is organized into chapters on each of the various species groups of fish, shellfish, other invertebrates, seaweeds and freshwater plants. Where appropriate, the chapters are subdivided into the different parts of the world where these animals are grown. At the end of each chapter there is an extensive list of reference and/or personal contacts.

"In the developing world the predominant problem is one of producing additional animal proteins, which may be so scarce that any meat, unless excessively cheap, is a luxury commodity available only to the wealthy few. The corollary here is that especially in developing nations herbivores or plankton filter feeders are most suitable for aquaculture, producing the most per surface or volume of water from the more-or-less natural amenities, such as solar energy, existing standing or flowing waters, and natural or man-enhanced fertility."

Fish Culture, book, 317 pages, C.F. Hickling, 1971, (Faber & Faber, Publishers, U.K.), out of print in June 1980, try to see a copy owned by another fish culturist or a library.

The information in this book is particularly suited to tropical climates, where the biology and chemistry of fish ponds requires different management from that usually recommended in the literature by Americans and Europeans. It is based on conditions at Malacca, Malaysia, where the author did much valuable research with minimum dependence on sophisticated technology. Hickling presents the biological basis for pondfish culture clearly, and terms broad enough to be useful in other locations and with other fish than those he studied. His approach to fish culture emphasizes locally available natural materials, and assumes that abundant labor but little equipment is available. He expects the reader to know some general biology as he considers the ecological and chemical relationships of the soil, water, plants and animals in a pond that affect fish production.

The most valuable aspect of this book for the user might be the orientation toward nutritious natural fish foods made of wild or easily-cultured plants. Nutritional data on many tropical plants and agricultural wastes, such as rice hulls, are presented. The fish genetics chapter is somewhat obsolete, but there is good treatment of biologically-significant aspects of water, soil, fertilizers (both natural and synthetic) and feeding. Hickling stresses low-cost methods (where appropriate) throughout the rest of the book as well, in chapters on pond construction, pond management, stocking rates, fish diseases, use of brackish and flowing water, mixed rice-fish culture, and public health. There is an excellent bibliography and index. Highly recommended for those with some background in biology.

Freshwater Fish Pond Culture and Management, book, 191 pages, by Marilyn Chakroff, 1976, free to Peace Corps volunteers and development organizations, from Office of Information Collection and Exchange, Peace Corps, 806 Connecticut Ave., N.W., Washington D.C. 20525, USA; others may obtain it for $6.50 from VITA.

The Peace Corps has done a lot of aquaculture work in different parts of the world. Several good local manuals were written in the past by their Indian branch. This latest publication integrates all the freshwater aquaculture projects, with the emphasis on warmwater species of fish. It is introductory in nature, aimed at an audience which did not like math or science in high school. Good illustrations complement the clearly written text. The contents cover the basic subjects important to a fish farmer: why grow fish, pond site selection,

Pushing eggs or sperm out of a fish, to be mixed in a dish

planning, construction and sealing, water chemistry and fertilization, fish spawning, stocking, feeding, harvesting, preserving and diseases.

"**Freshwater Fish Pond Culture and Management** is a how-to manual. It is designed as a working and teaching tool for extension agents. It is for their use as they establish and/or maintain local fish pond operations. The information is presented here to 1) facilitate technology transfer and 2) provide a clear guide for warmwater fish pond construction and management. A valuable listing of resources at the end of this manual will give further direction for those wishing more information on various aspects of fish pond operation." In fact, the resources section is practically impossible to use. It is not at all integrated into the subjects in the text, and most of the references are only available to those with access to excellent libraries. But this is the only failing of an otherwise good book.

Fish Culture in Central East Africa, book, 158 pages, by A. Marr, M.A.E. Mortimer and I. Van der Lingen, 1966, $6.00 from META.

Of these general "how-to" manuals, this is our favorite. More than any of the other manuals it emphasizes the economic and ecological constraints which

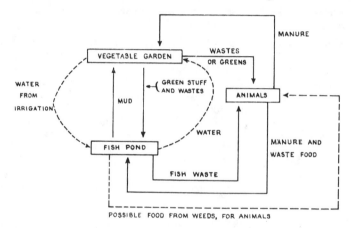

FIGURE 28. Combined fish, crop and livestock farming. The figure shows the relation between ponds, crops, and animals.

FIGURE 21. Making the walls of a contour pond. Soil to make the walls is dug first from the top of the pond. It is dug to a depth of 1 ft. Nearer the bottom wall, soil is dug less and less deep. As the walls are built up, the width is made less and less so that when the top of the posts is reached, the width is only 2 or 3 ft.

demand flexibility by fish farmers. Besides a good summary of the essentials of fish pond management, there is an excellent chapter on growing fish in lakes, reservoirs and seasonal farm ponds. Combining fish farming with other branches of agriculture is also stressed.

The book has a definite regional focus. Tilapias native to Africa are the only fish considered. Nevertheless, many of the ideas would be applicable in other areas. For instance, three aquacultural practices of wide interest are discussed: the polyculture of ducks and fish, the use of other animal manures as fish food and fertilizer, and the culture of fish in irrigated paddies. Fish such as carp or nonherbivorous tilapia can be stocked any time after the rice has rooted in paddies which have a water depth of 30-40 cm, or in paddies with 10-15 cm of water over the central rice growing area surrounded by a 1½ m-wide trench, 70 cm deep.

A glossary and 73 illustrations make the text easier to follow, but the last two chapters on biological production and fish biology are a bit dense.

Elementary Guide to Fish Culture in Nepal, book, 131 pages, by E. Woynarovich, 1975, FAO, $5.50 from META.

"**Elementary Guide to Fish Culture in Nepal** is designed for practical use in the training of extension workers and progressive fish farmers in the techniques of fish culture." This FAO publication emphasizes the culture and polyculture of the common carp, **Cyprinus carpio**, Chinese carps and Indian carps. In addition to the usual discussion of pond construction, management and harvest, there is a chapter on the biological background to fish production which includes information more sophisticated than is necessary for the successful management of a pond. There is also a good section on the food value of various feeds made from agricultural by-products.

Don't be fooled by the word "Elementary." While the book includes many simple illustrations, the author's attempt to cover all ecological processes very briefly can be confusing to someone who does not already know about them.

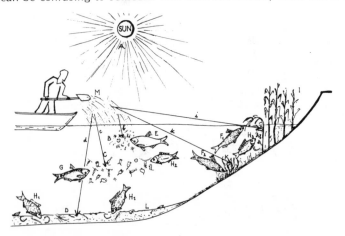

Food relationship in a pond stocked with a mixed fish population
A—solar energy, B—microphytoplankton, C—zooplankton, D—bottom animals, E—phytoplankton feeder (silver carp), F-1—grass carp as consumer of emerging (hard) plants, F-2—grass carp as consumer of submerged (soft) plants, G—zooplankton and bigger phytoplankton feeder (bighead), H-1—bottom feeder (common carp).

Freshwater Fisheries and Aquaculture in China, book, 84 pages, D.D. Tapiador, H.F. Henderson, M.N. Delmendo, H. Tsutsui, 1977, $6.00 from UNIPUB.

China accounts for more than half the world's fish production; yet very little was known about its aquaculture efforts until the visit of this FAO mission in 1976. "The particular forms of fish culture practised in China may not be directly applicable in many countries, particularly outside Asia. But the perspectives of Chinese fish farmers on self-reliance and on the inter-dependence of aquaculture, agriculture and animal husbandry, and their familiarity with fish and fish behavior under conditions of intensive culture, make their experience most valuable elsewhere."

"It seemed particularly significant that all of the major inputs, such as feed, fertiliser and fish seed, are produced within the farm...The use of organic fertilisers and locally-produced feed materials is especially to be recommended for most of the developing countries. Unfortunately, the latter have often elected to adopt commercial fertilisers and feeds simply because it is the practice in the developed countries."

Many people talk about making full use of water resources, recycling wastes and decentralizing planning decisions. In China they do it, and on a scale that produces four million tons of fish annually.

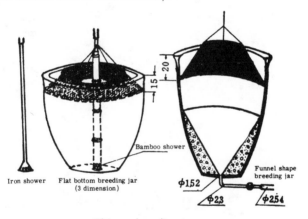

Chinese breeding jars

Making Aquatic Weeds Useful: Some Perspectives for Developing Countries, book, 175 pages, National Academy of Sciences, 1976, free from Commission on International Relations, JH 215, National Academy of Sciences, 2101 Constitution Avenue, N.W., Washington D.C. 20418, USA.

Aquatic weeds present serious problems to public health, fisheries production, water quality and navigation in the tropics where they grow most prolifically. "This report examines methods for controlling aquatic weeds and using them to best advantage, especially those methods that show promise for less-developed countries. It emphasizes techniques for converting weeds for feed, food, fertilizer and energy production. It examines, for example, biological control techniques in which herbivorous tropical animals (fish, waterfowl, rodents and other mammals) convert the troublesome plants directly to meat."

The major sections of this book focus on harvesting aquatic weeds either by herbivores which themselves can be harvested (e.g. grass carp, manatees, crayfish, ducks and geese) or by machines with additional treatment and processing. Throughout the book, the emphasis is on aquatic weeds as a resource rather than a nuisance. No information is given on growing the various plants, which might prove useful in certain wastewater treatment methods.

There is a good list of research contacts in the appendix.

Profitable Cage Culture, booklet, 30 pages, by Gregor Neff and Paul Barrett, 1979, $3.00 to U.S., Canada or Mexico, $4.00 elsewhere, from Inqua Corporation, P.O. Box 86, Dobbs Ferry, New York 10522, USA.

This is the most in-depth summary of the hows and whys of cage culture available—in part a publicity promotion for their brand of plastic mesh cages. If you keep in mind that indigenous materials can be used for the cage mesh and framing, and you have access to alternative feeds, then cage culture can add to your fish raising options. **Profitable Cage Culture** can give you hints on stocking, harvesting and managing your cages.

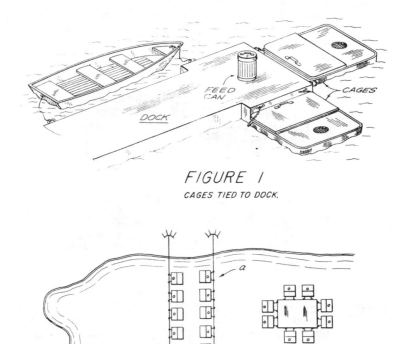

FIGURE I

CAGES TIED TO DOCK.

FIGURE 2

(a) CAGES FASTENED TO CABLE ANCHORED AT BOTH ENDS TO OPPOSITE SHORES.

(b) CAGES FASTENED TO FLOATING RAFT.

(c) CAGES CONNECTED IN PARALLEL TO CABLES.

Aquaculture publications from New Alchemy Institute (P.O. Box 432, Woods Hole, Massachusetts 02543, USA). While this group's publications are reviewed elsewhere (see pages 48, 144, 379), certain articles merit special attention by prospective fish farmers. (Note that Journal 3 is out of print.)

1) ''Midge Culture,'' by W. McLarney, J. Levine and M. Sherman, Journal 3, on pages 80-84.

Providing cheap sources of natural protein for fish feed is one way to speed their growth. Bug lights will collect protein-rich insects. An alternative is to raise the larvae yourself. This article tells you how.

2) ''A New Low-cost Method of Sealing Fish Pond Bottoms,'' by W. McLarney and R. Hunter, Journal 3, on page 85. Also found in **Book of the New Alchemists**, reviewed on page 379.

Various methods have been used to seal the bottoms of ponds to allow them to hold water, but most methods are expensive. Here the authors describe a virtually cost-free method using layers of manure and other farm wastes to create an anaerobic zone impenetrable to water.

3) ''Cultivo Experimental de Peces en Estanques,'' Journal 3, on pages 86-90. Also reprinted in **Book of the New Alchemists**.

This is a translated excerpt from a paper by Professor Anibal Patino R. which presents a plan for tropical aquaculture. For information on obtaining the original paper (in Spanish), write Cespedesia, Jardin Botanica del Valle, Apartado Aereo 5660, Cali, Colombia.

4) ''Cage Culture,'' by William McLarney, Journal 4, on pages 77-82.

Growing fish in floating cages is a traditional technique in Southeast Asia and of recent interest in the U.S. This article describes the reasons and methods for building and stocking cages. It also describes some of the pitfalls.

Raising Fresh Fish in Your Home Waters, pamphlet, 34 pages, B. Bortz, J. Ruttle and M. Podems, 1977, $1.00 from Barebo, Inc., 576 North St., Emmaus, Pennsylvania 18049, USA.

For the reader new to the idea of raising fish, this booklet is useful as a brief introduction to many of the topics of concern to a fish farmer. The central section, ''A Catalog of Fish,'' is a nice collection of the water quality tolerances

Position the net so that it covers the pond from bank to bank and crosses the very bottom.

and preferences of the major fish cultured in the U.S. While the pamphlet is written for the North American fish farmer, the sections on pond management are of general interest.

Fish Catching Methods of the World, book, 240 pages, Andres von Brandt, 1972, Ł7.75 from Fishing News Books Ltd., 1 Long Garden Walk, Farnham, Surrey, England.

This book is a testimony to the ingenuity of fishermen in their invention of an astounding variety of fishing gear and techniques to meet different environmental, economic and social requirements. While basically a scholarly treatment of the principles of fishing technology, its comprehensive discussions of both commercial and subsistence technologies present a fascinating tale to the lay reader. Of particular interest are the chapters on fish hooks, traps and nets. The potential utility of this book lies in the large number and variety of methods it presents.

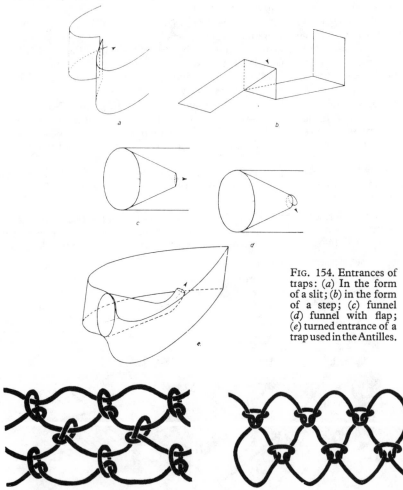

FIG. 154. Entrances of traps: (*a*) In the form of a slit; (*b*) in the form of a step; (*c*) funnel (*d*) funnel with flap; (*e*) turned entrance of a trap used in the Antilles.

FIG. 196. This illustrates the net-making technique used by lake-dwelling fishermen in Switzerland.

FIG. 198. Netting made with the well-known reef-knot.

Aquaculture Practices in Taiwan, book, 161 pages, by T.P. Chen, 1976, £5.00 from Fishing News Books Ltd., 1 Long Garden Walk, Farnham, Surrey, England.

T.P. Chen has provided a sampling of Taiwan's aquacultural practices for 29 species of animals including turtles, frogs, fresh and saltwater clams, shrimps and eighteen species of fish. The major emphasis is on milkfish, eels and Chinese carps with a wealth of production statistics and economics. This book is the best source of information on the culture of snakehead (Ophiocephalus), walking catfish (Clarias), mud skipper (Boleophthalmus), Corbicula clams, and the seaweed, Gracilaria.

Practical Shellfish Farming, book, 91 pages, by Phil Schwind, 1977, $8.95 from International Marine Publishing Company, Camden, Maine 04843, USA.

This easy-to-read book tells how to grow shellfish along the northeast coast of the United States. Consequently, it looks at the question of local regulations more deeply than would be appreciated by those living in other areas. The lively writing makes these few dull pages barely noticeable. The book's strength is its discussion of the management of bottom areas or rafts for maximizing shellfish growth. It lacks a good review of the physical characteristics which influence where shellfish larvae will settle and imaginative ways to collect these settled larvae.

Bamidgeh, quarterly magazine, since 1948, $7.50 per year (surface mail), make checks payable to The Fish Breeders Association in Israel, from Bamidgeh Editorial Office, Nir-David 19150, Israel.

Now published in a separate English edition, **Bamidgeh** presents the results of Israeli research in pondfish culture. In recent years, their semitropical systems have centered on carp, mullet and tilapia. Although some of the articles are extremely complex, using sophisticated engineering and ecological analysis, many show practical techniques and equipment that can be useful in other countries. The Israeli communes have never been wealthy, so the scientific approach is based on intermediate-level technology, using a maximum of ingenuity and energetic labor.

The topics covered in this magazine are usually strategies for efficient fish

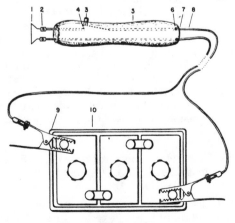

Electric fish brander connected to a car battery

production, such as pest control, safe use of manure fertilizers/feeds, and various species combinations in ponds. This is a magazine for those actually engaged in fish culture, who may want to use agricultural wastes for fish food, culture insect larvae for the same purpose, or start a selective-breeding program to reduce the number of bones in a fish. Particularly interesting recent research has been in all-male hybrids of tilapia, cage culture of carp, and comparison of low-cost aeration systems. If for no other reason, practicing fish culturists might consult **Bamidgeh** to appreciate the value of good record-keeping journals. Subject indexes are published each year for easy reference.

U.S. Government Publications on Aquaculture, order by title and number, from U.S. Government Printing Office, Superintendent of Documents, Washington D.C. 20402, USA.

A great deal of aquacultural work has been done by a few universities and the government. While the focus is often towards large scale, labor-extensive productions, much of the biological information is of great value to fish farmers of all types. This information comes in the form of technical papers, popular articles, how-to pamphlets, manuals, etc.

The following are available from the U.S. Government Printing Office: **Trout Ponds for Recreation**, Farmers' Bulletin No. 2249 (SN001-000-01455-2), $0.80; **Warm Water Fish Ponds**, Farmers' Bulletin No. 2250 (SN001-000-01533-8), $0.35; and **Ponds for Water Supply and Recreation**, Agricultural Handbook No. 387, $0.70 (good pond construction hints).

Publications, Department of Fisheries and Allied Aquacultures, Swingle Hall, Auburn University, Auburn, Alabama 36830, USA.

This is the most comprehensive collection of publications concerning the culture of freshwater, warmwater fishes. While many of these articles have appeared in scientific or trade journals, most of them would not be difficult for the non-professional to understand. Topics include aquatic ecology and marine biology, aquatic plants, baits and minnows, commercial fish production and aquaculture, farm pond management, fishery biology and population dynamics, fish feeds, fish food habits and nutrition, and water quality and waste management.

Foremost among the 300 available papers is H.S. Swingle's classic ''Biological Means of Increasing Productivity in Ponds'' in which he discusses use of efficient pond fish, polyculture species combinations and stocking rates, control of reproduction, and increasing production of fish food organisms. Swingle's successful combination of pond ecology with aquaculture has been unmatched and will be the key to future development of appropriate technology in aquaculture.

Auburn also has an international branch for aquacultural development in conjunction with USAID. Some of the publications from these groups would be of local interest to those people living near the various projects in El Salvador, Brazil, Colombia and the Philippines.

Aquaculture and Related Publications, of the School of Forestry and Wildlife Management, Publications Clerk, 249 Ag Center, Louisiana State University, Baton Rouge, Louisiana 70803, USA.

These reprints are divided into four major categories: crawfish, catfish, frogs and mariculture. Within these categories, most of the emphasis is on pond management and pond ecology. These are folksier than the Auburn reprints. A sample of some of the 100+ titles: "Agricultural By-Products as Supplemental Feed for Crawfish", "How to Keep 'em Alive in a Pond", "Polyculture of Channel Catfish and Hybrid Grass Carp".

Salmon Rancher's Manual, book, 95 pages, William McNeil and Jack Bailey, 1975, free from Alaska Sea Grant Program, University of Alaska, Fairbanks, Alaska 99701, USA.

"The North Pacific Ocean is a vast nursery ground for the Pacific salmon that spawn in streams and lakes in North America and Asia. These salmon reproduce in fresh water, but most of their growth occurs at sea. When mature they return to their freshwater ancestral spawning grounds, where tens of thousands of genetically-separate stocks segregate for reproduction."

In the past, growing salmonid fishes has required a capital investment beyond the means and interests of most groups. But the chemical imprinting of young fish so that they will return to the hatchery area to spawn after several years in the wild reduces the obstacle of their carnivorous habits. Fish can be spawned, hatched, raised and released from a hatchery to increase their chances of survival. While feeding at sea (or, in the case of steelhead rainbow trout, in a lake), the fish convert protein unavailable to man into animal flesh; hence the comparison to range cattle and the term "ocean ranching."

"Production of healthy fry is the 'core' of any salmon aquaculture system because the success of ocean ranching will depend largely upon the quality of juvenile fish released into the ocean. The primary purpose of this manual is to assist salmon ranchers with planning, constructing, and operating systems for artificial propagation of salmon fry."

Artificial Salmon Spawning, pamphlet, 21 pages, William Smoker and Curtis Kerns, 1978, free from Marine Advisory Program, University of Alaska, 3211 Providence Avenue, Anchorage, Alaska 99504, USA.

"This manual is designed primarily for the aquaculturist who is just getting started...(It includes) procedures that are least likely to go wrong for the novice egg-taker. Not all have been scientifically tested. But where they have been used, incubators have been filled with live eggs."

Aquacultural Planning in Asia, workshop report, 154 pages, FAO, 1976, $8.75 from UNIPUB.

Over the past few years, FAO has been holding aquaculture planning sessions in different regions of the world (proceedings are also available for Africa and Latin America). The bulk of these books contain the national aquaculture development plans for selected countries. Included here are Bangladesh, Hong Kong, India, Indonesia, Malaysia, Nepal, Philippines, Singapore, Sri Lanka and Thailand.

The problem which seems common to most of the countries is one of effective transfer of scientific knowledge and financial resources to potential producers—the few people who do receive training often decide to become bureaucrats rather than producers. Where fish culture has been long practiced, the results can be quite promising. In Thailand, for example, the average income of a fish farmer is six times that of an agricultural farmer and much higher than that of a coastal fisherman.

If you live in these countries, these reports will help you find out what programs your fishery officers are promoting—often they emphasize reaching small farmers—though it is probably cheaper to write to them directly.

Aquaculture for the Developing Countries: A Feasibility Study, book, 266 pages, F.W. Bell and E.R. Canterbery, 1976, $20.00 from Ballinger Publishing Company, Cambridge, Massachusetts 02138, USA.

This is an economic computer model with severe problems. It contains lots of irrelevant statistics glued together with too few ideas. Of possible interest to planners if they realize the limitations of the data base.

It does, however, have a well-annotated 94-page bibliography of the aquaculture literature, which can be an excellent tool for those with a serious interest.

ADDITIONAL REFERENCES ON AQUACULTURE

The Book of the New Alchemists describes the Ark, with its indoor fish tanks; see review on page 379.

Permaculture II includes fish ponds in its plan for ecologically-sound development; see page 434.

"Technology for the Masses" includes an article on aquaculture in India; see review on page 368.

WATER SUPPLY
AND SANITATION

Water Supply and Sanitation

After sufficient food stocks, a good clean water supply and adequate sanitation system are considered to be the most important factors in ensuring good health in a community. Improved water supply and sanitation systems were major elements of the public health measures that drastically cut death rates and improved health levels in the industrialized countries. Though it is not generally appreciated, these measures have been considerably more important than curative medicine in contributing to good health, long life expectancy and low infant mortality. Infant diarrhea, the largest killer in developing countries, is closely related to poor water quality.

This chapter has been subdivided to help the reader find his or her way through the large number of entries on these important topics.

The first three books provide a context for discussion of water supplies—the social and ecological effects of water systems (including large dams and irrigation projects in addition to community water supplies), and the nature of water supply needs, constraints, and possibilities for Third World communities.

Due to their great potential benefits, village level water supply systems have been favorite development projects of government and international agencies for several decades. They make a revealing topic of study for appropriate technology advocates, as they represent one task for which small scale technology has been widely promoted. A basic conclusion: a water supply or sanitation project that is imposed on a community, without community involvement in determining the need for and nature of the system, or without an effort to train some community members to do maintenance and repair, is very likely to fail. **Participation and Education in Community Water Supply and Sanitation Programmes: A Literature Review** offers valuable insights into the requirements for successful programs that fully involve the community. With 20-50% of handpumps in rural areas of the Third World broken down at any one time, the appropriate technology solutions seem to depend on local people and institutional arrangements that can ensure good maintenance and rapid repair. This also implies the use of equipment that can be repaired at the local level.

Research needs for new hardware are summarized in **Global Workshop on Appropriate Water and Waste Treatment Technology for Developing Countries.**. This is followed by two references that contain many ideas for water supply and waste disposal equipment.

Seven entries are manuals for the planning and installation of small water supply systems, including wells, pipelines, storage tanks, and drainage. Another seven publications on pumps and water lifters—from Thailand, Papua New Guinea, Nepal, The Netherlands and the U.S.—range from broad inventories of water lifting devices to construction plans for particular pumps.

Two additional entries describe the construction and use of ferrocement water tanks (sometimes used in roof rainwater catchment) and open ponds for water storage. These are followed by six publications on water filtration and treatment.

The bibliography **Low Cost Technology Options for Sanitation: A State of the Art Review** *offers an excellent summary of the technologies relevant to urban and rural settings in developing countries, and is a guide to the technical literature (mostly hard-to-get research reports).* **Small Excreta Disposal** *is a valuable small reference manual on the range of waste disposal alternatives that can be used in small communities. The next three books describe dry toilets in which human waste can be safely treated through composting, as an alternative to expensive water-borne sewage systems.* **Sanitation Without Water** *is most relevant to Third World conditions, while* **Compost Toilets: A Guide for Owner-Builders** *and* **Goodbye to the Flush Toilet** *were written for North American audiences. Natural treatment of water-borne sewage in a marsh pond is a relatively low technology approach that seems to have potential for some communities in North America and perhaps other places as well;* **Natural Sewage Recycling Systems** *describes work done on this technique in the United States.*

With several drought years and greater demands on existing water supplies in the western United States, there has been much recent interest in the re-use of household wash water in gardens and yards. **Residential Water Re-use** *is an excellent compendium of technology ideas and the basic technical considerations of such "greywater" systems. The last entry in this chapter,* **Management of Solid Wastes in Developing Countries**, *discusses refuse collection and transport, sanitary landfills, and composting of urban wastes.*

In Volume I:

Water Treatment and Sanitation: Simple Methods for Rural Areas, by H. Mann and D. Williamson, 1973, page 200.

Water Supply for Rural Areas and Small Communities, by E. Wagner and J. Lanoix, 1959, page 201.

Slow Sand Filtration, by L. Huisman and W. Wood, 1974, page 202.

Water Wells Manual, by U. Gibson and R. Singer, 1969, page 202.

Hand Pumps for Village Wells, by C. Spangler, 1975, page 203.

Chinese Chain and Water Pumps, by S. Watt, 1976, page 203.

Shinyanga Lift Pump, 1973, page 204.

The Salawe Pump, page 204.

Solar Distillation as a Means of Meeting Small Scale Water Demands, 1970, page 204.

How to Make a Solar Still (Plastic Covered), BRACE, 1965, page 205.

Simple Solar Still for the Production of Distilled Water, by T. Lawand, 1965, page 206.

Plans for a Glass and Concrete Solar Still, by T. Lawand and R. Alward, 1972, page 206.

Installation of a Solar Distillation Plant on Ile de la Gonave, Haiti, by R. Alward, 1970, page 206.

A Manual on the Hydraulic Ram Pump, by S. Watt, 1974, page 206.

A. Water Supply: General Considerations

The Social and Ecological Effects of Water Development in Developing Countries, book, 127 pages, edited by Carl Widstrand, 1978, $26.00 from Pergamon Press Ltd., Headington Hill Hall, Oxford, OX3 OBW, England; or Pergamon Press Inc., Maxwell House, Fairview Park, Elmsford, New York 10523, USA.

"During recent years it has been shown quite clearly that the expected social benefits from drinking water supplies have not been realized and that irrigation projects have created more problems than they solve. None has fulfilled the expectations of planners and government and most projects are used only to 50% of their capacity. This means that 100 million ha. of land with available irrigation are not used and that millions of rural people who are provided with pumps, pipes and installations cannot get any water out of them. This book is concerned with why this has happened and what can be done about it." These articles are by 10 people experienced in water systems work in developing countries.

Most irrigation schemes "appear to create subsidized income elites; contribute to food production only at high cost; facilitate preconditions for inappropriate mechanization and thus a disappointing employment creation record; and they lead to various aspects of environmental degradation. Public health considerations are typically ignored."

Much of the poor performance of water development schemes is attributed by two of the authors to structural problems in the way research and planning is conducted. David Henry notes that a poor learning situation for planners has prevented them from learning from the mistakes of the past two decades. Robert Chambers points to the problem of research priorities that are determined more by the need for recognition among professional colleagues than by the the real needs of rural people: ".. the primary criterion for good research should be that it is likely to mitigate poverty and hardship among rural people, especially the poorer rural people, and to enhance the qualilty of their lives in ways which they will welcome; that in short, priorities should be...grounded in the reality of the rural situation. Starting with rural people, their world view, their problems and their opportunities, will give a different perspective. To be able to capture that perspective requires a revolution in professional values and in working styles; it requires humility and a readiness to innovate which may not come easily in many reasearch establishments."

Some of the lessons for planners and donors: "more funds and more resources into public health training and education (with local teachers), more funds into training programs for operators and maintenance personnel — not producing full-scale engineers, but, instead, small-scale mechanics with some basic skills directly applicable to the water system — and more thought about the involvement of locals in the planning of water schemes."

Highly recommended.

Drawers of Water, book, 306 pages, by Gilbert White, David Bradley and Anne White, 1972, $14.00 from University of Chicago Press, 11030 S. Langley, Chicago, Illinois 60628, USA.

Combining engineering, economics, health and sociology, **Drawers of Water** is a "broad view of domestic water supply in the developing tropics". Using examples and studies from East Africa, the authors discuss: traditional water supplies and use in urban and rural communities; the range of attempted and possible improvements; the health costs and benefits of improved water systems; individual and social "costs"; and the successes and problems of standard economic and technical planning methods.

Drawers of Water is intended for "decision-makers" in developing countries — much of the book contains technical discussions of data from sociological and economic studies. The book's strength, however, is that it recognizes that "accepted" planning methods must be altered to account for local physical and social conditions.

Water for the Thousand Millions, book, 58 pages, edited by Arnold Pacey, written by the Water Panel of ITDG, 1977, £2.90 from ITDG.

This short volume differs from the other general water supply books in that it is explicitly about "appropriate water supplies," including the consideration of economic, social, environmental and health factors in determining "appropriateness". The authors have concentrated on how these factors are combined in a variety of low cost water systems that could be used by the thousand million people currently without clean drinking water.

"Water supplies are not an all-or-nothing phenomena. Almost every situa-

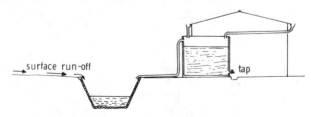

Rainwater catchment from roof and ground surface. With precautions to prevent dust and bird droppings being washed off the roof into the tank, and with a suitable cover, the roof tank can provide high quality water suitable for drinking, while the excavated tank may provide water suitable for washing or watering gardens (after Farrar, 1974).

tion lends itself to some improvement, even if funds and skills are severely limited." The key is matching people's needs and cultural patterns with the given water supply potential and much broader technical choice than is usually offered by governments or development agencies.

B. Participation: A Key to Successful Systems

Participation and Education in Community Water Supply and Sanitation Programmes: A Literature Review, book, 204 pages, by Christine Van Wijk-Sijbesma, March 1979, Technical Paper No. 12, from International Reference Centre for Community Water Supply, Information Section, P.O. Box 140, 2260 AC Leidschendam, The Netherlands.

An extraordinary review of conclusions from a wide literature on the participation of communities in water supply and sanitation programmes, this should be required reading for people working in these fields. For more information on specific concepts, the reader can refer to the original studies. This volume offers planners and community organizers the opportunity to avoid many of the common mistakes of the past, and create programs with a maximum of community participation. It is also a good general guide for involving the community in any kind of appropriate technology activity, stressing their own perceptions of problems and solutions. An annotated bibliography, published as No. 13 in the Technical Papers Series, contains detailed abstracts of the 145 most relevant works on which the literature review is based.

Community participation in decision-making and implementation brings a number of rewards. It is a more democratic approach than imposition of projects from outside. It provides good opportunities for the growth of skills and competence at the grass-roots level—increasingly recognized as the most central goal of development. And it is more likely to be successful in solving problems.

Some of the authors have noted that handpumps are broken down 20 to 70 percent of the time, and that in some countries village water systems are breaking down faster than they are being built. "A community is more likely to cooperate in the implementation, operation and maintenance of new systems if it has had a say in the preparation of plans."

In some countries, water supply programs have been divided into three categories. In communities where water supply and sanitation problems are felt by the entire population, the government agency offers assistance with forming a local committee and planning a work program. If problems are felt only by the village leadership, these people are supported with media and locally-planned primary school educational programs to generate broader motivation to solve the problems. If problems are felt only by the water supply specialist, "various surveys are carried out with the involvement of the villagers, a motivation and education campaign is set up, and assistance is provided in solving other, more deeply felt village problems."

Many observers have "stressed the importance of presenting the community with the various technological solutions which are feasible, ranging from simple source protection and pit latrines to multiple house connections... Community choice should include the possibility of rejection of any immediate source improvement...Although this may seem a negative outcome...each community has its own criteria for calculating sets of trade-offs, so that their perceptions of the usefulness and effects of improvements may differ considerably from those of the agency. Besides, self-made choices will ensure a greater commitment than solutions presented from outside.''

Recently, some authors have emphasized the need for participatory research, "because it is a process which is part of the total educational experience, serving to identify community needs and to effect increased awareness and commitment.'' Two of the more innovative information-gathering and educational approaches briefly discussed are the ''environmental sanitation walk'' with a group of villagers, and the ''community self-survey''.

Highly recommended.

Hand Pump Maintenance, booklet, 38 pages, Arnold Pacey, 1977, revised 1980, Ł1.25 from ITDG.

This booklet is part of an Oxfam series on socially appropriate technology. The author looks at community well projects in developing countries, and examines why over 60% of these break down and/or are not used. The reason for this is not due to faulty design of the pumps themselves, but because ''...the community or village has not been adequately involved in the project in the first place, and has not accepted the social responsibility for the task of maintaining the pump.''

''An effective pump system is not simply a technological object but a conglomerate of technology, institutions and people.'' With this in mind, three approaches are possible: 1) total village self reliance, where a pump is manufactured using only those materials and skills available locally, 2) partial self reliance, where a pump may be made outside the village, but the responsibility for maintenance lies within the village, and 3) elimination of village responsibility, usually by use of a manufactured imported pump which requires no maintenance (the most expensive option).

The conclusion is that the partial self reliance path is most applicable in a variety of situations. Locally-made pumps will also work for low-lift applications.

This booklet is most valuable as a reminder that local people and their institutions are at least as important as the hardware in the introduction of any community-level technology.

Recommended.

C. Research Needs and References

Global Workshop on Appropriate Water and Waste Water Treatment Technology for Developing Countries, report, 54 pages, 1977, on request from the International Center for Community Water Supply, P.O. Box 140, Leidschendam, The Netherlands.

These are the results of a workshop held in 1975, concerned with water supply and sanitation in the developing world. The objectives were: ''to assess the state of the art and to identify the role of appropriate technology in the development of water supply and sanitation...to formulate recommendations and agree upon priorities...to discuss the development of international programs to implement the activities recommended.''

The report is divided into three parts: water supply, waste disposal, and related socio-economic aspects. Each section contains specific recommendations for further areas of study needed. The water supply section is mostly on water treatment and quality, while the socio-economic section deals with methods for information and technical change, training, planning, on both the national and local levels.

A good summary of research needs in water supply and waste water treatment technologies for rural areas.

Using Water Resources, book, 143 pages, VITA, 1970 (reprinted 1977), $5.50 from VITA.

Due to the international demand for information on water supplies, VITA has reprinted this from its **Village Technology Handbook** (see review on page 41). Subjects are:

a) Developing water resources: basic well-drilling and digging information, including how to make various hand drilling tools;

b) Water lifting and transport: measuring water flows, bamboo piping systems, chain pump and inertia pump for irrigation, hydraulic ram pump;

c) Water storage and water power: springs, cisterns (tanks), dams, power transmission;

d) Water purification: boiler for drinking water, chlorination methods, sand filter.

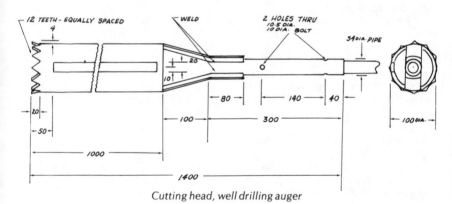

Cutting head, well drilling auger

Contributions to a Mail Survey on Practical Solutions in Drinking Water Supply and Waste Disposal for Developing Countries, large loose-leaf book, 150 pages, WHO International Reference publication, 1977, free from WHO International Reference Centre for Community Water Supply, P.O. Box 140, 2260 AC, Leidschendam, The Netherlands.

This valuable book supplements other reference books (such as **Water Supply for Rural Areas**, by Wagner and Lanoix, reviewed on page 201) which are becoming very expensive. It is a "picture bibliography" of rural technology. The 150 entries were gathered from healthworkers around the world.

The entries are indexed according to subject, including: rainwater collection, wells; surface water; treatment, filtration and disinfection; pumps and hydraulic rams; solar and wind energy; waste collection and disposal. Among the interesting entries are: a rotary filter with bamboo strainers from Japan; and a simple disinfectant device from India; a bamboo water pump from Laos; and a "comfort station" from Nigeria, with a combined unit providing toilet, bathing and washing facilities for groups of families (200-600 people) which can be constructed on a self-help basis.

The first section illustrates designs with brief descriptions, and lists contact people for further details. The second section provides references and contacts for more detailed information without illustrations. The use of pictures makes this a valuable source of ideas.

Rotary filter with bamboo strainers (Japan)

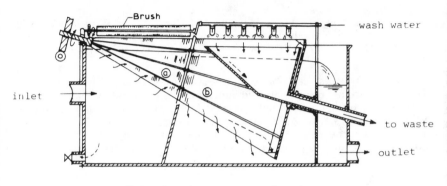

ⓐ Strainer (made of bamboo)
ⓑ Micro-strainer

D. Small Water Supply Systems

Village Water Systems, book, 100 pages, by Carl Johnson, free to serious groups from UNICEF, Box 1187, Kathmandu, Nepal; to be replaced in 1980 by **Gravity Flow Systems**, 150 pages, by Thomas Jordan Jr., which will be available from UNICEF at the same address.

This is written as a reference for designing water distribution systems, with an emphasis on conditions found in the mountain regions of Nepal. It does not describe how to pump or otherwise obtain water; rather, after you have the water supply, it tells you how to distribute the water through pipes to an entire community or village.

"Design criteria are presented where standard designs cannot be practically used, while the standard designs that are included are for guide purposes only..."

With charts and sample calculations, the author covers initial surveys, intake

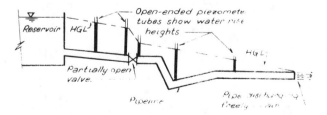

Physical interpretation of the hydraulic grade line

works, pipeline sizing, break pressure and reservoir tanks, and public taps. Water quality and/or treatment, windmills and hydraulic rams are mentioned briefly, and the reader is referred to other publications in the bibliography (the book assumes you have access to the UNICEF or WHO libraries in Nepal). An appendix presents a sample design, with calculations, for a rural water supply system, and the cable suspension of a flexible pipe (a frequent need in mountainous areas).

Recommended.

Small Water Supplies, book, 78 pages, S. Cairncross and R. Feachem, 1978, ₤1.50 from The Ross Institute, London School of Hygiene and Tropical Medicine, Keppel Street (Gower Street), London WC1E 7HT, England; or ITDG.

This handbook is for ''someone who wishes to build only a few water supplies (systems) using simple equipment easily available to him; typically a rural health worker.'' It is not intended for those working with large-scale water supply systems. All aspects of designing a water supply system are presented for the novice. The subjects include preliminary design, water sources (wells and boreholes), raising water (how to choose pumps of all

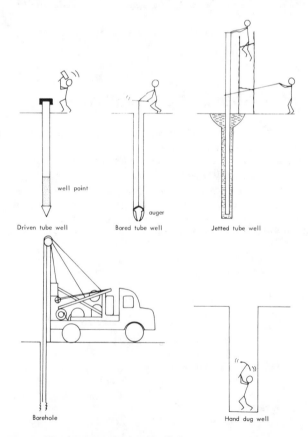

Five basic methods of well digging and drilling

types), water treatment, storage (dams and tanks), pipes, and distribution. There is an extra chapter on purification on an individual scale.

An appendix describes bacterial analysis of water using simple equipment and MacConkey broth (available from Oxoid Ltd., Wade Road, Basingstoke, Hampshire, England; or Difco Laboratories, Detroit, Michigan 48201, USA).

This book covers material similar to that in **Water Supply for Rural Areas** by WHO (see review on page 201). **Small Water Supplies** is significantly less expensive, and while less detailed it does include a wide range of material on water purification and general aspects of small water systems. For these reasons, we highly recommend it.

Manual for Rural Water Supply, book, 175 pages plus fold out drawings, by Swiss Center for Appropriate Technology, 1980, SwFr 34 or US$20.00 from Swiss Center for Appropriate Technology (SKAT), Varnbuelstrasse 14, CH-9000 St. Gall, Switzerland.

A very thorough book on small community water supply systems based on 15 years experience of the Swiss Association for Technical Assistance and the Community Development Dept. in Cameroun. The basic elements of a distribution system are presented—wells, springs, stream diversions, storage tanks, distribution pipelines and standpipes. There is a brief maintenance checklist.

This book is unusually broad in scope, beginning with the yearly water cycle (rain to groundwater) and then discussing standards for water quality that are realistic and affordable in rural Cameroun. Also unusual: coverage of the corrosive effects of water flowing through a variety of piping materials, and what can be done about this. Emphasizes the planning of distribution systems for expansion with expected population growth.

Widely relevant.

Manual for Water Systems and Pipe Work, 37 pages, by Andreas Bachmann and Nir Man Joshi, 1977, free to serious groups from SATA-Director, Box 113, Kathmandu, Nepal.

Subtitled "A Brief Introductory Course for the Establishment of Rural Water Supplies in Nepal", this engineering manual is for those involved in design, construction and plumbing of water supplies in rural areas. The concepts are presented with dimensional drawings and simple English explanations.

The first section provides introductory design information for natural gravity and hydraulic ram distribution systems and water conduits (pipes and valves). The rest of the manual covers the use of the three types of pipe available in Nepal: galvanized iron, HDP (high density polyethylene) and PVC (polyvinyl-

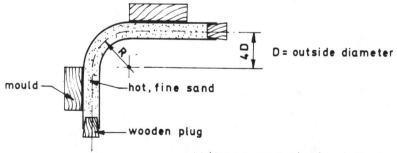

Making a permanent bend in plastic pipe

chloride). Included is information on laying pipe and making many different kinds of joints.

Excellent as a field training manual; some of the information is contained in Rural Water Supply in Nepal, Manual #4, reviewed below. Highly recommended.

Rural Water Supply in Nepal: Technical Training Manuals 1-5, 5 short booklets, 1978, by Local Development Department of Nepal, free in limited quantities only, from UNICEF, Box 1187, Kathmandu, Nepal.

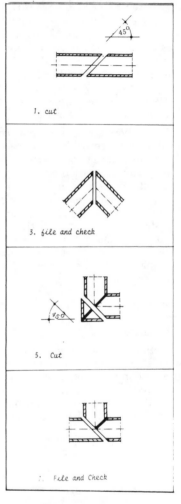

1. cut

3. file and check

5. Cut

7. File and Check

Making a 90-degree branch from PVC pipe

These four manuals are very simple training materials for those working in rural water supply development in Nepal. They range from 13 to 30 pages in length and give brief introductions in the following areas:

1) Hydrology and Water Cycle—the climatology and general water availability in Nepal.

2) Stone Masonry—how to build tanks and basins for water storage, using stones and cement.

3) Concrete—mixing and using concrete; also for water storage.

4) Pipes and Fittings—an introduction to pipes and fittings commonly used in Nepal; both galvanized iron and HDP (high density polyethylene). This last manual is useful in showing welding techniques for joining pipes. The advantages and disadvantages of each material are presented: iron is heavier but stronger and easily available; HDP is an easily damaged synthetic that is cheaper, easily connected, and doesn't require special joints.

5) Construction Design Course—Source protection, water treatment, storage tanks, pressure reducing units, pipelines, and public water tanks.

These manuals are useful examples of how simple but necessary skills for a set of local conditions can be communicated.

Drinking Water Installations and Drainage Requirements in Buildings in Nepal, book, 141 pages, 1976 and 1978, Andreas Bachmann, free to serious groups from SATA-Director, Box 113, Kathmandu, Nepal.

This is an engineering handbook for designing water installations in Nepalese buildings (assuming a low-pressure supply of piped water already

exists). Four sections include:
1) Design criteria and tables for making design calculations for low pressure systems
2) Examples of water system designs, including several solar water heating, natural gas and biogas installations
3) Drainage requirements for drains and sewers.
4) Legal requirements for installing water systems in buildings in Nepal

Published in cooperation with the Nepal Water Supply and Sewering Board, this is definitely a book for engineers, with many detailed system design drawings and design tables. It is also meant for buildings in a city like Kathmandu, where water distribution systems already exist, or will exist in the near future.

The special value of this manual is that it represents an adaptation of standard plumbing design and practice to some of the materials, tools, and building construction practices found locally in Nepal.

Self-Help Wells, FAO Irrigation and Drainage Paper #30, 78 pages, 1978, by R. G. Koegel, from Distribution and Sales Section, FAO, Via delle Terme di Caracalla, 00100 Rome, Italy.

This is a good survey of self-help well drilling and digging techniques. The emphasis is on local materials and labor, not on imported technologies. Techniques are described for both small (15 cm) and large diameter wells. Drilling methods discussed include boring, percussion and rotary drilling for small diameter wells. Excavation techniques for larger diameter wells are presented.

The amount of detail in descriptions varies, but there are many good drawings. Even though this is not a construction manual, these drawings are very useful in explaining the ideas. The materials range from simple wooden tools to metal drill bits, making these techniques adaptable to a wide variety of local conditions. Labor intensive techniques are emphasized.

Also included are sections on health aspects of drilling, how to find water that is likely to be uncontaminated, safety precautions while drilling, and non-vertical wells (e.g. Qanats—horizontal tunnels that intercept a sloping water table).

A useful book for those working with local water projects. See also the excellent volume **Hand Dug Wells**, reviewed in Volume I.

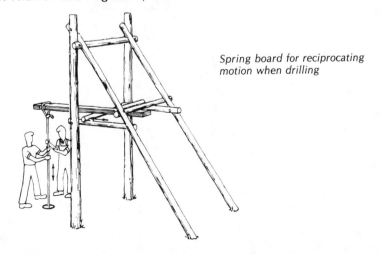

Spring board for reciprocating motion when drilling

E. Pumps and Water Lifters

Water Lifters and Pumps for the Developing World, book, 317 pages, Alan D. Wood, 1976, $9.00 postpaid in U.S., $10.00 postpaid elsewhere, from Civil Engineering Department, Engineering Research Center Publications, Foothills Campus, Colorado State University, Fort Collins, Colorado 80523, USA.

A survey of simple water-lifting mechanisms in use around the world, from simple buckets to hydraulic rams and centrifugal pumps. The emphasis is on those that are built locally in developing areas, using local materials.

For each type of device, drawings, operating principles, and most appropriate conditions are given. Also included: discussions of historical uses of water pumps, prime movers (the energy source that powers a pump, from animal power and falling water to electric motors), criteria for choosing pumps for particular applications, general pumping principles with sample calculations, and how to read performance curves.

This is the best 'encyclopedia' of simple water lifters available. It is comparable to the out-of-print FAO book **Water Lifting Devices for Irrigation**. No detailed construction drawings are given. There are 123 clear illustrations, making this an excellent idea book for those who want to know what types of pumps are in use all over the world. It could serve as a starting point for someone to design and build his or her own pump for a particular application.

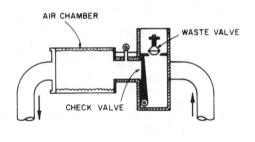

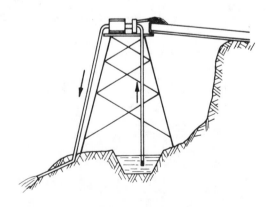

*Siphon elevator has been claimed
to work, using some hydraulic ram principles*

Hand Pumps, book, 230 pages, by F. Eugene McJunkin, 1977, from WHO International Reference Center for Community Water Supply, P.O. Box 140, 2260 AC Leidschendam, The Netherlands.

This is a reference book on hand pumps, describing various types of pumps and principles of pump operation. It is "intended to serve public health officials, engineers and field staff who are planning and implementing water supply programs with hand pumps." Fewer pump types are described than in **Water Lifters and Pumps** (see review). The section on pump principles is detailed and technical, and includes design principles for each part of a pump assembly (plunger, stand, suction pipe, seals, valves, cylinders). It is a useful reference for someone interested in detailed engineering design of simple pumps; many examples are given. There are two sections describing recent research in hand pumps, using wood, bamboo, plastic and steel, and local manufacturing methods for steel parts (such as casting, machining and welding).

A very complete, detailed handbook.

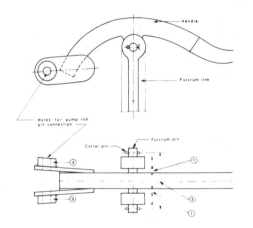

Examples of common defects in pump handle link assembly:

1. *Excessive tolerance between fulcrum link and handle*
2. *Distance between cotter pins too great*
3. *Misaligned bushings*

Design of Simple and Inexpensive Pumps for Village Water Supply Systems, research report, 41 pages, by N.C. Thanh, M.B. Pescod, and T.H. Venkitachalam, 1977, $8.00 in Thailand, $10.00 to developing countries, and $15.00 to developed countries, from Library and Regional Documentation Center, Asian Institute of Technology, P.O. Box 2754, Bangkok, Thailand.

The final report of a project to develop two pumps for rural water supplies that can be built and maintained locally, describing the design and testing of a bicycle-driven inertia pump and a bellows pump (also pedal powered). The bellows design achieved a pumping rate of 89 liters/minute with a 1.5 meter lift, and the inertia pump in some cases pumped as much as 300 liters/minute with the same lift. No tests were made to determine how much water could be lifted over a longer period of time, considering the pedalling endurance of one man. Construction details are not given but dimensional drawings illustrate the pumping principles adequately. Materials needed: sheet metal, cast iron piping, some canvas and rubber, and miscellaneous hardware.

When this report was printed further development was not planned. The designs appear viable, however, and should be pursued.

Pedal-powered inertia pump

Pitcher Pump, Technical Bulletin No. 15, 8 pages, $1.00 from VITA.

Drawings and text for a simple hand pump that lifts 8-10 gallons per minute about 10-15 feet. Construction requires some welding equipment and basic metal working tools. (In some parts of the world pitcher pumps are made out of clay and fired like pottery.) Some of the drawings are not very clear and may present problems to readers unfamiliar with simple pumps.

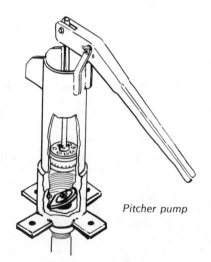

Pitcher pump

Diaphragm Pump, Technical Bulletin No. 16, 13 pages, by Dr. Richard Koegel, $2.00 from VITA.

Drawings and text for the construction of a simple pump from wood, auto inner tube, and pipe fittings. "Two to three liters per stroke can be pumped to a height of three to four meters. The pump can be operated by one or two men and can be easily adapted for use with animal or wind power." Diaphragm pumps could be valuable when combined with sail windmills for low lift irrigation of small plots. They are advantageous for use when the water being pumped contains particles that might jam other pumps.

Once the principles of a diaphragm pump are understood, the design can be adapted to use a wide range of materials and construction techniques; only old auto inner tubes are essential.

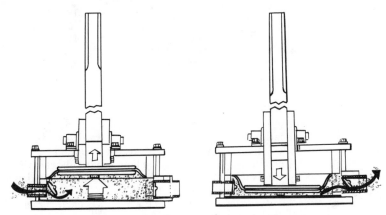

Diaphragm pump

The Construction of a Hydraulic Ram Pump, booklet, 36 pages, by Allen Inversin, 1978, exchange your publication with the Appropriate Technology Development Unit, P.O. Box 793, Lae, Papua New Guinea; also available from VITA for $3.00 (ask for VITA Technical Bulletin No. 32).

Allen Inversin has designed and tested the first working hydraulic ram design we've seen that requires no welding or special skills. "Both VITA and ITDG have long had available designs for a low-cost ram pump. These designs, however, require special skills and tools to construct, have not been tested to any extent over a range of operating conditions, and consequently provide little data on the performance of the actual rams described."

Starting from the basic ITDG ram design, Inversin has redesigned the parts so that studs, nuts and bolts carry the loads, using epoxy adhesive only as a sealant. The valves have been simplified, eliminating much of the machining and/or special tools that are required in the ITDG and VITA designs (see reviews on pages 206 and 208).

The author intends this to be built by someone with very little machining experience; however, "...it is quite probable that those who have had machine shop experience will prefer alternative means of construction (rather than the simple ones described in the manual)." Very simple yet detailed instructions, easily the clearest we've seen, are given, along with clear illustrations. Construction is simple, as only handtools and a drill press are necessary. In addition to commercially available pipe fittings, small strips of scrap steel, nuts, bolts and epoxy are the only materials required.

A significant part of this manual is the 8 pages devoted to performance information, based on a year of testing and improvement in PNG. The pump

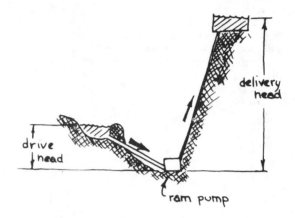

has been tested at drive heads (heights) of .5 to 4 meters. For delivery heads of 10 times the drive head, it can deliver 3600 liters of water/day. Graphs for predicting performance at various operating heads are included.

There is purposely no information provided on how a hydraulic ram works, or how to install, operate and maintain a ram, since these subjects are adequately covered in other publications, such as the ITDG and VITA booklets, or the UNICEF/Nepal booklet (see review in this section). However, this manual is a welcome improvement in simple ram design, and would be valuable to use in combination with any of the three books mentioned above.

Use of Hydraulic Rams in Nepal: a Guide to Manufacturing and Installation, booklet, 46 pages, Mitchell Silver, free to serious users from UNICEF, Box 1187, Kathmandu, Nepal.

This book includes what **Construction of a Hydraulic Ram Pump** (see review in this section) does not: how to install, use and maintain a hydraulic ram after you've built or bought one. It is written in plain English, rather than technical language.

Although aimed at local conditions in Nepal (for example, the availability of supplies in Kathmandu hardware stores is discussed), this book is useful for anyone with little mechanical background who wants to use a hydraulic ram in any location.

Included are chapters on: how a hydraulic ram works, surveying a site, descriptions and design considerations for intake tanks, reservoirs, operating a ram pump, maintenance and repair. Conversion tables are included.

There are 14 pages on building a ram pump using standard pipe fittings. The valves can be either bought or made. The instructions and drawings are not as detailed as in other hydraulic ram publications, but they could be used by someone with mechanical experience. The most useful part of this booklet, however, is that it shows how hydraulic rams can be simply adapted for use in differing local conditions.

Recommended for use with a hydraulic ram construction manual.

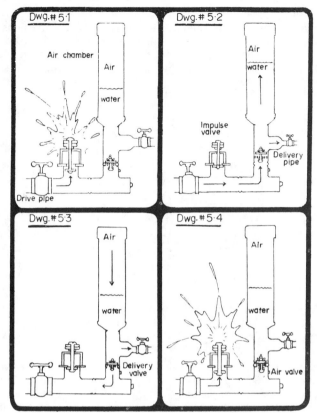

Opening and closing of the valves and the flow of water during one pumping cycle of a hydraulic ram (Nepal)

F. Tanks and Ponds for Storage

Ferrocement Water Tanks and Their Construction, book, 118 pages, by S.B. Watt, 1978, Ł2.95 from ITDG.

The book covers a number of different ways to use ferrocement (wire-reinforced cement mortar) to construct water storage tanks and jars of many shapes and sizes. The author has worked with ITDG for years and has written a book that thoroughly covers design, construction and use of ferrocement water storage containers. The book is an excellent construction manual, but it also covers catching and using rainwater from roofs and land surfaces, health aspects of water storage, use of ferrocement linings for earthen tanks, and sources of further ferrocement information. There are many detailed photos and illustrations.

''The Dogon people of Mali, living in the Sahelian drought zone with a rainfall of only about 40 cm. (16 in.) per year, suffer greatly from water shortages during the dry season. The method of storing water described in this chapter was devised to provide a cheap tank for water collected from the flat roofs of the houses. The water tanks, which consist of traditional (adobe) grain bins, lined with a thin layer of reinforced (ferrocement) mortar, are readily acceptable to the users and fit well into their social and cultural visions of life.''

This book seems to contain all the information one would need to build ferrocement water storage tanks. The water tank construction techniques would also be useful to someone building other types of ferrocement structures such as grain bins or houses.

An excellent book, highly recommended.

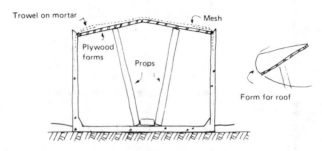

Constructing the roof of a ferrocement water tank

The Village Tank as a Source of Drinking Water, paper, 17 pages, 1969, free on request (paper #WHO/CWS/RD/69.1, limited supplies available), from Environmental Health Technology and Support Unit, World Health Organization, 1211 Geneva 27, Switzerland.

Open water storage ponds and pools (sometimes called 'tanks'), are sources of water for many villages around the world. They are generally highly polluted and spread disease. This paper is an outline for needed developments identified in 1969 to make this type of water supply safer. Research needs are suggested for these parts of such water storage systems: intake, filter, disinfection (for short term emergencies), handpumps, prevention of erosion in excavated slopes, using charcoal waste from lumber industries, and preventing contamination of water supplies.

Many valuable suggestions and ideas are presented (some that are, as yet, untested). Drawings help explain these ideas, though they are not intended as

River to pond to handpump system

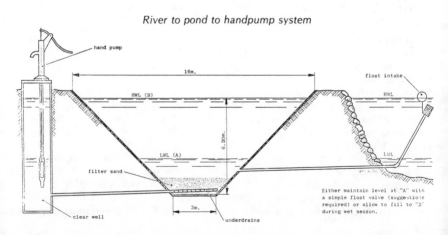

guides for construction. These ideas include placement of a well next to the storage pond, handpump/filter setups and drip-feed chlorinators.

Although written in 1969, this paper will still be useful reading for those working with water supplies in rural areas. Those with field experience to share should contact the Community Water Supply Unit at the above address.

G. Filtration and Treatment

Treatment Methods for Water Supplies in Rural Areas of Developing Countries, paper, 38 pages, L. Huisman, 1975, from WHO International Reference Center for Community Water Supply, P. O. Box 140, 2260 AC, Leidschendam, The Netherlands.

This paper is a summary of rural water supply sources and treatment (purification) methods. Making no attempt to provide details, Huisman briefly describes these treatment methods: aeration, plain sedimentation, chemical coagulation and settling, slow sand filtration, rapid filtration, activated carbon filtration and disinfection. He also compares and contrasts water sources, such as rain water, ground water, surface water and spring water. (For an introduction to wells, pumps and water lifters, see **Village Water Supplies**.)

Of particular interest is an appendix with drawings illustrating water catchments, wells, and treatment methods from around the world.

While much of the information here is available elsewhere, this paper is useful as a non-technical introduction to rural water treatment methods.

The Purification of Water on a Small Scale, Technical Paper #3, booklet, 19 pages, 1973, on request from WHO International Reference Center for Community Water Supply, P.O. Box 140, 2260 AC Leidschendam, The Netherlands.

This short booklet contains practical instructions for water purification by boiling, chemical disinfection and filtration. Chemical disinfection includes chlorine, iodine and potassium permanganate (not recommended). Filtration

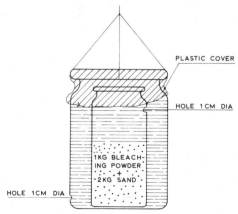

FIG. 1   CHLORINATION POT FOR HOUSEHOLD WELLS
CAP. 400 LITRES PER DAY (100 GPD) OF DRINKING WATER;
CHEMICAL RECHARGE EVERY 3 WEEKS
(SOURCE: CPHERI, NAGPUR, INDIA)

methods include sand filters for coarse filtration, and ceramic filters for finer filtration.

A simple filter design is given. This sand filter removes visible dirt and large organisms such as ova and cysts; the authors say that it won't work against bacteria.

The importance of storage in preventing re-contamination of water is also emphasized. Unfortunately, there is no discussion of the relative advantages of chlorine versus iodine as disinfectants, including the dangers of overuse.

Recommended.

Slow Sand Filtration for Community Water Supply in Developing Countries, Bulletin No. 9, annotated bibliography, 40 pages, 1977, free to serious groups, from WHO International Reference Center for Community Water Supply, P.O. Box 140, 2260 AC, Leidschendam, The Netherlands.

Slow sand filters have been in use for almost 150 years in public water supplies. Even though some European cities use them on a large scale, they tend to be dismissed as old-fashioned or out of date. However, many people are once again concluding that slow sand filtration for purifying drinking water is very effective, and deserves more widespread use, particularly in rural and urban fringe areas of developing countries.

This bibliography contains 79 entries on the technical aspects of slow sand filtration. Each entry is summarized and indexed according to author and keywords. Included is a list of institutions around the world active in developing low-cost slow sand filtration methods.

L. Huisman notes in the preface: "In developing countries...slow sand filtration is often applied as a single treatment process, only where necessary preceded by a simple pre-treatment for turbidity removal. Optimal use can be made of locally available materials such as bricks, mud blocks and mass concrete, while also filter sand of good specifications is readily available in most countries. Operation and maintenance are relatively easy and can be done by semi-skilled operators. Operational costs are minimal, the more so as no chemicals are required. Slow sand filtration may be regarded as an appropriate water treatment process and its wider application may considerably contribute to an improved provision of safe drinking water in developing countries."

An extensive, excellent bibliography.

Application of Slow Filtration for Surface Water Treatment in Tropical Developing Countries, technical report, 75 pages, by N.C. Thanh and M.B. Pescod, 1976, $8.00 in Thailand, $10.00 in developing countries, $15.00 in developed countries, from Library and Regional Documentation Center, Asian Institute of Technology, P.O. Box 2754, Bangkok, Thailand.

This is a report on performance tests of 3 different slow filtration systems for purifying water using various combinations of sand, burnt rice husks, and coconut fibers. Data are given on the influence of turbidity (cloudiness) and filtration rates on the quality of filtered water.

The goal of these studies is to provide relatively inexpensive treatment for surface waters, using materials common to Southeast Asia. The systems tested are for a village water supply, and would require an operator. The construction cost of such a system is estimated at 15,000-20,000 Baht ($750-$1,000).

Very technical; useful only for researchers in slow filtration techniques. For a study of **prefilters** to be used with slow sand filters, see review below.

Horizontal-Flow Coarse-Material Prefiltration, research report, 46 pages, by N.C. Thanh and E.A.R. Ouano, 1977, $8.00 in Thailand, $10.00 to developing countries, and $15.00 to developed countries, from Library and Regional Documentation Center, Asian Institute of Technology, P.O Box 2754, Bangkok, Thailand.

Slow sand filters are accepted means of treating water for drinking; they can, however, be clogged by inorganic materials. The authors have been working at the Asian Institute of Technology on prefilters which are horizontal and use coarse crushed stone to filter out large inorganic material before the water goes through the main sand filter. Having the prefilter horizontal allows both the crushed stone and gravity to help settle out the suspended inorganic and large organic materials.

This report describes and gives results from research on coarse prefilters used with both sand filters and burnt rice-husk filters. The authors found that the horizontal prefilter was effective, removing up to 60-70% of the solid matter in water before it reached the regular filter. It was found that the sand filter produced a better tasting water, although the rice-husk filter performed better in terms of water flow. The filter/prefilter combinations also removed much, but not all, of the non-fecal coliform organisms present in the unfiltered water. This indicates that a high degree of health protection is possible with such filter combinations.

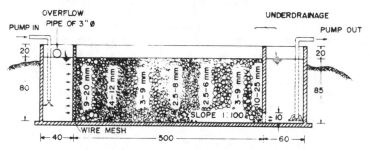

Pilot scale horizontal flow prefilter
(all dimensions are in centimeters)

Much of the report is technical, with graphs, photos and drawings of the experimental setups. Design drawings of the horizontal prefilter and slow sand filter are particularly useul. Of interest to researchers working with slow filtration methods.

Simplified Procedures for Water Examination, Manual M12, laboratory manual, 190 pages, 1978, $14.00 from American Water Works Association, 6666 West Quincy Avenue, Denver, Colorado 80235, USA.

A detailed, step-by-step laboratory manual for testing water quality, **using modern laboratory equipment**. Chemical, bacteriological and biological (microscope) examinations are all discussed.

Chemical tests can be used to determine whether too much or too little disinfectant is being added to water. Bacteriological tests determine the presence of coliform bacteria, which can make drinking water unsafe.

Useful **only** in areas where laboratory facilities already exist.

H. Sanitation: Low-Cost Options

Low-Cost Technology Options for Sanitation: A State of the Art Review and Bibliography, book, 184 pages, by Witold Rybczynski, Chongrak Polprasert, and Michael McGarry, 1978, free to local groups in developing countries, $10.00 to others, from IDRC.

This is a sourcebook on sanitation alternatives. The first quarter of the book is a review of the existing methods for collection, treatment, reuse and disposal of human wastes.

"There exists a wide range of effective alternatives between the unhygienic pit privy and the Western waterborne sewerage system. These systems are generally far cheaper. Most of them do not demand a heavy use of water. And many make creative use of the nutrients in human waste to fertilize fields and fish ponds or to contribute to biogas production—and they can do this without serious risk of returning pathogens to human food or drinking water."

The bibliography fills the rest of the book. Thousands of pieces of literature were examined in the process of choosing 530 documents summarized here. "Emphasis has been placed on technological issues, but institutional, behavioural, and health-related aspects of excreta disposal were also considered." Most of this literature is available only in English, but some of it is also available in Spanish, French, Norwegian or Swedish.

Very little documentation of indigenous excreta disposal practices exists, the authors note. "It has been assumed that these (practices) are of little importance as they will eventually be replaced by sewerage. As a result, the potential for upgrading existing practices has been largely ignored...Once existing conditions are understood as a starting point, certain solutions will be more compatible with resources available. Particular options will integrate reuse possibilities that reflect energy, food, or agricultural needs of the particular community. Whether or not such solutions lead to waterborne sanitation is less important than the fact that they will be the beginning of a dynamic process of development."

Unfortunately there are no prices listed, and most source addresses given are incomplete. This will make acquisition of the documents difficult.

There is a thorough index and an excellent glossary of sanitation terms.

Highly recommended.

Fertilizer pits being used in rotation

Small Excreta Disposal Systems, booklet, 54 pages, R. Feachem and S. Cairncross, 1978, £1.30 from The Ross Institute, London School of Hygiene and Tropical Medicine, Keppel Street (Gower Street), London WC1E 7HT, England; or ITDG.

This booklet presents "the range of technologies available for excreta disposal in small communities and describes each system in simple terms. Design formulas are included where appropriate and (for experienced people) it is possible using this booklet to design the main elements of the systems."

Two chapters cover individual components and complete waste systems, from bucket latrines to water seal privies and septic tanks. The information here is largely descriptive. A third chapter covers the design and construction of squatting slabs, pit latrines, water privies, septic tanks, soakways (leach-lines or drainfields), and waste stabilization ponds.

A good mix of description and design details is presented. The book's strongest point is that it concentrates on rural, tropical areas in developing countries, analyzing what is and isn't appropriate. (For example: the Clivus Multrum self-composting toilet "has not yet proved its worth in the tropics or among low income communities.")

Recommended.

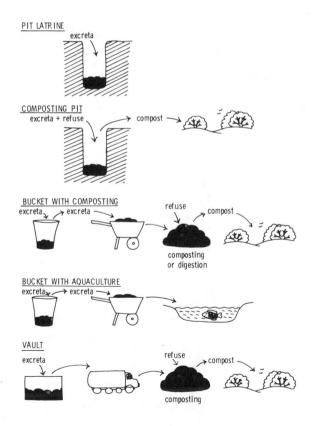

A variety of excreta disposal systems

Compost Toilets: A Guide for Owner-Builders, book, 51 pages, NCAT, 1979, $3.00 from National Center for Appropriate Technology, P.O. Box 3838, Butte, Montana 59701, USA.

Water-borne sewage systems are the standard technology for safe disposal of human wastes in urban areas in the industrialized countries. These systems are extremely expensive to build and operate. By 1978, the capital investment necessary to connect one house to such a system in the U.S. was reported to be a minimum of $4000. A substantial amount of water, 25-60 gallons per person per day, is required to flush the toilets. Treatment of the sewage is also costly, energy-intensive, and represents the destruction of a potential resource.

In many areas both inside and outside the United States, a dry composting toilet appears to be a much more economically and environmentally-sound alternative. **Compost Toilets** is a well-organized introduction to the subject. It explains the composting problems associated with the major designs. Draining systems and simple jar fly traps are shown that have been successfully developed to overcome the common problems of liquid buildup and flies. There are three complete slant-bottom designs presented for owner-builders, with discussion of the pros and cons of each.

''A compost toilet...can be a safe and efficient, sanitary human waste treatment system. The main question facing owner/builders is whether they are prepared to take the time to **manage** the system efficiently.'' The right liquid and carbon/nitrogen balances must be kept or the toilet will not function properly. In most designs, the composted material must be removed approximately every 3 months after the initial start-up period.

Costs of commercially-made compost toilets are given as $850 to $2500. Systems built at the site are expected to cost about $450 for materials and $500

Composting toilet

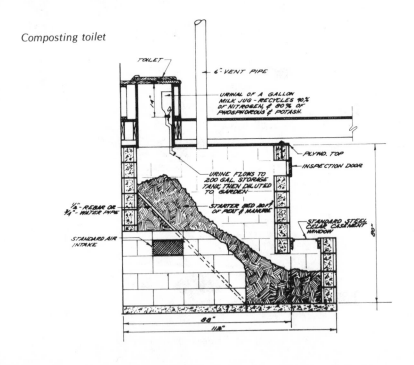

for labor (U.S. prices). Compost toilet installation must also be accompanied by some kind of greywater recycling system. It should be possible to bring these costs down considerably in the Third World, near the $55 (1976) cost of the Minimus built in the Philippines (documented in **Low Cost Technology Options for Rural Sanitation** and **Goodbye to the Flush Toilet**, reviewed in this section). In rural areas of the Third World, the Vietnamese composting pit privy may be a lower-cost solution (see review of **Stop the Five Gallon Flush**, page 210, and **Health in the Third World—Stories from Vietnam**, reviewed in this section).

In addition to the problems of cost in developing countries, this report does not take into account the different kinds of problems likely to be faced under tropical conditions. We do, however, recommend this book to all readers, as it certainly offers a good starting point for adaptation work.

Sanitation Without Water, book, 116 pages, Uno Winblad, Wen Kilama, Kjell Torstensson, 1978 (revised edition available soon), price unknown, from Swedish International Development Authority, S 105 25, Stockholm, Sweden.

"The flush toilet cannot solve the problems of excreta disposal in the poor countries. Nor has it indeed solved those problems in the rich part of the world." To document some alternatives, the authors have written this practical manual on water-less waste treatment systems, primarily compost privies and pit latrines for individual households. These units do not require water-borne sewage disposal lines, and are thus less expensive and complex.

The information presented here is intended for health officers, medical workers, and village technicians in East Africa, although the authors feel the information is adaptable elsewhere. The book should also be useful as a guide in training programs. There are four parts:

1) The relationship between sanitation and disease; the digestion and composting processes; 2) a description of 10 dry sanitation systems around the world (for a more extensive summary, see **Stop the Five Gallon Flush**, reviewed on page 210); 3) a latrine manual, including explanations of latrine components, design information, operating instructions, and proper location for latrines; 4) Appendix: fly control in dry latrines.

The authors also add a note of caution: "There are no miracle solutions to the problem of excreta disposal …The methods and systems described can work and some do work very well indeed. But when not fully understood by the users or constructed in the wrong way or in the wrong place, they may fail completely."

The best (and only) book we've seen that covers all aspects of dry sanitation systems on an easily understood level. The 97 drawings are very helpful in understanding these simple methods of sanitation that can be applied with limited resources.

Highly recommended.

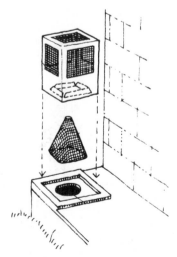

Fly trap for a composting toilet

Goodbye to the Flush Toilet, book, 296 pages, edited by Carol Hupping Stoner, 1977, $6.95 from RODALE.

''A waste is a resource out of place.'' This book describes the best biological systems developed over the last few years in North America for on-site recycling of human wastes and waste water (greywater). Numerous designs are presented, with text and drawings that would allow readers to build or maintain such devices. The Farallones Institute vault and drum privies, Ken Kern's solar shower and privy, the Clivus Multrum and other manufactured units are featured. This book is written primarily from a North American context, and is probably most valuable in temperate regions and industrialized countries.

An essay by Witold Rybczynski describes dry toilets in the Third World: ''The first installation in a developing country was completed in Magsaysay Village in the Tondo area of Manila, Philippines. The Tondo is a squatter settlement of 160,000 people, three-quarters of whom have no waste disposal facilities at all. The prevalent system is euphemistically called ''wrap and throw''! The Minimus (a low cost composting toilet design) chamber is built out of cement blocks, plastered inside, and has a concrete bottom. The vent pipe is galvanized metal, and the air ducts are PVC. The total material costs (not including labor) was US$55. The construction time was six man-days. Half a dozen of these have been built, both in the Tondo and in a resettlement area

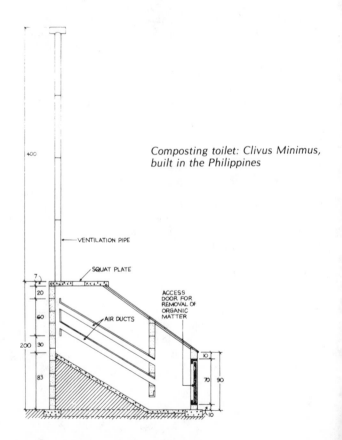

Composting toilet: Clivus Minimus, built in the Philippines

outside Manila."

"The operation of the Minimus in hot climates remains to be seen, and the Manila models will show the problems, if any, of ventilation in hot/humid situations. It is too early yet to tell, though the fact that hot temperatures will certainly speed up the composting process itself is promising."

"It should be stressed that there is no one 'design' for the Minimus. It must be adapted to meet local climatic conditions, available building materials, local skills, and conditions. The application of composting sanitation technology to developing countries cannot be on a piecemeal basis. It must be done on a community (not individual) scale and integrated with social and educational development. It was precisely in such a way that rural composting toilets were introduced to North Vietnam during the years 1961-1965."

Natural Sewage Recycling Systems, report, 36 pages, by Maxwell Small, 1977, $6.00 (Accession No. BNL 50630/LL) from NTIS.

"This paper reviews the work done at Brookhaven National Laboratory in the development of natural systems which produce potable water from sewage." A pond is constructed that is marsh at one end and open water at the other. Human wastes are pumped into this pond after passing through an aeration pond. The first pond adds oxygen to the sewage to reduce nitrogen levels and eliminate odors. "Conventional treatment plant hardware beyond aeration is not used...and no sludge is generated."

The marsh/pond system of natural treatment is much less costly than traditional Western systems of sewage treatment. Human wastes could be brought to the system in pipes, buckets or tanks. The aeration pond requires a floating aerator pump with .3 hp of capacity for each 1000 gallons per day of sewage passing through it. A pump is also recommended to move the aerated human wastes to the second pond.

"Experiments with two prototype systems are described and performance data are presented in detail for the marsh/pond. Empirical interpretations of results achieved to date are suggested for use in the design of marsh/ponds as natural sewage recycling systems."

Plants and aquatic life can be harvested to provide food, fiber and energy. The marsh plants are of great importance and must be harvested regularly to prevent overcrowding.

I. Water Re-use and Solid Waste Management

Residential Water Re-Use, book, 533 pages, by Murray Milne, 1979, California Water Resources Center Report No. 46, $10.00 from Water Resources Center, University of California, Davis, California 95616, USA.

This is the best book available on the re-use of greywater (household waste water, not contaminated with human excreta). A good illustrated summary of proven and experimental household greywater reuse systems is followed by an explanation of the necessary components. The major recommended use for greywater is in garden and tree crop irrigation; a lengthy section on this topic summarizes the technical considerations involved. A limited amount of information is offered on developing sources of fresh groundwater and surface water for household use.

"The conclusion of this study is that residential on-site water reuse systems are already technically feasible and environmentally sound, and are becoming more economically attractive everyday, due primarily to the rapidly increasing cost of energy required for pumping and treatment by centralized water and sewage systems. The objective of this book is to help homeowners, builders, developers, architects, planners and lawmakers understand the design and installation of small on-site residential water reuse systems.''

Two briefly treated topics of special interest are roof rainwater catchment tanks (some interesting ideas for filters) and a dry toilet called the ''earth closet''. This dry toilet worked ''on the principle that powdered, dry earth, which contains clay and natural soil bacteria, will absorb and retain all offensive odors, fecal matter, and urine. It required a supply of dry and sifted earth, or a mixture of two parts earth and one part ashes. After the user got up from the seat a sufficient amount of dry earth was discharged to entirely cover the solid wastes and to absorb the urine. The wastes and the dry earth fell into a pail that was easily removed for emptying, or into an existing holding pit. The pit did not have to be emptied for up to a year.'' (See drawing, opposite page.)

A directory of manufacturers of special equipment and an extensive annotated bibliography on greywater are also included. Though developed for conditions in the State of California, this book has much that will be of interest anywhere in the world.

Recommended.

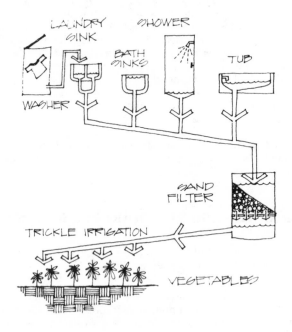

GREYWATER RECYCLING
HAWAIIAN ENERGY HOUSE

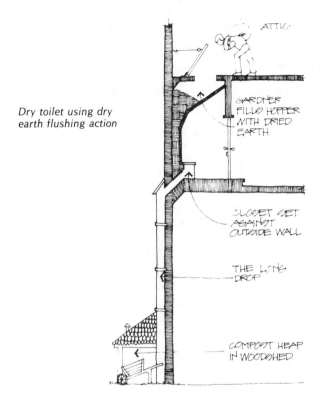

Dry toilet using dry
earth flushing action

Management of Solid Wastes in Developing Countries, book, 242 pages, by Frank Flintoff, 1976, Sw. fr. 20 ($12.00), half price to health workers and institutions other than commercial concerns within the South-East Asia Region of WHO, available from World Health Organization, Regional Office for South-East Asia, Indraprastha Estate, New Delhi 110 002, India.

"Wherever people live, wastes, both liquid and solid, are produced. While the disposal of liquid wastes more often receives priority attention, the management of solid wastes has generally been a neglected field. The aim in developing countries must not be to mimic the technology of the industrialized countries, but rather to employ the technology appropriate to their own situations, while still meeting the basic needs of public health."

This book covers methods for disposal and recycling of solid wastes (both household and commercial). It is intended "to provide a reference source for engineers, municipal officers, administrators and other interested persons, and to fill a need for a training manual for technicians in a field of universal and growing importance." Major topic areas include economic and other aspects of refuse collection and storage; sanitary landfills; and composting of urban wastes.

The coverage of composting programs is quite comprehensive. "Transport between the compost plant and the farm is an important cost element; in most situations this cost limits the marketing range to about 25 kms. If the potential marketing area for compost is a circle 25 kms. diameter and if the plant is in a very large city, much of that circle will be occupied by urban areas; therefore, the larger the city the smaller the potential market for compost. The larger the city, however, the greater the quantity of wastes. Thus composting as a policy suffers from the paradox that the potential market is in inverse ratio to potential wastes production...The most successful composting plants have been those which serve small towns in agricultural areas.''

In the large cities of the People's Republic of China, such as Shanghai, enormous amounts of compost are generated but used in agriculture **within** the cities. Some changes in the land use and ownership patterns of many countries—both developed and developing—would allow agriculture to re-enter the cities in the form of neighborhood gardens, tree crops, and container gardening on rooftops and balconies. This would help solve solid waste and malnutrition problems in many cities.

Though this book is primarily devoted to large-scale solid waste operations, there is much that would be useful in small-scale projects.

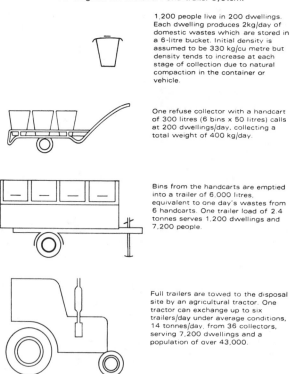

SHORT-RANGE TRANSFER
Flow diagram for handcart and trailer system.

1,200 people live in 200 dwellings. Each dwelling produces 2kg/day of domestic wastes which are stored in a 6-litre bucket. Initial density is assumed to be 330 kg/cu metre but density tends to increase at each stage of collection due to natural compaction in the container or vehicle.

One refuse collector with a handcart of 300 litres (6 bins x 50 litres) calls at 200 dwellings/day, collecting a total weight of 400 kg/day.

Bins from the handcarts are emptied into a trailer of 6,000 litres, equivalent to one day's wastes from 6 handcarts. One trailer load of 2.4 tonnes serves 1,200 dwellings and 7,200 people.

Full trailers are towed to the disposal site by an agricultural tractor. One tractor can exchange up to six trailers/day under average conditions, 14 tonnes/day, from 36 collectors, serving 7,200 dwellings and a population of over 43,000.

ADDITIONAL REFERENCES ON WATER SUPPLY AND SANITATION

Appropriate Technologies for Semi-Arid Areas: Wind and Solar Energy for Water Supply, reviewed on page 566.

The use of human wastes in biogas plants is described in **Compost, Fertilizer, and Biogas Production from human and Farm Wastes in the People's Republic of China**, page 650.

Repairs, Reuse and Recycling discusses strategies for reducing the flow of solid waste, page 364.

ENERGY: GENERAL

Energy: General

"It is the experience of most developing countries that energy produced through centralized thermal, hydroelectric and nuclear power stations rarely flows to rural areas where the bulk of the population lives. A typical distribution for such centralized power production is about 80% for urban industry (based on energy-intensive Western technology), about 10% for urban domestic consumption, and only about 10% for rural areas."
— **CERES: The FAO Review on Development**, March-April 1976

With each price increase in the world's diminishing oil supply, renewable energy sources are made more attractive. The decentralized supply of these renewable energy sources—wind power, solar energy, water power and biofuels—matches the decentralized settlements of the rural Third World. Planners and program administrators are increasingly convinced that these technologies have a major role to play in the energy supplies of rural communities.

These same people, however, have been slower to accept the idea that urban communities could also be largely powered with renewable energy technologies. **Soft Energy Paths** *is for that reason a landmark document, making the case that a conservation-oriented, decentralized industrial society can be operated entirely with "soft" (renewable) energy technologies. Not only is this technically feasible, argues author Amory Lovins, but it is also necessary to avoid the disastrous political, economic, and environmental consequences of the alternative "hard" path, with its reliance on dangerous nuclear power stations, accelerating environmental disruption, and the nuclear weapons proliferation associated with nuclear power. Lovins has crafted probably the most tightly documented, academically respectable analysis of the political, economic, social and environmental consequences of energy technology choice, and in so doing has provided helpful direction and ammunition to appropriate technology advocates.*

Soft Energy Paths *and* **Rays of Hope** *together make the arguments that the exponential increases in energy consumption characteristic of industrial societies cannot continue, and therefore industrial development in all countries will have to shift towards decentralization, conservation, improved energy conversion efficiency, and a better matching of energy quality to end use needs.*

The next two books in this section review the most attractive renewable energy technologies likely to fit the circumstances in the rural Third World, where they may help slow urban migration. **Renewable Energy Resources and Rural Applications in the Developing World** *also notes the domestic and foreign policy implications that come with choice of energy strategy.* **Energy for Development: Third World Options** *points specifically to reforestation programs for fuelwood and soil conservation as high priorities in energy planning.*

Appropriate Technologies for Semi-Arid Areas: Wind and Solar Energy for Water Supply and **Small Scale Renewable Energy Resources in Nepal** *also document a variety of energy technologies. A catalog of commercially available small scale power generating equipment, entitled* **The Power Guide,** *has been introduced by ITDG. This book includes both renewable energy devices and diesel and gasoline engines.*

The increasing acceptance of an important role for renewable energy systems, noted earlier, has lead to a proliferation of pilot projects. Economic feasibility has not been properly considered in many of these projects, a fault perhaps most common in the efforts of large international and bilateral aid agencies, who should know better. **The Economics of Renewable Energy Systems for Developing Countries** offers three case studies illustrating this problem, and a methodology for evaluating the economic appeal of any renewable energy project. Author David French notes that in particular, large agencies seem to have forgotten that most of the rural poor do not use commercial fuels and thus cannot simply switch cash payments towards the purchase of new equipment:

''Most renewable energy devices now tend to be attractive primarily to people already using costly commercial power. Just as is happening in the United States, for example, some Third World city-dwellers are discovering that solar energy may be cheaper than electricity for heating water...Such systems will be of greatest use to the wealthy; there is little reason to suppose they will be of comparable interest to the poor.''

In the rural Third World, most of the energy used is in the form of firewood and crop residues gathered and burned in cooking fires. Low cost locally-built cooking stoves can greatly increase the efficiency of cooking, reducing the demand for firewood by up to one-half. This would both slow the rate of deforestation and lighten the burden of long distance wood hauling. Technologies such as improved stoves and village woodlots, that use local materials and skills, are more likely to be immediately affordable than expensive devices such as solar pumps, photovoltaic systems, and even biogas plants in many cases.

Other renewable energy technologies relevant to the rural Third World include locally-built waterpumping windmills (increasingly attractive for small plot irrigation and community water supplies) and windgenerators (more expensive but of interest where a small amount of electricity production has a high value); both can be found in the ENERGY: WIND chapter. Waterwheels and water turbines could play a greatly expanded role in rural crop processing and in supplying the energy needs of rural industries, as has been the case in China (see ENERGY: WATER chapter). Direct solar technologies for crop drying, currently probably the most important area for solar energy use, are examined in CROP DRYING, PRESERVATION, AND STORAGE. Solar home heating and cooling, water heating, cooking, refrigeration, water pumping, and electricity production are all reviewed in the ENERGY: SOLAR chapter. The anaerobic fermentation of animal manures and crop residues to produce biogas and fertilizer is covered in the ENERGY: BIOGAS chapter, including two new translations of manuals on the remarkably successful Chinese biogas plants.

A significant locally available energy source is animal power, which could be made more efficient through the design of better pumps, crop processing equipment, harnesses, carts, and agricultural implements. **The Management of Animal Energy Resources and the Modernization of the Bullock Cart System** examines this largely neglected topic, noting that for agricultural activities on the Indian farm, two-thirds of all energy is provided by animals, while humans contribute 23% and electricity and fossil fuels together amount to only 10%.

Alcohol fuels have received a great deal of attention recently, bringing the hope that they could replace increasingly expensive and scarce gasoline. The technical requirements of alcohol production are presented in **Fuel from Farms** and **Makin' It On The Farm**. Alcohol fuels can be made from crop residues and tree crops. However, large scale alcohol fuel programs ignore these feedstocks in favor of more economically attractive grain, cassava, and sugar cane. In **Food or Fuel: New Competition for the World's Croplands**, Lester Brown argues that this will tend to reduce food supplies, as the major exporting nations could easily consume all their surplus crop in alcohol conversion programs. The first people to be affected will be the urban poor in developing countries.

Pedal power offers some possibilities for use in small tasks that otherwise would require hand-cranking or high payments to the owners of engine-driven equipment. The small amount of power that a healthy person can produce (75 watts—1/10 hp—continuously, 200 or more watts for brief periods) is best suited to short, intermittent tasks, such as the operation of workshop equipment. While pedal powered agricultural processing equipment, particularly pedal threshers, can be much more efficient in labor use than many traditional techniques, larger quantities of crops require the power available from draft animals, waterwheels, or small engines. **The Use of Pedal Power in Agriculture and Transport in Developing Countries** summarizes the potential applications, and is followed by two entries that present construction details for pedal powered equipment. **Bicycling Science** (see review in TRANSPORT chapter) reviews the performance of human beings operating stationary pedal power units.

Steam engines can be used for a wide variety of rural power requirements, and can be fired with agricultural residues; they are reportedly built in many small workshops in Bangkok. Five of the entries in this chapter are concerned with the design and construction of very small scale steam engines and boilers, mostly in the range of 1-2 hp—equivalent to the pedalling power of 10-20 men. Steam engines are inefficient in conversion of fuel energy into work, and they are heavier and require more materials and space than small gasoline or diesel engines of similar power. Yet they are less technically demanding to make, with larger acceptable tolerances when fitting parts. They are probably most commonly found in small sawmills in the Third World, where a ready supply of sawdust, bark, and wood scraps is available. (For repair and maintenance of small gasoline engines, see **How to Repair Briggs and Stratton Engines** and **Small Gas Engines**, in the AGRICULTURAL TOOLS chapter.)

The final two entries in this chapter are on electrical systems. One is a reference on installing electrical lines in small communities. The other describes synchronous inverters that allow the direct linking of wind-generators or small hydroelectric units to the electrical grid, thereby avoiding the need for costly battery systems.

In Volume I:

An additional 15 cross-references on pedal powered equipment are listed on 196.

Soft Energy Paths: Toward a Durable Peace, book, 231 pages, by Amory Lovins, 1977, reprinted by Harper and Row 1979, $3.95 from Harper and Row, Publishers, Inc., Keystone Industrial Park, Scranton, Pennsylvania 18512, USA.

In this book Amory Lovins demonstrates the feasibility of an advanced industrial society fueled entirely by renewable energy sources. Two strategies for energy development are compared and contrasted.

The conventional strategy or "hard path" is aimed at sustaining a high growth rate in energy consumption in the face of increasingly inaccessible and expensive petroleum supplies. This path would rely on massive expansion of coal and nuclear-based power generation. The latter in particular threatens world peace, as "nuclear power is considered by most informed observers today to be the main driving force behind the proliferation of nuclear weapons."

The "soft path", by contrast, would rely on a variety of sources carefully matched in scale and energy quality to specific tasks. "The laws of physics require, broadly speaking, that a power station change three units of fuel into two units of almost useless waste heat plus one unit of electricity. This electricity can do more difficult kinds of work than can the original fuel, but unless this extra quality and versatility are used to advantage, the costly process of upgrading the fuel—and losing two thirds of it—is all for naught... Where we want only to create temperature differences of tens of degrees, we should meet the need with sources whose potential is tens or hundreds of degrees, not with a flame temperature of thousands or a nuclear reaction temperature equivalent to trillions—like cutting butter with a chainsaw." This approach would allow conservation of scarce non-renewable fuels during a transition to renewable sources.

In his comparison of the hard and soft paths, Lovins argues that political and environmental considerations should lead us to choose a soft path. Furthermore, he shows that the hard path's centralized plant and distribution systems would require a huge percentage of all available investment capital.

In the United States, this book has helped to trigger a widespread re-evaluation of our energy strategies. The facts and arguments are also relevant to energy strategies in the Third World, particularly in urbanindustrial areas.

This volume is probably the best synthesis of the technical, macroeconomic, and humanistic arguments for a serious political and social commitment to renewable energy alternatives.

Rays of Hope: The Transition to a Post-Petroleum World, book, 233 pages, by Denis Hayes, 1977, $3.95 from W.W. Norton & Co., 500 Fifth Avenue, New York, New York 10036, USA.

Beginning with an overview of patterns of petroleum resources depletion, Denis Hayes shows that our planet cannot continue to support the way of life now characteristic of advanced industrial societies. With costs of delivering increasingly scarce petroleum skyrocketing, it will ultimately require more energy to deliver dispersed fuel in marginal deposits (e.g. tar sands and shales) than is available in the fuel itself. And while coal reserves are relatively plentiful, their use is limited by the capacity of the environment to process waste carbon dioxide. Public opposition to nuclear power is growing because of waste problems, rapidly rising capital costs, and vulnerability to accident and misuse (in weapons production around the world).

Following these arguments, most of this book explores more attractive, sustainable energy consumption paterns based upon direct solar, wind, water, and biomass resources. An analysis of energy use patterns in space heating and food and transportation systems shows that conservation efforts using technical approaches could save great quantities of energy: ''A $500 billion investment in conservation would save the U.S. twice as much energy as a comparable investment in new supplies could produce.''

Looking to the medium- and long-term future, Hayes also stresses the importance of policy and the pivotal nature of decisions made today: ''Oil and natural gas are our principal means of bridging today and tomorrow, and we are burning our bridges. Twenty years ago, humankind had some flexibility; today, the options are more constrained. All our possible choices have long lead times...inefficient buildings constructed today will still be wasting fuel fifty years from now; oversized cars sold today will still be wasting fuel ten years down the road...''

An important perspective on the global energy situation.

Energy: The Solar Prospect, Worldwatch Paper 11, 40 pages, 1977, and **The Solar Energy Timetable**, Worldwatch Paper 19, 78 pages, 1978, both booklets by Denis Hayes, $2.00 each from Worldwatch Institute, 1776 Massachusetts Avenue N.W., Washington D.C. 20036, USA.

In these two papers, Denis Hayes outlines possible strategies for a global transition to a ''solar-powered world'' within 50 years, an ambitious but necessary goal. In discussing solar resources, he notes that the possibilities range from the simple to the complex, and from the decentralized to the centralized. This makes solar applications adaptable in many different societies, rich and poor, urban and rural.

Hayes argues forcefully that shifting to renewable (primarily solar) energy sources is technologically feasible—the barriers are economic and political. He suggests that by taking the lead in developing renewable energy supplies, Third World nations could achieve a level of development independent of the western oil-based economies.

These thoughtful papers provide an important analysis of the potential for solar energy use. Recommended.

Renewable Energy Resources and Rural Applications in the Developing World, book, 168 pages, edited by Norman Brown, 1978, American Association for the Advancement of Science, $19.00 from Westview Press, 5500 Central Ave., Boulder, Colorado 80301, USA.

This set of papers offers a valuable look at the potential for use of renewable energy resources in rural areas of developing countries. Norman Brown provides a thoughtful introduction to the topic:

"Choosing conventional large-scale capital-intensive technologies implies **a priori** decisions, conscious or not, about many important policies. These include the course of urban development, expanding industrialization, environmental impact, large-scale borrowing (or foreign investment) with long-term indebtedness and problems of debt servicing, and last but not least, the foreign policy stance dictated by these requirements."

"On the other hand, the choice of small-scale decentralized power systems (e.g., solar heating, cooling, and generation of electricity; windmills; small scale hydroelectric plants) implies a different set of **a priori** decisions. These include, for example, de-emphasis of western-style industrialization as the sole or primary immediate goal of development; dispersal of industry and, perhaps, changes in financial mechanisms; and a shift from western agricultural techniques to emphasis on improvement of indigenous agricultural practices, with consequent reduced demand for energy-consuming nitrogenous fertilizers. All of these factors could contribute significantly to a slowing down of migration to the cities and urban growth, with important effects on the rate of growth of dependence on commercial energy supplies."

The other papers include a general introduction to rural energy requirements, a description of the U.S. photovoltaic program, a look at the potential for solar energy use, an evaluation of wood waste as an energy source in Ghana, a summary of the process of methane production, and discussion of a wide variety of alternative energy technologies for Brazil. An article on wind energy conversion in India argues that waterpumping windmills seem to be the most promising new energy technology in the rural areas. There is also a good historical summary of the development of waterwheels and water turbines, and their importance in the growth of the rural economies of the United States and Europe (and China today).

"The history of small scale hydro-power development provides sound suggestions of how to aid...rural areas of developing countries (that have water power potential) in achieving an improved standard of life...The major role in this development was played by the simple waterwheel...Later the small turbine provided more power at a given site than was feasible with the waterwheel...(In the rural areas of the United States water turbine production) was rapidly taken over by blacksmiths and foundrymen who found it easy to make, in great demand, and an extremely profitable business...By the middle of the 19th century the French turbine had been so radically altered by rural American craftsmen that American turbines began to take the names of their

many improvers.''
    These mills showed the potential for rural industry based on decentralized
power sources. They ''turned out such household products as cutlery and edge
tools, brooms and brushes...furniture, paper...pencil lead...needles and
pins...watches and clocks, and even washing machines.''
    ''For the farm they turned out fertilizers, gunpowder, axles, agricultural
implements, barrels, ax handles, wheels, carriages. There were woolen,
cotton, flax and linen mills;...tannery, boot and shoe mills...and mills turning
out surgical appliances...and scientific instruments.''

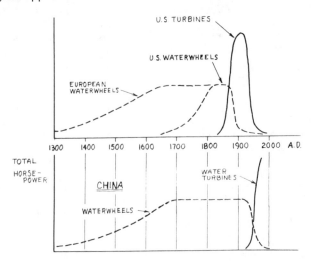

Comparison of the number of small waterwheels
and water turbines in Europe, the United States,
and China over time

Energy for Development: Third World Options, Worldwatch Paper 15,
booklet, 43 pages, by Denis Hayes, December 1977, $2.00 from Worldwatch
Institute, 1776 Massachusetts Avenue, N.W., Washington D.C. 20036, USA.

    This is a short overview of the energy prospects for the Third World, pre-
senting the arguments for decentralized renewable energy technologies.
The relatively dispersed population living in the rural Third World matches the
decentralized energy available from sun, wind, water and plant sources.
These energy sources also offer the possibility of helping to slow urban
migration by making the villages more appealing places to live.
    After promotion of improved cook stoves (not discussed here), reforestation
is probably the cheapest way to invest in energy production. Two of the
generally less discussed but significant technologies are small scale hydropower and passive solar housing. The potential for small scale hydropower has
never been fully explored. Passive solar buildings could be very important in
cold regions such as the Andes, the Himalayas, northern China, and south of
the Tropic of Capricorn; indeed, traditional housing in these regions often
already incorporates some passive solar principles.
    The disadvantages of nuclear plants include safety hazards, technical
complexity, waste disposal dilemmas and centralization, as in the industrialized countries. In addition, they represent a large drain on foreign

exchange for plants and fuel, and in all but five developing nations plant size is too large to prevent a complete grid shutdown when a plant is shut down.

Soft Energy Notes, journal, 6 issues per year, $25 to individuals ($15 to Friends of the Earth members) in the U.S., $35 overseas, ask for prices for local governments and businesses, small organizations who cannot afford these subscription rates can exchange other information for the journal, from International Project for Soft Energy Paths, Friends of the Earth Foundation, 124 Spear St., San Francisco, California 94105, USA.

This thoughtful international journal, which has grown out of Amory Lovins' book **Soft Energy Paths**, reports on studies of "soft energy" technologies and strategies. "Soft energy paths are routes to renewable energy futures... away from an exclusive reliance on fossil fuels and nuclear power, leading to a pattern of energy sources and uses that is renewable, diverse, efficient, and environmentally benign."

"**Soft Energy Notes** reports to a network based in more than 70 countries on soft energy technologies, efficiency studies, new research methods, and policy issues under development worldwide." Perhaps a quarter of the articles and references are on Third World (mostly rural) energy topics. For example, the June/July 1980 issue includes an article on energy use and social structure in a village of Bangladesh, and an article examining the policy measures that can support indigenous R&D on decentralized rural energy technologies.

A good source of hard information that can be quite valuable in persuading policy makers and project planners of the technical and socio-economic merits of soft energy strategies.

The Economics of Renewable Energy Systems for Developing Countries, report, 67 pages, by David French, 1979, available on request from Office of Development Information and Utilization, Development Support Bureau, USAID, Washington D.C. 20523, USA.

The author examines three projects employing some of the more sophisticated renewable energy technologies: solar pumps in Senegal, biogas plants in India, and solar-electric pumps in Chad. He presents a careful economic analysis and concludes that none of these technologies is now a good investment, nor does any of them appear likely to become a good investment in the next decade.

"Most renewable energy devices now tend to be attractive primarily to people already using costly commercial power. Just as is happening in the United States, for example, some Third World city-dwellers are discovering that solar energy may be cheaper than electricity for heating water...Such systems will be of greatest use to the wealthy; there is little reason to suppose they will be of comparable interest to the poor."

"Rather than concentrating on devices of the sort described above, organizations concerned with the poor might seek to meet basic energy needs through simpler systems: village woodlots, improved wood stoves, hand or pedal pumps and grinders, hydraulic ram pumps, and so on. Emphasis would be on systems whose benefits were likely to be commensurate with their costs, and whose costs were likely to be within reach of the poor. Given this approach, ways might be found to make energy widely available to people most in need of it."

In addition to pointing out the dubious appeal of the higher-cost group of alternative technologies, the methods of economic analysis clearly presented here can be used to help evaluate other renewable energy technologies. This report will also be helpful to people who need to understand the methods and concepts of analysis often used by major aid agencies.

The Management of Animal Energy Resources and the Modernization of the Bullock Cart System, book, 137 pages, by N.S. Ramaswamy, 1979, available on request to serious groups in the Third World, from Prof. N.S. Ramaswamy, Director, Indian Institute of Management, 33, Langford Road, Bangalore 560 027, India.

''For short hauls, small loads, versatile movement over any available surface and low freight charges, the cart has no peer either in the rural areas, or, for that matter, in the towns and cities. It is still cheap, readily available, and safe.''

The author presents statistics to convincingly demonstrate the importance of animal power and carts in the Indian economy. Discussing deficiencies in design that need to be overcome, he offers evidence that the improvement of harnessing devices, agricultural implements, and carts should be given a high priority by the Indian government. To accomplish these objectives, he proposes the establishment of an Animal Energy Development Corporation, and outlines a program of activities. He also argues for less cruelty to the animals both in general use and through promotion of improved slaughterhouse facilities.

Animal power inputs on the Indian farm are even greater than in the transport sector. Two-thirds of all farm energy is provided by animals, while human energy provides 23% and electricity/fossil fuels only 10%.

Carts are used in moving 15-18 billion tonne-km of freight per year in India. But ''the traditional cart is defective in design. The draught power of the animal is wasted due to friction resulting from rough bearings and crude and inefficient harnessing, etc. The wobbling rim cuts into the road surface and damages it...Weights run high. Traditional carts can be easily improved by: smooth bearings, lower weight, the introduction of a log-brake, better harnessing, the use of pneumatic tires on paved roads'' and the use of hard rubber tires in rural areas.

As this is a compilation of papers, much of the text is repetitive. Photos of old and new cart designs are included.

Appropriate Technologies for Semi-Arid Areas: Wind and Solar Energy for Water Supply, book, 334 pages, 1975 conference report, DM 20 ($11.20) from German Foundation for International Development, Endenicher Strasse 41, 53 Bonn 1, Federal Republic of Germany.

Participants from developing and industrialized countries contributed to this report. It is divided into sections on wind energy, solar pumping and electricity, solar distillation, research and development, and conditions and restraints for wind and solar energy in water supply. The articles cover both social and economic aspects of water supply projects around the world as well as summaries of various technologies that have been successfully applied. For example, there are several excellent articles on low-cost waterpumping sail windmills for small-plot irrigation.

A set of recommendations is addressed to international organizations, suggesting ways in which their work can best encourage local development in rural areas. These include acceptance of a need for both complex and simple machines, depending on circumstances, and an emphasis on more local control and maintenance of any water supply. The importance of a water supply to the development of any community is recognized, and it is felt that future research into solar and wind energy should reflect the need for adequate water. Areas for specific development are identified: biogas digesters (to provide gas for engine-driven pumps); better, lower cost concentrating and flat plate solar collectors; locally available wind measuring equipment; and low speed, reliable wind pumps that can generate 1-2 hp at 15 km/hr wind speeds. (There is, however, no agreement on how this type of research should best be carried out.)

A useful summary of water supply efforts based on renewable energy sources, with examples drawn from around the world.

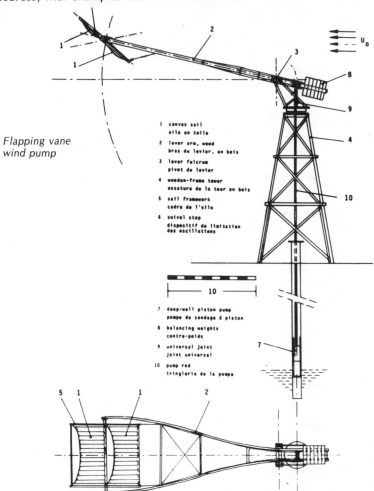

Flapping vane wind pump

Small Scale Renewable Energy Resources in Nepal, booklet, 13 pages, 1979, free from Swiss Association for Technical Assistance, Jawalakhel, P.O. Box 113, Kathmandu, Nepal.

A short booklet filled with pictures of traditional and new technologies that can be seen in Nepal. 62 photos of stoves, solar dryers, watermills, microhydroelectric turbines, solar water heaters, bio-gas plants, and passive solar buildings.

The Power Guide: A Catalogue of Small Scale Power Equipment, book, 240 pages, compiled by Peter Fraenkel, 1979, ₤7.50 from ITDG.

This is a guide to commercially available equipment. This book differs from the **Energy Primer** in that less attention is given to explanation of the different energy systems—but the text and the equipment listed is more directly relevant to the Third World.

The guide provides criteria that should be used in selection, but makes no specific recommendations of equipment, as ITDG has no facilities for testing machinery. No prices are given, as they are rapidly changing. Names and addresses are listed for manufacturers and their agents, information services, and organizations doing R&D work on small scale power production.

Major topics are: solar electric cells, solar engines, solar space and water heaters, windgenerators, windpumps, hydroelectric units, hydraulic ram pumps, iron and steel cooking and heating stoves, methane digesters, steam boilers, wood gas producers, diesel engines, gasoline (petrol) engines, steam engines, alternators and generators, electric generating units, and batteries.

The lowest cost small scale energy devices, however, are rarely available from commercial sources, and cannot be found in this catalogue. Such devices are built within many developing countries. These include solar dryers for crop preservation, low wind-speed water-pumping windmills for small plot irrigation and for village water supply, waterwheels for direct mechanical power, efficient cooking stoves built on the spot out of clay and sand, and Chinese biogas plants. Efficient hydroelectric units, hydraulic ram pumps, and steam engines are all being built within developing countries, and at much lower cost than the imported alternatives. Plans for all of these can be found in the energy chapters of the **Appropriate Technology Sourcebook**.

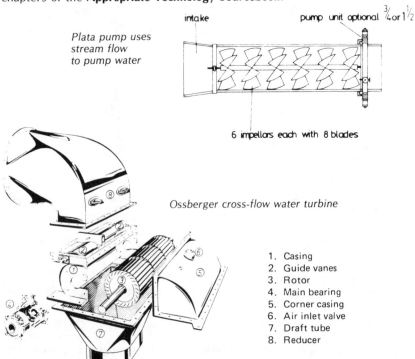

intake   pump unit optional $\frac{3}{4}$ or $1\frac{1}{2}$

Plata pump uses
stream flow
to pump water

6 impellors each with 8 blades

Ossberger cross-flow water turbine

1. Casing
2. Guide vanes
3. Rotor
4. Main bearing
5. Corner casing
6. Air inlet valve
7. Draft tube
8. Reducer

Food or Fuel: New Competition for the World's Cropland, Worldwatch Paper 35, 43 pages, by Lester Brown, 1980, $2.00 from Worldwatch Institute, 1776 Massachusetts Avenue N.W., Washington D.C. 20036, USA.

This well-documented paper indicates that if present trends continue, the world's alcohol fuel programs will be taking food away from the poor, especially the urban poor in developing countries.

The major grain exporters are all launching alcohol fuel programs that could easily consume all of their surplus crop. New Zealand figures for complete reliance on alcohol as a liquid fuel by the year 2000 require an area equal to the entire current area under crops. In Australia, by 1985 15-20% of liquid fuel could come from wheat—requiring roughly the entire current wheat crop. Converting the entire U.S. grain harvest into alcohol fuels would replace 30% of current gasoline consumption.

Operating a typical American car 10,000 miles/year at 15 mpg would require almost 8 acres of cropland for the alcohol fuel—enough to feed 39 people in the developing countries or 9 people in the United States. (These figures do not include the amount of liquid fuel needed to produce the grain in the first place.)

In Brazil, "the decision to turn to energy crops to fuel the country's rapidly growing fleet of automobiles is certain to drive food prices upward, thus leading to more severe malnutrition among the poor. In effect, the more affluent one-fifth of the population who own most of the automobiles will dramatically increase their individual claims on crop-land from roughly one to at least three acres, further squeezing the millions who are at the low end of the Brazilian economic ladder."

"Brazilian officials claim that the production of energy crops will be in addition to rather than in competition with that of food crops. Yet energy crops compete not only for land but also for agricultural investment capital, water, fertilizer, farm management skills, farm-to-market roads, agricultural credit, and technical advisory services."

"A carefully designed alcohol fuel program based on forest products and cellulosic materials of agricultural origin could become an important source of fuel, one that would not compete with food production."

Fuel from Farms—A Guide to Small Scale Ethanol Production, book, 157 pages, by the Solar Energy Research Institute, 1980, limited number of copies, free from Technical Information Center, U.S. Department of energy, Post Office Box 62, Oak Ridge, Tennessee 37830, USA. Additional copies $4.50 from the Superintendent of Government Documents, U.S. Government Printing Office, Washington, D.C. 20402, USA.

This Solar Energy Research Institute publication is a part of the Department of Energy's effort to encourage the production of alcohol for fuel in the United States. It "...presents the current status of on-farm fermentation ethanol production as well as an overview of some of the technical and economic factors. Tools such as decision and planning worksheets and a sample business plan for use in exploring whether or not to go into ethanol production are given. Specifics in production including information on the raw materials, system components, and operational requirements are also provided. Recommendation of any particular process is deliberately avoided because the choice must be tailored to the needs of each individual producer. The emphasis is on providing the facts necessary to make informed judgements."

This analysis is aimed at the American farmer, demonstrating how an ethanol plant can support and complement other farm activities. Surplus grains and spoiled or marginal crops can provide the feedstock base for ethanol production, which can then be mixed with gasoline or burned directly in farm vehicles. The remainder can be sold to blenders of ''gasohol'' (a commercially available fuel mixture composed of 90% gasoline and 40% ethanol). The solid by-product can be fed directly to animals or mixed with other fodder as an animal feed supplement.

While ethanol can be produced from a wide variety of crops and agricultural byproducts and wastes, profitable production depends on many factors such as current prices of feedstocks, availability of a low-cost source of fuel for the distillery apparatus, and how much of the ethanol produced would be used directly by the farmer. Another key factor is high initial equipment investment for the cookers, wells, pumps, still, condensers, and storage tanks. The sample business plan includes a feasibility analysis of a 25 gallon/hour installation on a 1,200 acre farm and feedlot operation, and assumes an initial plant and equipment investment of nearly $125,000. This is hardly small scale in most countries.

It is clear that farm-based production of ethanol is not the whole solution to increasing scarcity and cost of petroleum fuels. Even if all the crop land in the United States were devoted to cultivation of ethanol feedstocks, the fuel produced would not meet current demand from the transportation sector alone. A move toward, renewable sources of energy is, however, crucial to a sustainable agriculture. Decentralized ethanol production may be a way for some American farmers to reduce their dependence on traditional fuel sources.

Makin' It On The Farm: Alcohol Fuel is the Road to Energy Independence, book, 87 pages, by Micki Nellis, 1979, $2.95 plus $0.75 postage from American Agriculture Movement, P.O. Box 100, Iredell, Texas 76649, USA.

This low-cost report has been published in the belief that small-scale ethanol production can be an important step toward viability for the American family farm. ''Community size alcohol plants using locally grown farm products could make towns independent of Big Oil almost overnight. Alcohol fuel plants would...provide fuel, use up those 'burdensome surpluses' that are blamed for depressing farm prices, and provide a high-protein feed byproduct for local use that is oftentimes as valuable as the raw commodity that went into the alcohol...The farmer has the advantage over large alcohol plants because he can put up a small plant in a few weeks, compared to a two-year lead time on large plants. The farmer can use his own wastes when they are available...He can use the crops he grows best, produce as much fuel as he needs, adjust his livestock numbers for the amount of high-protein feed he will produce, and work at making fuel during slack times.''

The core of this book is a discussion of how ethanol is produced by fermentation and distillation of mash made from grains and other crops. Several brief case studies document construction and performance of privately and cooperatively-financed plants producing up to 30 gallons of fuel-grade alcohol per hour, fired with crop residues and/or steam injection. All the plants described involve an airtight vat or tank for fermentation of the grain mixture, a vertical column through which alcohol vapors rise, and a condensing apparatus. Drawings and parts lists are provided for construction of a plant with three fermentation tanks and distillation columns. (How to generate the dry steam which vaporizes the alcohol in the columns, is not explained.) The authors claim that this plant can produce about 30 gallons/hour of 192 proof alcohol (96% ethanol, 4% water).

Other chapters briefly cover simple solar ethanol stills, ideas for using waste irrigation pump heat in alcohol plants, and modifications to improve performance of engines burning alcohol fuels. Appendices include lists of alcohol plant manufacturers, informative papers, farm products that can be used to make alcohol, and a glossary.

Fuel from Farms (see review in this section), gives a better cost and technical analysis of ethanol plants which produce a relatively high volume of high-quality alcohol. **Makin' it on the Farm** stresses a low-investment, do-it-yourself approach. For this reason it should be a more useful resource for many family farmers and Third World groups interested in alcohol fuels. These are the only two alcohol fuel books we are aware of that are not simply high-priced booklets full of exaggerated claims.

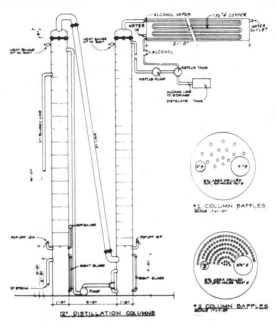

Some of the equipment used in farm-scale alcohol production

Solargas: How to Easily Make Your Own Auto and Heating Fuel for Pennies a Gallon, book, 202 pages, by David Hoye, 1979, $7.95 from International Publishing of Los Angeles, 309 South Sepulveda, Manhattan Beach, California 90266, USA.

One of a flurry of high-priced books on alcohol fuels, this is full of unsubstantiated claims and inaccuracies. We include it only because it does have a lot of interesting ideas for backyard alcohol fuel production (though most of these are probably untested). For example, the author suggests the use of sealed plastic garbage cans as fermentation vessels, and aspirators (attached to water lines) to create partial vacuums and thus reduce heat requirements for distillation in solar panels or other apparatus. A corncob-fired still with steel link chain filling the distillation column (instead of specially cut and drilled plates, the conventional design) is also described. Other proposals: automotive fuel system modifications for adequate performance on homemade ethanol; and commercial-scale production using swimming pools and gasoline storage tanks for sprouting grains and fermentation.

These ideas may inspire countless dead-end experiments. Because the technical content is unreliable, we urge the reader to handle this with great caution.

Methanol and Other Ways Around the Gas Pump, book, 140 pages, by John Ware Lincoln, 1976, $4.95 from EARS.

This is a good general introduction to methods of making fuels from forest and agricultural wastes. The fuels covered include methanol, alcohol, wood gas (produced in a 'gasogen'), hydrogen, methane, and gasoline blends. There is not enough detailed information for actual construction of devices, but there is a bibliography with further sources of information. The author briefly discusses alternative types of engines, including the Stirling (hot air) engine, steam engine, and gas turbine.

The 'gasogen' "is no more than a stove, usually turned upside down, and in its elemental form, the gasogen has no moving parts. At worst, it has a moving grate, and an electric or hand-driven blower to get it started. At the top there is a fuel loading door; in about the middle, there is a small door to light the fire; at the bottom another door, for removal of ashes."

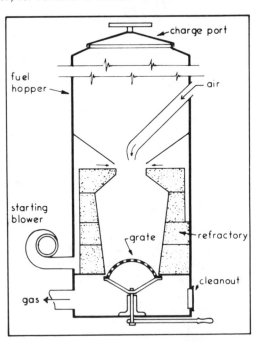

Early and heavy gasogen design

The Use of Pedal Power for Agriculture and Transport in Developing Countries, report for ITDG Transport Panel, 22 pages, by David Weightman, May 1976, small charge for postage, from David Weightman, Faculty of Art and Design, Lanchester Polytechnic, Gosford St., Coventry CV1 5RZ, England.

"This report examines the existing and potential applications of pedal power for simple agricultural machinery and transport devices in developing countries." For these uses, pedal power is compared to other power sources, such as draft animals, electric motors, biogas plants, wind machines, and internal combustion engines. Built-in pedal and treadle mechanisms, separate pedal drive units, and bicycle-connected drive systems are all compared.

"Both treadle and pedal actions are used to drive machinery. The treadle action is commonly operated by one leg, only using half the available power, but enabling the operator to support himself on the other leg and load the machine. Treadle mechanisms are commonly inefficient and much higher power outputs are obtained from the pedal crank arrangement." However, pedal drive systems restrict the operator's freedom of movement, making lathes, potter's wheels, and sewing machines better suited to a different foot-powered approach.

"If more widely used, the lower speeds and lower axle loadings (of pedal powered vehicles) could enable savings in rural road construction costs to be made...Certainly the Chinese transport system, being based on the bicycle rather than the car or lorry, indicates the particular suitability of pedal power to the transport requirements of developing countries."

No photos or drawings included.

Design for a Pedal Driven Power Unit for Transport and Machine Uses in Developing Countries, report, 27 pages, by David Weightman for the ITDG Transport Panel, 1976, small charge for postage, from David Weightman, Faculty of Art and Design, Lanchester Polytechnic, Gosford St., Coventry CV1 5RZ, England.

This report describes a proposed pedal powered unit for use in rural areas of developing countries. The author discusses the need for and the desired characteristics of such a device, and the range of human power outputs that is possible. Ten photographs and a number of drawings are included.

An attached bicycle, an integral pedal drive mechanism, and a dynapod are compared as alternative ways to use pedal power efficiently for different machines and circumstances. The author lists a wide variety of agricultural and workshop equipment that is suitable for pedal power, including winnowers, threshers, grain mills, cassava grinders, maize shellers, winch plows, coffee pulpers, winches, blowers, air compressors, bandsaws, drills, grindstones, lathes, and potter's wheels.

For maximum utility and lowest cost, the author proposes a design for a

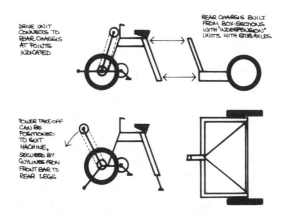

Pedal power concept combining transport and equipment operation

one-wheeled basic unit. It could be used to power equipment, and could have a two-wheeled trailer attached to allow use as a tricyle for transport. It is not a proven design, but suggests an interesting avenue for further investigation.

The Dynapod: A Pedal Power Unit, booklet, 32 pages, by Alex Weir, 1979, $3.95 from VITA.

This two-person unit can be connected to any small mill or machine. Maximum power to be expected over an extended period (an hour or more) is about 150 watts; for shorter periods higher power output is possible. Very little information is offered on actual arrangements for powering different kinds of machines.

The unit has a frame made of wood, and uses wooden bearings. However, also required are 6 bicycle cranks, two saddle seats, a bicycle wheel, a main drive gear and two chains. Five other parts require the use of a metal-turning lathe. Saddle supports are made of steel tubing and plate, and require welding. It appears that many of these parts could be replaced with wooden or simple metal ones, leaving a much simpler and cheaper design with similar performance.

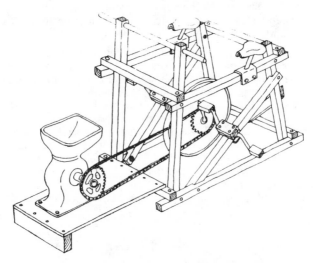

DYNAPOD WITH CHAINDRIVE
&
MILL ON MOUNTING PLATFORM

Steam Power, quarterly magazine, 60 pages, $17.00 for one year, from Steam Power, 106A Derby Road, Loughborough, Leics., LE11 0AG, England.

This magazine reports on the latest advances and most useful technologies in steam power. It covers many different steam power topics, such as vehicles, boats, sawmills, engine plans, history, solar steam power, boilers, fuels, machine shop techniques for making equipment, and conferences.

Letters from interested individuals and many advertisements for kits and plans for fully functional steam engines help make this a useful magazine. Well-illustrated. Back issues are available.

Model Stationary and Marine Steam Engines, book, 168 pages, by K.N. Harris, 1964, $5.25 from META.

A detailed treatment of the design of small steam engines for many different uses is provided in this book. Readers who would like to build small-scale steam engines should find it valuable. The model engines shown are 1 hp and smaller in rated power.

There are no steam engine construction plans contained in this book, but the design principles should be valuable when building a steam engine from plans acquired elsewhere. Knowledge of metalworking, including lathe techniques, is an important part of steam engine construction and necessary for effective use of this book. Many illustrations.

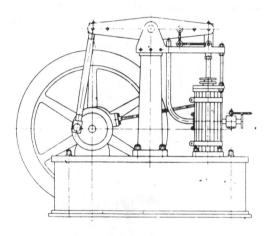

Beam engine
driven by steam

Model Boilers and Boilermaking, book, 185 pages, by K.N. Harris, 1967, $5.25 from META.

Steam engines cannot operate without steam, and the steam boiler is where steam is produced and fuel is consumed. This book presents the theory and construction of small steam boilers. These boilers can be used to power engines of up to several horsepower. There are illustrations covering many types of

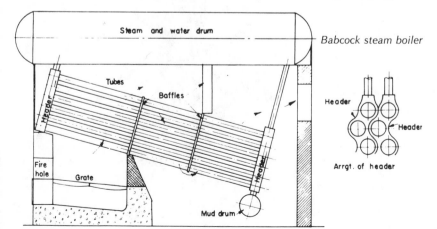

Babcock steam boiler

boilers and fuel systems. Safety considerations are carefully outlined for both boiler construction and operation.

"Fig. 3-32 shows the Babcock, probably one of the most extensively used boilers in full-size pattern for both land and marine work."

A very useful book for anyone interested in producing steam boilers for steam engines or other uses such as heating.

Craftsmanship Catalog, 62 pages, $1.00 from Caldwell Industries, 603-909 E. Davis St., Box 591, Luling, Texas 78648, USA.

This catalog includes kits and plans for a range of small steam engines, most of them models but some of them large enough to be used as a power source.

"The Clarkson 2x2 engines were designed to meet the demands for a larger more robust engine capable of real work...2″ bore and 2″ stroke cylinder. The bearings have been made stronger in proportion to the rest of the engine...The Verticle Compound is capable of about 2 horsepower...care has been taken in the design for the ease of construction in the home shop...These engines do take a lot of steam...The 2x2H and the 2x2V are robust, single cylinder, double acting steam engines fully capable of sustained high speed operation under load...fairly easy to make...The Verticle Compound is the engine for you if you have in mind something like a steam powered row boat or perhaps a motorcycle."

"The 5a engine...is rated at 1½ hp per cylinder...The steam boat crowd

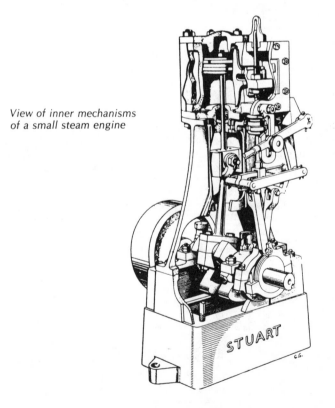

*View of inner mechanisms
of a small steam engine*

often use this engine on small boats, others have used it to pull electric generators at remote sites.''

The kits of castings for these engines ranged from about US$90 to $200 in 1978. The buyer is expected to do a substantial amount of the work to finish the engines, before assembling them. Drawings can be purchased alone for $12 to $15. To operate these engines you will need small boilers; these could use agricultural residues to produce the steam.

Live Steam Magazine, monthly, 68 pages average length, $20.00 per year in USA, $22.00 per year surface mail outside the USA, from Live Steam Magazine, P.O. Box 581, Traverse City, Michigan 49684, USA.

This magazine is primarily devoted to the building of model steam railroad engines. These devices are powerful enough to pull a dozen or more human passengers. While the tiny locomotives are far too small to be of practical use for transport, the detailed information presented can be applied to other small steam engine uses.

Live Steam is full of illustrations and plans for steam engines and boilers. The kind of metalworking techniques required to make efficient small steam machinery is a major topic. Construction of valves, descriptions of other types of steam power equipment (such as water pumps and steamboats) and news of steam railroading are among the other topics covered.

People actually building small steam power equipment will find many useful construction details and design ideas in this magazine. Back issues are also available.

The Planning, Installation and Maintenance of Low-Voltage Rural Electrification Systems and Subsystems, book, 151 pages, 1979, $5.95 from VITA.

Originally compiled in 1969 by American electricians, this manual was written for the training of Peace Corps Volunteers who were to be assigned to rural electrification projects in developing countries. An introductory chapter provides a simplified introduction to electrical theory. Succeeding sections discuss wiring of houses, wiring for distributing power to houses, and connection of a village-scale electrification system to a generation plant or other power source.

Some attention is given to the differences between electrical installations in rural areas of developing countries and in the U.S. For example, the authors suggest that the trainees should wire two homes, one with standard techniques

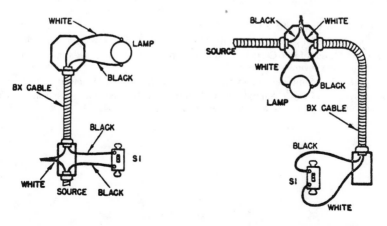

and one with "techniques applicable to mud construction". Yet in fact, the subsequent discussion assumes availability of commercial U.S. cable, connectors, meters, fuseboxes, switches, and other components. And the reader is informed that "(convenience outlets) are normally located about 12" above floor level. They should be placed near (2 or 3 feet) corners of rooms rather than in the center of the wall to lessen the chance that they will be blocked by large pieces of furniture." With such an unrealistic orientation, this book misses an opportunity to present information on simpler systems (providing 50-100 watts per house) commonly found in developing countries.

A brief section on "Planning Requirements" explains how to calculate materials, labor, and overhead costs of an electrification project. There is no discussion of the needs for electricity in rural areas, how new electric installations might complement existing sources of power, or whether rural people will be able to afford their new electricity.

This manual is part of an approach oriented toward purchase and installation of an expensive energy technology with little regard for whether or not it fulfills basic human needs. Still, it provides a good overview and bibliography on the fundamentals of electricity and electrification.

Gemini Synchronous Inverter Systems, 11-page booklet, 1978, on request from Windworks, Inc., Rt. 3, Box 44A, Mukwonago, Wisconsin 53149, USA.

A synchronous inverter is a device that allows the connection of small alternative energy systems to the large electric distribution networks (grids). All of the available power from the alternative energy system is converted to alternating current (AC) for normal use. Any extra electricity need is supplied by the grid, and any surplus electricity generated is fed back to the grid, where it can be used by other consumers on the same grid. In this way, the owner of a small windgenerator, microhydroelectric system, or solar photovoltaic array can use the grid in place of a costly battery storage system, or in place of no storage system at all. A synchronous inverter could be used with a combination 10 kw diesel generator set and windgenerator, for example, if the windgenerator never exceeds ⅓ the diesel generator capacity.

Windworks has a booklet and three similar papers describing the theory behind synchronous inverters and the possible applications. Synchronous inverters are available for single phase and three phase equipment, and cost (in early 1976) $160 per kw capacity for small units, down to $40 per kw capacity for 1000 kw installations.

ADDITIONAL REFERENCES ON ENERGY
Several kinds of pedal-powered agricultural equipment are reviewed on pages 469-471.
Construction Manual for a Cretan Windmill contains plans for a pedal-powered wood turning lathe; see review on page 598.
Two books on the maintenance and repair of small engines are reviewed on page 467.
The Employment of Draft Animals in Agriculture reviews the power associated with draft animals and the use of animal power gears; see page 464.

ENERGY:
IMPROVED COOKSTOVES
AND CHARCOAL PRODUCTION

Energy:
Improved Cookstoves
and Charcoal Production

While the major forces threatening the world's forests include clear-cutting in agricultural and lumbering activities, observers calculate that when all tree-cutting in the Third World is considered, up to 80% of the wood is used as fuel for cooking, mostly with open fires. Such observations indicate that current patterns of daily firewood consumption around the world are key factors in an advancing environmental crisis which affects us all. When wood is simply too expensive or too far away, animal manures and crop residues formerly returned to the soil as fertilizers frequently are burned as fuel instead. This practice, increasingly common in many parts of Africa and South Asia, adds to the downward spiral in soil fertility. Once the trees and vegetation on hillsides are removed, soil erosion proceeds rapidly with rain water runoff and flooding, and the land can be turned into a desert.

"At current rates the world will lose about 1/3 of its arable land between now and the end of the century—a period in which the human population will continue rapidly to increase...The Third World must double food production before the century is out just to keep pace with present inadequate standards. And, of course, on less land. By a strange and mocking coincidence the projected figure for cultivated land in 2000 (940 million ha.) is roughly equal to the amount of land already lost to desertification. It is an area the size of the United States. This is not enough to feed the world on. It is far too much to have thrown away."

—Ray Dasmann, "Planet Earth—1980", 1980

Though a few oil-producing nations have been able to temporarily slow the rate of deforestation by instituting massive subsidies (e.g. kerosene is sold in Indonesia at less than a third of world market prices), for most communities there is no alternative to gathered fuels. It is in this context that improved cooking stoves with lower fuel consumption hold so much promise. Simple stoves can save up to half the wood fuel that would be consumed cooking over open fires.

Community-wide conversion to improved stoves alone might halt deforestation in many areas, by allowing regrowth to overtake a reduced firewood cutting rate. (The establishment of village woodlots and vigorous reforestation efforts are important complementary activities to stove promotion; see discussion in the FORESTRY chapter. Improved efficiency stoves make both woodlots and reforestation efforts more effective.) Yet efforts to promote the use of improved stoves have typically met with little support at decision-making levels and less success at the grass roots. Most rural people gather (rather than pay for) firewood, and cannot afford a manufactured stove. Some observers have calculated that a $10 investment in an improved cooking stove is probably a reasonable upper limit in most communities. One particularly promising approach that stays within these cost constraints is the sand-soil block technique described in **Lorena Owner-Built Stoves**. Stoves of a great variety of sizes and designs can be built using this method. Lorena stoves can achieve high efficiencies, can be made by local craftsmen, and cost little, requiring only local materials except for the chimney or stovepipe (which costs an additional $1.50 to $7.00, depending on the materials used). Because Lorena stove building can be easily learned, the technology has become the basis of large stove promotion programs in several countries where firewood and deforestation problems are most acute. In theory, master stove craftsmen could train more craftsmen who could train yet another group and so on, such that tens of thousands of stoves could be functioning in a region within a year's time. Experiences we are aware of, however, show that efficient, acceptable stoves for any given community depend on a locally-determined sand-soil mixture of sufficient strength, and a tailoring of stove design to suit local cooking patterns and preferences. To use an improved stove to best advantage, the cook must understand how it works; otherwise, little fuel may be saved. And the stove builders themselves must thoroughly understand the operating principles; promoters are finding that poorly built stoves can have efficiencies lower than traditional cooking methods.

The secondary effects of existing cooking systems must also be understood. In many places, smoke from indoor cooking fires is a significant contributor to lung and eye disease. Yet this smoke also serves to protect thatched roofs from insect damage, and dry crops hung over the cooking area. Successful stove promotion efforts may depend on the availability of effective alternatives for low cost crop preservation and roofing technologies (see the CROP DRYING, PRESERVATION AND STORAGE and CONSTRUCTION chapters). In highland regions and other colder areas, the space heating function of the indoor cooking fire may need to be included in cook stove design.

These experiences suggest that despite the clear and critical need for the immediate implementation of massive woodsaving schemes, adoption of Lorena technology cannot occur "all at once" for an entire nation or region. It will, instead, depend on involvement of local people in careful, systematic training which emphasizes design, testing, and cooking methods. These issues are discussed from the point of view of the development worker in the manual **Helping People in Poor Countries Develop Fuel Saving Cookstoves** (this section).

Woodburner's Encyclopedia is a good reference book on the basic physics of woodburning. **Lorena Owner-Built Stoves, Wood Stoves: How to Make and Use Them** and **Cookers** all provide ideas and construction principles for a variety of possible low cost cookstove designs.

In the affluent countries wood stoves for home heating are gaining popularity with the increasing prices of electricity and heating oil. At the time of this writing, for example, the price of softwood seldom exceeds $60 per cord (3' x 3' x6') in many of the sparsely populated heavily forested areas of the northern United States. With these rates, even elaborate hot water baseboard stove systems costing $1000 and more are cheaper to own and operate than the conventional systems. The wood stove industry in the United States is now selling a remarkable 1-2 million units a year. **Woodburner's Encyclopedia** is an excellent reference for North Americans interested in wood heating systems.

Charcoal has a high energy content per unit of weight and is thus easier than wood to transport long distances. When fuel wood hauling becomes a serious problem in Third World communities, charcoal production tends to increase *significantly, so that more energy resource can be transported in a single load. Although many kinds of biomass can be converted into charcoal, almost all charcoal produced in the rural Third World is made from wood or woody shrubs. Because of the energy required to fire charcoal kilns and the removal of volatile gasses that is an essential feature of the charcoal making process, the final product contains less cooking or heating energy than the wood from which it is made. Included in this chapter are several publications that should be helpful in making and using improved kilns which produce more charcoal from the same amount of wood than traditional kilns and pit-fired techniques. This appears to be the crucial point at which charcoal production and use can be made more efficient. Indonesian charcoal cookers are reportedly several times as efficient as wood stoves in getting heat to the cooking pot. Thus even considering the substantial initial losses in charcoal production, the net efficiency of the whole system may be similar to traditional wood-burning stoves. It is, however, likely to be much easier to improve the low efficiency of traditional wood cooking systems than to improve the already higher efficiency of charcoal cookers.*

This chapter concludes with a review of **Rice Husk Conversion to Energy***. In the Third World, rice husks (or hulls) not consumed as fuel are usually returned to the soil or used as a binder in building materials such as bricks. The FAO report reviewed here notes that globally most rice hulls are already being used in one way or another, and that only about half of the remainder could be used. There is much work to be done, though, in the search for more efficient rice hull burning methods. (***Lorena Owner-Built Stoves** includes an illustrated appendix on rice hull stoves in Java.)*

In Volume I:

Rice Hull Stove, set of drawings, page 176.

The Woodburner's Handbook, by David Havens, 1973, page 177.

The Working Woodburner, by D. Dahlin, 1976, page 178.

The Complete Book of Heating with Wood, by Larry Gay, 1974, page 178.

How to Build an Oil Barrel Stove, by Ole Wik, page 178.

Oil Drum Handicraft, by Gary Brooks, 1973, page 179.

Oil Drum Stove, by John McGeorge, 1976, page 179.

The $1.50 woodburning stove, photos and drawing, page 179.

The Chula Stove, by D. Brett, 1970, page 180.

Double Drum Sawdust Stove, by J. Wartluft, 1975, page 180.

Sawdust-Burning Space Heater Stove, by D. Huntington, 1975, page 180.

Wood Burner's Encyclopedia, book, 155 pages, by Jay Shelton and Andrew Shapiro, 1976, Vermont Crossroads Press, $8.00 from META.

Here is an excellent reference book on 'wood as energy.' The chapter on energy, temperature and heat nicely provides an understanding of the basic physics involved in burning wood—in an open fire, a heater, or a cookstove. Although this book is primarily intended for use by North Americans wanting to use wood to heat their homes, the considerable amount of background discussion is often relevant to cook stove design. There are valuable sections on fuelwood, combustion, energy efficiency (wood heating), safety considerations,

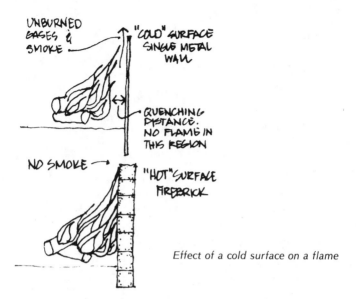

Effect of a cold surface on a flame

and creosote and chimney fires.

There is a list of manufacturers and some product information (though no designs detail or efficiency claims) for 33 different cook stoves made by 12 different companies. While these stoves cost so much ($400-500) that they are not realistic models for developing countries, some lessons about the design of efficient cook stoves might be learned through careful analysis and testing.

Recommended for cook stove experimenters as a reference.

Lorena Owner-Built Stoves, book, 90 pages, by Ianto Evans, 1979, (revised, expanded edition to be published in early 1981), current edition $3.00 plus $0.59 postage, from Appropriate Technology Project, Volunteers in Asia, P.O. Box 4543, Stanford, California 94305, USA.

"What is the Lorena stove? It is a permanent cookstove made with a mixture of sand and clay. Almost anyone can build it, without special tools, at almost no cost and with only this book or a few days training." This inexpensive stove, originally developed in highland Guatemala, was designed for high efficiency using a variety of organic waste fuels in addition to wood. Hot gases from the firebox are directed through passages to the bottoms and sides of a series of pots, increasing the total amount of heat available for cooking and heating water. The snug fit of pots in their holes prevents heat losses and keeps smoke out of the kitchen. Dampers made of scrap sheet metal allow control of draft and cooking speed.

In illustrated step-by-step fashion, this manual explains Lorena stove construction: how to test for suitable sand-clay mixtures, design the stove, and build and carve out the sand-clay block. Cooking methods and possible design modifications are suggested. A final section describing research on acceptance and use of the stove by Guatemalans shows that builders continually alter the designs. These innovations improve (rather than reduce) fuel efficiency only when builders and users fully understand how the stoves work. Training courses must therefore communicate the operating principles in addition to the

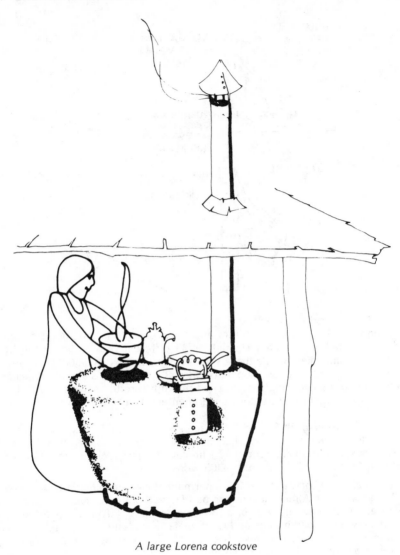

A large Lorena cookstove

construction techniques. When these principles are mastered, a stove made of local materials can evolve rapidly to meet local cooking and heating needs.

An appendix describes and illustrates rice hull burning stoves from Java.

Helping People in Poor Countries Develop Fuel Saving Cookstoves, book, 148 pages, by Aprovecho Institute, 1980, free to serious groups, from German Agency for Technical Cooperation (GTZ), P.O. Box 5180, 6236 Eschborn 1, Federal Republic of Germany; or, from Aprovecho Institute, 359 Polk Street, Eugene, Oregon 97402, USA.

Aprovecho's involvement in the development of fuel-saving stoves in Guatemala resulted in publication of **Lorena Owner Built Stoves** (see review in this section). Since that experience, Aprovecho has carried out further research on how Lorena and other low-cost stoves might be improved, and continues to provide assistance to cookstove popularization efforts in other Third World countries.

This book is about such efforts, written for field workers (such as volunteers and extension agents), administrators and planners (especially those responsible for forestry and soil conservation programs), and researchers. The purpose of the manual is not to present construction methods in detail for specific stoves. Instead, the emphasis is on how to encourage poor people to develop solutions to their problems, with the focus on cooking technologies. Topics covered include important background information on how deforestation, declining agricultural production, and stagnating rural economies are related; working with villagers to design stoves; and systems for spreading information and training stove builders. "If improved stoves are to make a major impact on fuel demand, work will need to be immediate and broadscale. At current deforestation rates, it may be too late to merely seed a good idea, then go away, leaving it to grow naturally in its own time. Stove projects must be extensive, well-organized and adequately funded. Yet if heavy-handed methods are used for distributing or developing this technology, it may never be accepted. Involvement of the people is essential to effective dissemination.

"There are as many ways of going about dissemination as there are cultures, but (several points covered here are) raising public awareness; setting up an approach for dissemination; where to go for help in distributing information; promotion: ideas to try; where and how to start dissemination; setting up stove centers; training; involving women; evaluation and follow-up; use training; sponsoring and advising small businesses."

Three final chapters discuss how woodstoves work; how to design simple comparitive stove testing procedures; and brief illustrated instructions for building a variety of lorena, clay, metal, and other stoves.

Historically, efforts to introduce "appropriate technology" have relied on convincing people that they need a manufactured product. This valuable book is a down-to-earth discussion of how development workers can help people make use of their own ideas about what they need, to develop an improved technology for themselves.

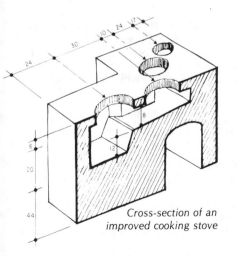

Cross-section of an improved cooking stove

Cookers, in **Food Preparation, Series 1, Rural Home Techniques, Volume 3,** United Nations Food and Agriculture Organization (FAO), $7.00 from UNIPUB, or from local FAO offices, or from Chief, Home Economics and Social Programmes Service, Human Resources, Institutions and Agrarian Reofrm Division, FAO, Via delle Terme di Caracalla, 00l00-Rome, Italy.

This is a complete set of drawings for construction of several cooking stoves with improved fuel efficiency. This material has been taken from a 1961 report to FAO written by Hans Singer, working in Indonesia, in which he did a detailed analysis of the fuel efficiency of existing Indonesian stoves and charcoal cookers. Unfortunately, only the construction details for his proposed improved stoves are reprinted here; the rest of the report (which we believe is out of print) would be valuable for people working on stove improvement programs.

"Cookers" is one of three sections in the FAO publication **Food Preparation: Series 1, Rural Home Techniques, Volume 3**. The other two sections contain brief instructions and drawings for making cleaning brushes and storage cupboards. Text for all three is in English, Spanish, and French.

Wood Stoves: How to Make and Use Them, book, 194 pages, by Ole Wik, 1977, $5.95 from Alaska Northwest Publishing Company, Box 4-EEE, Anchorage, Alaska 99509, USA.

Unlike most North American books on woodstoves, this one is concerned with making stoves. It also contains many ideas on design and construction of cooking stoves, which tend to be ignored in the literature. Those people experimenting with the design of improved-efficiency cookstoves will certainly want to read this book.

Only metal stoves, requiring purchased metal stovepipe and made primarily from discarded oil drums, are discussed. In the Third World, these stoves are

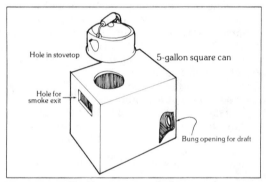

Hole in stovetop

5-gallon square can

Hole for smoke exit

Bung opening for draft

Cooking stove made from a can

expensive to build and corrode quickly. In addition, this book is based on years of experience in a very cold climate where wood is abundant and efficiency of combustion is not as important as in most semi-deforested regions. Also, protecting the cook and kitchen from excess heat is of little concern to the author. Designers using this book in the Third World will want to keep these differences in mind.

Cooking with Wood, booklet, 23 pages, SKAT, 1980, SwFr 7 from Swiss Center for Appropriate Technology (SKAT), Varnbuelstrasse 14, CH-9000 St. Gall, Switzerland.

Here are photos of stove designs (many of them experimental) from all over the world, assembled for an exhibition in 1980 in Geneva. A short text is in German, English, and Spanish.

The photos of stove exteriors do not reveal important internal design features. No price or efficiency information is provided, and some of the other information given is inaccurate. Most of the stoves are made of sheet metal or bricks, and thus possibly too expensive for many rural people who most need them.

The collection, while not exhaustive, is worth looking at for adaptable ideas.

Cash in on Charcoal, booklet, 39 pages, by Jose B. Blando, 1976, AGRIX How To Series No. 26, Asia Edition, $1.25 from Agrix Publishing Corporation, 79 Dona Hemady cor. 13th Avenue, Quezon City, Philippines.

Pictures and scale drawings of several different charcoal kilns in the Philippines. These are: variations of the simple pit method, an oil drum kiln, a multiple oil drum kiln, a 4-drum vertical retort, and 3 larger masonry kilns. Instructions are provided for the construction and operation of each of these kilns. Wood, coconut shells and other materials are used to make the charcoal.

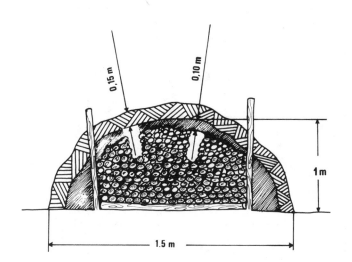

Earth kiln

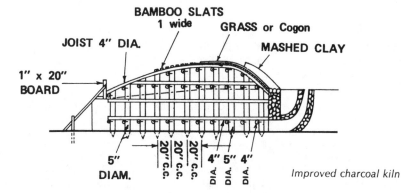

Improved charcoal kiln

Charcoal Production Using a Transportable Metal Kiln, Rural Technology Guide 12, booklet, 18 pages, by A.R. Paddon and A.P. Harker, free to public bodies in countries eligible for British aid, Ł1.00 to others, from Tropical Products Institute, 56/62 Gray's Inn Road, London WC1X 8LU, England.

Photos and text show how to properly load and operate a lightweight sheet metal charcoal kiln that can be rolled from place to place. When properly operated, this kiln is more efficient than traditional pit systems. Production is ½ to ¾ ton of charcoal per batch. With two kilns, two men can produce 2-3 tons of charcoal per week. Construction details for the kiln will be published separately as Rural Technology Guide 13.

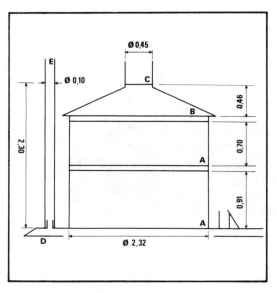

Portable steel charcoal kiln

Charcoal Making for Small Scale Enterprises: An Illustrated Training Manual, 26 pages, International Labour Office, 1975, 6 Sw. frs. ($3.60) from International Labour Office, CH 1211 Geneva 32, Switzerland.

This short, large format booklet is a good source of details of correct operation for two kinds of low cost charcoal making kilns: earth kilns (most common and virtually no-cost) and small portable steel kilns (approximately $2000 each). "Earth kilns are simple to construct and operate, and produce good results when managed by experienced people."

The language is simple and there are many drawings and photos. Notes on the preparation of wood, tools required, calculation of production costs, marketing, and charcoal-making cooperatives are included. Unfortunately, there are no rules given for estimating efficiency of a kiln, nor is the end-use efficiency of charcoal vs. direct wood burning discussed.

Rice Husk Conversion to Energy, FAO Agricultural Services Bulletin No. 31, book, 175 pages, by E. Beagle, 1978, $10.00 from UNIPUB.

This is an extensive reference book on the enormous variety of energy applications for rice husks around the world. About half of the world's 60 million tons of rice husks produced annually are currently used; another 20% (12 million tons) apparently could be used as well.

The author discusses the general processes for converting rice husks into energy along with existing technologies for doing this. Steam engines, producer-gas engines, paddy dryers, and domestic cooking stoves are among the topics considered. Where parboiling is done, small steam engines can effectively be used to power the mills and provide heat for parboiling. Where parboiling is not done, the best power choice for small (less than 5 tons per hour) mills would be "an engine fueled by gas produced from rice husk...This system of 'producer gas' is of proven technology, having been in continuous use for over 75 years." The great range of technologies discussed is unfortunately not supported by enough drawings.

The format allows the reader to go on to find more detailed information when relevant. For example, it is noted that the standard rice mill in Thailand is driven by a rice hull fired steam engine. A 224-entry list of contacts includes makers of such equipment in Thailand, and the 264-entry bibliography leads to further information on a wide variety of other topics.

The author concludes that rice husks are used far more extensively as an energy source than is generally recognized, that manufacturing capabilities for the related equipment are greater than realized, and that difficulties in information exchange prevent wider progress in applications. This book is a major step in overcoming the information exchange problem.

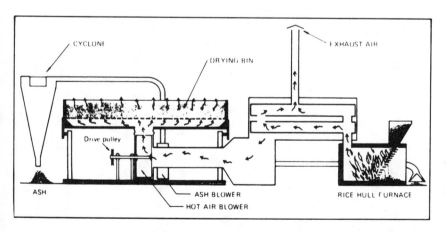

Grain dryer using rice husks as fuel

ADDITIONAL REFERENCES ON COOKSTOVES

Minitechnology includes several designs for cookers and ovens; reviewed on page 373.

Nepal Biogas Newsletter contains drawings of a burner for use of biogas in cooking; see drawings with review on page 654.

ENERGY: WIND

Energy: Wind

The wind has been a significant source of power for centuries. Early windmills in China and Southeast Asia lifted water into rice fields. In Europe the windmill developed into an enormous structure, nearly the size of a small sailing ship, developing power in the range of 25 hp and higher, for use in grain grinding, drainage, and a multitude of small industrial tasks. The first windmills in North America and the Caribbean were of this type. In the late 19th century, waterpumping windmills were manufactured by the thousands, and several million machines are estimated to have been in operation by the middle of this century. These were mostly lifting water for isolated livestock watering spots and farmhouses.

Windgenerators for electricity spread by the hundreds of thousands across rural North America in the 1930s, supplying the farmhouses with small amounts of power for radios and a few lights. Both the waterpumping windmills and the windgenerators went into decline with the arrival of rural electrification, which offered cheap electricity for running electric pumps and many more household uses. With the energy crisis, however, sales of waterpumping windmills and windgenerators have greatly increased in the United States.

It is the waterpumping windmill that appears to be the most immediately relevant for rural energy needs in the developing countries, both for high-value community water supplies and for irrigation pumping. Irrigation is the biggest single factor in improving farm yields, and there are many places where low lift irrigation on small plots could be accomplished with windmills. Thailand and Greece have both had large numbers of windmills in irrigation use; thousands are still operating today. Japan and Portugal are among the other nations where significant numbers of irrigation windmills have been used in recent times. In North America, farmers built thousands of scrap wood waterpumping windmills before the manufactured steel machines appeared. In all of these national experiences, local windmill designs were developed to fit pumping needs, wind conditions, and materials available. These machines were built in small workshops; this kept prices low and repair skills nearby. In other countries where manufactured windmills have been directly introduced, the high initial cost and lack of repair skills have greatly reduced their attractiveness.* Thus the historical record suggests that successful windmill promotion programs in developing countries will need to focus initially on locally-

---

*South Africa and Australia may be exceptions. In these industrialized countries, variations of the American fan-bladed windmill have been widely used to water livestock and supply isolated farmhouses. These are expensive, high performance machines requiring infrequent but skilled maintenance and repair.

adapted designs and craftsman-based production using local materials, with a limited number of manufactured parts (e.g. bearings). Promotion programs might include credit mechanisms whereby the windmill itself is both loan and collateral. Also of interest is the Las Gaviotas approach in which the buyer assembles and installs a metal windmill from a kit (see **Un Molino de Viento Tropical**).

Waterpumping windmills for irrigation purposes are most economically competitive in areas that do not already have electricity for powering irrigation pumps. (Only 12% of the rural Third World has electricity.) Low lift windmills for rice paddy irrigation appear to be economically attractive, compared to engine-driven pumps, for a wide area of South and Southeast Asia. Higher lift applications for high-value vegetable farming may be economically competitive in many parts of the world. The economic appeal of locally-built windmills is even greater when the savings of scarce foreign exchange from reduced oil imports, and the village-level economic multiplier effects are considered. Other advantages of locally-built windmills include village capital formation using local labor and materials, much lower initial cost, and avoidance of maintenance problems associated with engine-driven pumps. Such windmills appear to have more frequent but simpler maintenance requirements than manufactured windmills.

A small number of people are working on waterpumping windmill designs in developing countries. Most promising at present appear to be sail windmills—machines with cloth or bamboo sails rather than fixed blades. These include versions of the Greek (Cretan) sail windmill—brought to its peak in The Gambia where efficiency has been doubled and the windmill has been adapted to operate in relatively low wind conditions. (See **How to Build a Cretan Sail Windpump**.) Even cheaper are the cloth and bamboo sail windmills of Thailand (see **Proceedings of the Meeting on the Expert Working Group on the use of Solar and Wind Energy**, reviewed on page 135). The Thai designs in particular have the advantages of low cost and use of local materials and labor. They are well matched to the low lift small volume pumping needs of small plot rice agriculture. For a comparative summary of waterpumping windmills costs, production requirements, and performance, see the UN Economic Commission for Asia and the Pacific (ESCAP) paper **Report on the Practical Application of Windpowered Waterpumps.**

In the United States, isolated houses have become a major market for windgenerators for electricity. **Wind Power for Farms, Homes and Small Industry** is recommended for readers considering such an installation. Technical advances now also allow a windgenerator to feed surplus power back into a conventional electric grid, a practice which makes wind-generated electricity in urban and suburban settings much more attractive than before, as the substantial expense of a battery system can be avoided (see **Windmill Power for City People**). Large-scale windgenerators are being tested by utilities in several locations, and may prove to be a significant source of grid electric power.

For any wind machine, the choice of site is very important. Trees and buildings can greatly reduce the useful winds reaching a windmill. A small difference in wind speed can mean a big difference in power available, because the power in the wind varies with the cube of the wind speed. Thus a 12 mph wind has 8 times as much power as a 6 mph wind. Windgenerators operate in the highest range of windspeeds, and the user will usually want to find the windiest spot possible for such an installation. Waterpumping windmills, on the other hand, need greater protection from the extremes of high winds, and *are usually designed to operate in low and medium winds. We have included several publications on site selection for wind machines, including two on vegetative indicators of high average windspeeds at particular locations.*

In Volume I:

Is There a Place for the Windmill in the Less-Developed Countries?, by M. Merriam, 1972, page 139.

A Survey of the Possible Use of Windpower in Thailand and the Philippines, by W. Heronemus, 1974, page 140.

Considerations for the Use of Wind Power for Borehole Pumping, 1976, page 140.

Food from Windmills, by Peter Fraenkel, 1975, page 140.

The Homemade Windmills of Nebraska, by E. Barbour, 1898, page 142.

Low-Cost Windmill for Developing Nations, by H. Bossel, page 143.

Sahores Windmill Pump, by J. Sahores, 1975, page 143.

A Water-Pumping Windmill That Works, New Alchemy Institute, 1974, page 144.

Wind Power Poster, by Windworks, page 145.

How to Construct a Cheap Wind Machine for Pumping Water, by BRACE, 1975, page 146.

Performance Test of a Savonius Rotor, by M. Simonds and A. Bodek, 1964, page 146.

Simplified Wind Power Systems for Experimenters, by Jack Park, 1975, page 147.

Electric Power from the Wind, by Henry Clews, page 147.

Wind Power, by Syverson and Symons, 1974, page 147.

Wind and Windspinners, by Hackleman and House, 1974, page 148.

The Homebuilt Wind-Generated Electricity Handbook, by Michael Hackleman, 1975, page 149.

Homemade 6-Volt Wind-Electric Plants, by H. McColly and F. Buck, 1939, page 150.

Proceedings of the United Nations Conference on New Sources of Energy, 1961, Volume 7, page 150.

Energy from the Wind, by B. Burke and R. Meroney, 1975, page 151.

Wind Power Digest, page 151.

The Generation of Electricity by Wind Power, by E. Golding, 1955, page 152.

There is a list of 11 additional cross-references on page 152.

Feasibility Study of Windmills for Water Supply in Mara Region, Tanzania, book, 89 pages, by H. Beurskens, March 1978, $7.00 (single copies free to research institutes in developing countries), from SWD.

This is a very interesting example of a region-wide study of the potential for water pumping windmills. A wide variety of ''different sites were selected where water is needed for irrigation, domestic purposes or cattle farming. The water needs, heads and piping distances were determined for these sites...

Only the lake was considered as a water source...Although a number of good sites to install windmills along riversides were noticed, they are not mentioned in this report.''

The economics of diesel sets, central grid electricity use, commercial windmills and locally-built windmills are compared. Locally-built windmills appear to be most economical, in addition to creating employment, using locally available materials, saving foreign exchange, conserving fuel, and promoting self reliance in the villages.

A set of design considerations are proposed for such locally built windmills to match the region's winds, water needs, local production capability, and available skills. The elements and costs of a pilot testing phase are briefly described.

The possibility of financing windmills through the gains from a single irrigated harvest are described based on an existing irrigation project: ''In 1977 a group of 10 farmers grew onions on a diesel irrigated plot of 0.6 ha (1.5 acres) at Mugubya. The net income amounted to approximately shs 11,000 for one harvest...Total investment for a windmill for 0.6 ha (1.5 acres) irrigation would vary roughly from shs 3,400 to shs 6,800, depending on the type of windmill... So the income per harvest is even higher than the costs of a windmill. This seems a very reasonable proposition for windmill irrigation, at least at plots with relatively low heads.''

Highly recommended to anyone interested in the use of windmills for irrigation in the Third World.

Report on the Practical Application of Wind-Powered Water Pumps, 26 pages, by Marcus Sherman, 1977, available on request from Natural Resources Division, U.N. Economic and Social Commission for Asia and the Pacific (ESCAP), United Nations Building, Sala Santitham, Bangkok 2, Thailand; also reproduced in ''Proceedings of the workshop on biogas and other rural energy resources held at Suva, and the roving seminar on rural energy development, held at Bangkok, Manila, Tehran and Jakarta'', Energy Resources Development Series No. 19, ESCAP, 1979, U.N. Sales No. E.79.II.F.10, $10.00 from U.N. Sales Section, Rm LX 2300, New York, New York 10017, USA.

This short paper contains a unique and very useful set of tables comparing the operating characteristics, design features, and costs of a wide variety of water pumping windmills. The author's intent is to ''assist in the design and evaluation of future wind-powered water pumps projects for a wide range of environments. Water source, water use, local wind conditions, and availability of labor, capital, and materials are the major determinants of design selection.''

Most windmill types are covered. These include Greek (Cretan) sail, multi-vane, Savonius, Chinese vertical axis, Thai cloth, Thai bamboo, medium speed cloth, medium speed metal, and high speed windmills. Categories for comparison include rotor diameter, blade material, pumping rates, starting and rated wind velocity, initial capital cost, expected lifetime, maintenance costs and cost per cubic meter of water lifted a standard distance. The format allows quick comparisons of different windmill types, though for some (e.g. Chinese vertical axis windmills) little performance information is available.

''It appears that local design and construction of wind-powered water pumps is generally feasible...The selection of low capital cost, low technology, high labor input designs is usually preferred for agricultural applications unless farmer credit schemes can be used. Higher cost is tolerable for public drinking water supply because the initial cost can be amortized through a long term community budget.''

How to Build a 'Cretan Sail' Windpump for Use in Low Speed Wind Conditions, construction manual, 56 pages, by R.D. Mann, 1979, £2.95 from ITDG.

MATERIALS: flat steel, steel angle (used for the 23' tower legs), ½'' steel bar, 1'' outer diameter galvanized steel water pipe (and slightly larger and smaller sizes), 5/8'' inside diameter thin wall galvanized steel water pipe, 1/8'' galvanized steel wire, 3 sealed bearings, one self-aligning ball-bearing, hardwood for block bearings, 3'' diameter Dempster lift pump

PRODUCTION: cutting, drilling and welding steel; bending circular pieces out of 1'' x 1/4'' flat mild steel

This waterpumping windmill design was based on the low lift windmills which had been built on the Omo River in Ethiopia (see **Food from Windmills,** reviewed on page 140), which had themselves evolved from the sail windmills of Crete. The author adapted the design for the lighter winds of The Gambia, and succeeded in nearly doubling the efficiency of the Omo River design. He reports on field testing done in 1978, and provides complete drawings and text for the construction of the windmill.

This machine was developed for irrigation use on small farms. In this region of the Gambia, there is no wind 31% of the time, wind of more than 12 mph only 6 % of the time, and moderate winds to 12 mph 63% of the time. Needed is a windmill that will operate in winds of 5-10 mph. ''The wind speed required to start the windpump from rest was calculated to be between 5.2 and 5.6 mph, and once started the windwheel continued to run in a steady wind down to 4.5 mph.'' During a series of 9-hour pumping trials spread over four months, the windmill lifted 1700 to 3400 gallons of water a height of 13'4''; windspeed

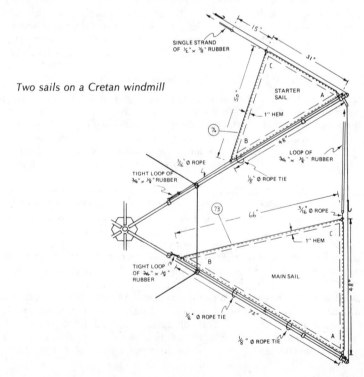

Two sails on a Cretan windmill

averaged 5.1 mph at the low end and 6.75 mph at the high end of this range.

The windmill has 6 sails, three full-sized and three smaller sails that help in starting. There is a 23-foot tower. Estimated cost of the windmill is Ł750 ($1650). As of this report, the windmill had only been used to operate a lift pump, with a 14' lift. Future tests will involve a force-pump and 45' head (lift).

The drawings are separated from the text, making the book a bit awkward to use. However, the drawings can be clipped from the book and spread out separately, and with study they become clear to the reader. There are also 12 photos.

The construction section that is most difficult to understand is the folding tail. This, however, is an unnecessary complication, and can be eliminated from the design.

There is no complete list of tools or materials required.

The author's design improvements establish this version of the Creten sail windmill as one of probably the two most cost-effective waterpumping windmills for low lift circumstances. The other would be the Thai sail windmill (see review of **Proceedings of the Meeting of the Expert Working Group on the Use of Solar and Wind Energy**, on page 135). Highly recommended.

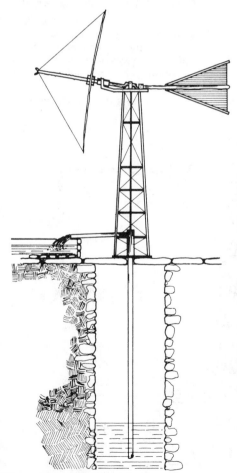

Windpower in Eastern Crete, booklet, 9 pages, by N. Calvert, 1971, 35 pence including postage, from The Society for the Protection of Ancient Buildings, 58 Great Ormond St., London WC1, United Kingdom.

This booklet provides a good description of the techniques and materials used to build the Cretan sail windmill. It is not a construction manual, and it does not provide precise dimensions.

These machines, thousands of which still operate in the plain of Lassithi, were evidently mostly built during the period 1900 to 1950. Many of them were constructed partially from military debris from the two world wars. There seem to be three basic types: 1) those which could have been made by a blacksmith-wheelwright using wood and metal and fastened with wedges and rivets; 2) those which could have been built by mechanics, using mostly metal parts welded or bolted together; and 3) those which have a stone tower instead of a steel one.

"Observations were made on a number of machines in the fully rigged state and in rotation, at wind speeds commencing at 2.2 m/sec. (5mph). A useful output of water appeared at a wind speed of 2.75 m/sec. (6 mph). When the wind rose to 3.5 m/sec. (8 mph), a four meter diameter machine would run at a speed of up to 25 revolutions per minute (the highest observed)." The author later built a similar water-pumping windmill for testing in Britain, and notes that a four meter machine under full sail would develop power of 220 watts in a wind of 3.5 m/sec. (8 mph). "There is no doubt that the Cretan Mill excels in its ability to utilize low wind-speeds. This is consistent with the maximum number of operating hours per year and, in an irrigation context, is probably a criterion of excellence...The efficiency of 30% noted in the author's tests compares satisfactorily with that recorded for any other type of windmill."

The Cretan Mill "can hardly be improved for the efficient use of material. Aerodynamically, the low speed efficiency is high and it has an inherent stability against accidental overspeed."

Construction Manual for a Cretan Windmill, book, 59 pages, by N. van de Ven, October 1977, available in English and Dutch, $5.00 (single copies free to research institutes in developing countries), from SWD.

MATERIALS: logs, rugged sawn wood, wood screws, sail cloth, water pipe.

PRODUCTION: simple hand tools.

This is a construction manual for a water pumping sail windmill similar to the ones found in Crete. This version was built at the Twente University of Technology in The Netherlands.

The low cost design shown here

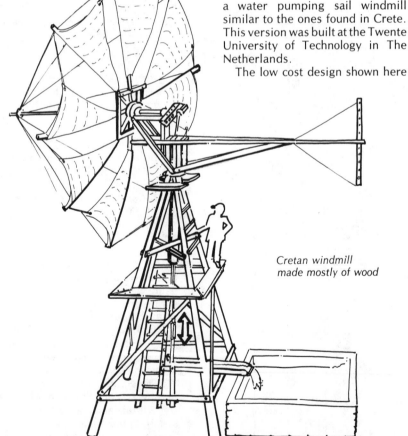

Cretan windmill
made mostly of wood

could be built almost anywhere in the world with mostly local materials. It is best suited for low lift pumping. The rotor diameter is 6 m, but could be made smaller. Sail windmills are especially interesting in areas where the winds are occasionally very high—the sails can be removed and the windmill protected under conditions that would destroy a commercial windmill.

Plans are also included for a pedal-powered wood-working lathe, which can be built with hand tools using wood and a few bicycle parts. The lathe is used in making some of the windmill parts. A shallow borehole, hand-drilling method using locally-made drill bits and augers is shown. A piston pump design is also provided.

The manual is well illustrated, with over 100 photos and drawings. The English edition contains some confusing wording, but this should not be a big problem.

Vertical Axis Sail Windmill Plans, 16 pages, 1976, reprinted 1979, $4.00 from Low Energy Systems, 3 Larkfield Gardens, Dublin 6, Ireland.

MATERIALS: wood for tower, cement and steel for foundation and anchors, steel road and plate for axle and hub, cables, metal fittings, electrical conduit pipe, bearing, canvas sails.

PRODUCTION: Some careful machine shop work is required for the axle and hub.

This design combines some of the principles of sail and sailwing rotors. "The rotor consists of two or more sailwings mounted vertically at equal distance from a vertical axis...Each sailwing is formed from a rigid spar...at the leading edge of the sail...The surface of the sailwing is made from a cloth envelope... When the wind impinges on the sailwing it takes up an airfoil shape with a concave surface facing into the wind...During one complete revolution of the rotor the sailwing switches the concave surface from one side to the other automatically...It is self starting, unlike the Darrieus rotor, to which it is similar in some other respects."

This small lightweight windmill is used by its designers to grind grain. It develops a maximum power of about ¼ hp in a 20 mph wind.

(This design should not be confused with the traditional Cretan sail windmill, which has a **horizontal** axis, and is used for irrigation water pumping in Crete.)

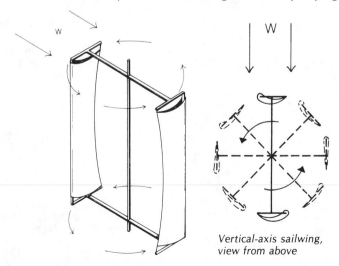

Vertical-axis sailwing,
view from above

Un Molino de Viento Tropical: Manual de Instalacion y Manejo, booklet, in Spanish, 45 pages, by Centro Las Gaviotas, 1980, available from Centro Las Gaviotas, Apartado Aereo 18261, Bogota, Colombia.

Presented in a popularized "foto-novela" (picture novel) format, this manual introduces a small water pumping windmill designed by the Colombian appropriate technology center Las Gaviotas. The large number of photos and drawings are intended to allow the buyer of a windmill kit to assemble and install it himself.

The windmill described is the production version of the latest in a series of 56 prototypes built by Las Gaviotas in their attempt to develop a low cost windmill that would operate in low windspeeds. This one is 1.9 meters in diameter, with a double acting piston pump, able to pump to a depth of 25 meters. From a 10 meter depth, this windmill will pump 2 cubic meters of water per day in a light and sporadic wind, and 4-5 cubic meters of water per day in a moderate continuous wind. These windmills are made in a well-equipped large workshop.

Optimization and Characteristics of a Sailwing Windmill Rotor, report, 82 pages, March 1976, by M. Maughmer of Princeton University, $8.00 (Accession No. PB259898) from NTIS.

This is the final report of the Princeton sailwing windmill project. "Through many years of extensive research, the sailwing has been found to provide a simple, light-weight and low-cost alternative to the conventional rigid wing, while not suffering any performance penalties throughout most low-speed applications."

This unusual wind rotor design uses a sail cloth sleeve over a spar and tension cable, instead of a solid blade.

Rapid evaluation of comparative performance of 8 different rotor shapes was made possible by using a test tower mounted on a jeep, and a homemade cup anemometer, demonstrating that effective testing can be carried out at low cost.

Many technical terms are used.

Savonius Rotor Construction; Vertical Axis Wind Machines From Oil Drums, booklet, 53 pages, by Jozef Kozlowski, 1977, $3.25 from VITA.

The author "has built two Savonius rotors—one in Wales and the other in rural Zambia. This manual details the construction of these machines...puts the rotors in a perspective which allows potential builders to judge the applicability of such machines for meeting their needs and then provides effective

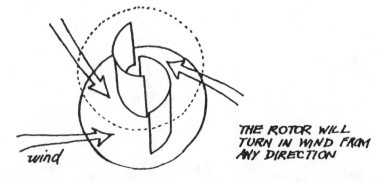

THE ROTOR WILL TURN IN WIND FROM ANY DIRECTION

wind

guidelines for constructing each.'' One of the rotors is for pumping water, and one is for charging automobile batteries.

The rotors are not very efficient compared to other low-cost windmills. For example, ''The data from Bodek and Simonds' experimental S-rotor in the West Indies shows that the useful energy from a 12 mph wind...means that one can pump 75 Imperial gallons/hour up to 30' above the water level (341 liters/hour up to 9.14 m). In an 8 mph wind...only 25 Imperial gallons/hour (104 liters/hour) can be pumped to the same height.'' (This compares unfavorably to the 5.4 m Cretan sail windmill, which is reported to pump as much as 15 times this volume of water in an 8 mph wind. Low cost sail or bamboo mat windmills in Thailand also appear to be considerably more productive.)

The summary of performance data on Savonius rotors and the reviews of other S-rotor publications are useful. The construction details are good, although many of the drawings are poorly reproduced.

Rotor Design for Horizontal Axis Windmills, book, 52 pages, by W. Jansen and P. Smulders, May 1977, $4.00 (single copies free to research institutes in developing countries), from SWD.

''This publication was written for those persons who are interested in the application of wind energy and who want to know how to design the blade shape of a windmill rotor...a lot of attention is given to explaining lift, drag, rotor characteristics, etc...In the selection of a rotor type, in terms of design spread and radius, the load characteristics and wind availability must be taken into account...The availability of certain materials and technologies can be taken into account in the earliest stages of design. We therefore hope that, with this book, the reader will be able to design a rotor that can be manufactured with the means and technologies as are locally available.''

The reader will need at least a good high school mathematics and physics background and familiarity with abstract technical presentations to be able to use this book.

airfoil name	geometrical description		$(C_d/C_1)_{1min}$	α^o	C_1
sail and pole	$c/10$	$c/3$	0.1	5	0.8
flat steel plate			0.1	4	0.4
arched steel plate		$f/c=0.07$	0.02	4	0.9
		$f/c=0.1$	0.02	3	1.25
arched steel plate with tube on concave side	$d<0.1c$	$f/c=0.07$	0.05	5	0.9
		$f/c=0.1$	0.05	4	1.1
arched steel plate with tube on convex side		$f/c=0.1$	0.2	14	1.25
sail wing	$c/10$ cloth or sail tube steel cable		0.05	2	1.0
sail trouser $f/c=0.1$ $d_{tube}=0.6f$	$c/4$ cloth or sail		0.1	4	1.0

Comparison of different blades

Horizontal Axis Fast Running Wind Turbines for Developing Countries, book, 91 pages, by W. Jansen, June 1976, $7.00 (single copies free to research institutes in developing countries), from SWD.

This is a highly technical report of some work on the design of rotors for high speed wind machines. The authors argue that ''in contrast with airplane propeller design, a maximum energy extraction is reached by enlarging the chords of the blades near the tips.''

''A simple method for manufacture of twisted, arched steel plates is given. Six rotors were built of blades that were manufactured with this method.''

This report will be of value to readers with an engineering background. ''Final conclusion is that with simple materials high power co-efficients are possible.''

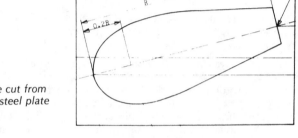

*A blade cut from
curved steel plate*

Matching of Wind Rotors to Low Power Electrical Generators, book, 85 pages, by H.J. Hengevold, E.H. Lysen, and L.M.M. Paulissen, December 1978, Serial Number SWD 78-3, free to research institutes in developing countries, $7.00 to others, from SWD.

Here is a much needed, good presentation of the design choices for the most likely application of windgenerators in the Third World: isolated, rural, low

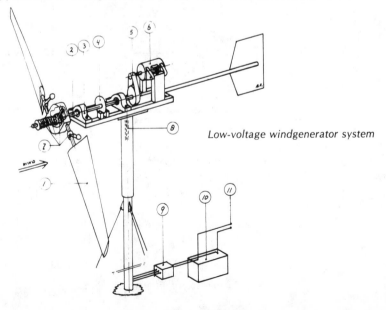

Low-voltage windgenerator system

voltage, small capacity systems with battery storage. The text explains a number of design "rules of thumb" for this kind of application, to maximize daily electricity output while minimizing cost. A good set of charts shows the important relationships between wind speed, power output, rotor diameter, and generator size. Readers will require some knowledge of basic physics, though an appendix explains the operation of a generator.

The authors begin by showing how to use information on the local wind conditions, and the computed energy demand, to calculate the necessary rotor diameter and rated power of the generator. "The emphasis (of the book) lies on the electrical part of the system and its optimum matching to the rotor...In the case of rural applications most windgenerators will be used to charge batteries for lighting purposes and to feed radio or TV equipment. Therefore we will limit ourselves here to DC loads, to avoid the complications of computing reactive loads." Particular attention is given to automobile generators and alternators. "These components are not the most suitable for our purpose, but since they are low priced and readily available they cannot be neglected."

Selecting Water Pumping Windmills, booklet, 14 pages, 1978, free from New Mexico Energy Institute, P.O. Box 3E1, Las Cruces, New Mexico 88003, USA.

This booklet is an introduction to the multi-blade windmill commonly seen on North American farms. It describes the parts of a windmill, tank sizes, and the lifting capacity of windmills of different sizes. 'Selecting' in the title refers to the size (diameter), not the type of windmill.

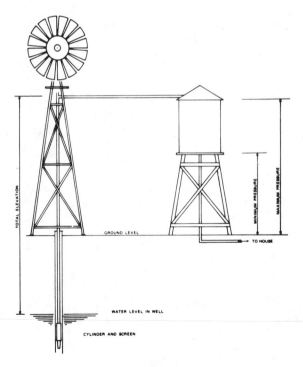

Pumping scheme for American farm windmill

Piston Water Pump, publication No. T-114, two pages of blueprints, 1977, $5.00 from Brace Research Institute, MacDonald College of McGill University, Ste. Anne de Bellevue, Quebec, Canada H9X 1CO.

The fabrication and assembly of a piston water pump for use with water pumping windmills is shown. Materials required include galvanized water pipe and steel rod. Some welding is required.

Wind Power for Farms, Homes and Small Industry, book, 277 pages, by Jack Park and Dick Schwind, 1978, Document No. RFP 2841/1270/78/4, $9.50 (foreign orders add $2.50 handling charge) from NTIS.

This is a no-nonsense introduction to windpower and windmachines for the North American. It is not a design manual, but a book to help the reader understand how to decide whether to buy a windmachine, considering needs, wind conditions, and other power options. The author discusses the different kinds of wind measuring equipment, different electrical systems, possible legal problems, and the routine tasks that come with owning a wind system. Monthly wind data for most of the United States is included. If the reader decides to get a windmachine, the book will help him decide what kind, what size, and what kind of energy storage system to use. Windgenerators and waterpumpers are considered. Highly recommended for North Americans considering installing a wind system.

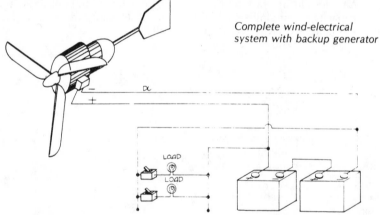

Complete wind-electrical system with backup generator

In the Third World, this book would be most useful as a guide in establishing the true costs (high) of a windgenerated electric system, for small groups or individuals who want to investigate their applicability as an alternative in their region.

Typical relative costs for small wind power systems in U.S.

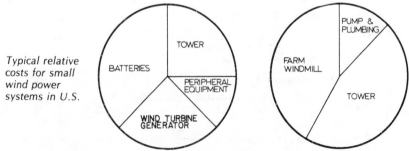

Windmill Power for City People, book, 65 pages, 1977, $2.60, Stock No. 059-000-00001-2, from Superintendent of Documents, U.S. Government Printing Office, Washington D.C. 20402, USA.

This book documents the experience with an urban wind energy system, on the roof of the cooperatively-owned 11th Street Movement building in New York City. The 2-kw rebuilt Jacobs windgenerator is linked to the electric grid by means of a synchronous inverter. With a moderate average windspeed of almost 9 mph, the windgenerator is expected to produce at least 2000 kwh of electricity per year, worth $200 given New York City's electric rates (10 cents per kwh, the highest in the nation). Labor was provided free by the tenant owners of the building, and total costs were $4134. This made the system economically attractive given anticipated increases in utility electric rates in the future and an expected 30-year system life.

In other locations such a system would be economically feasible only where higher average windspeeds exist to offset the lower value of the electricity saved.

An interesting example of A.T. for low-income people in the urban United States, and the kind of political confrontation with the big utility companies required to "legalize" decentralized energy production.

A Siting Handbook for Small Wind Energy Conversion Systems, book, 120 pages, by Harry L. Wegley, et al, 1978, Accession No. PNL 2521, $9.00 from NTIS, U.S. Dept. of Commerce, Springfield, Virginia 22161, USA.

"The primary purpose of this handbook is to provide siting guidelines for laymen who are considering the use of small wind energy conversion systems." This kind of information is essential in promoting the effective use of wind power in the best locations.

The choice of a site for a wind machine is very important because: 1) the energy in the wind is proportional to the cube of the windspeed, and thus small differences in windspeed mean large differences in wind power available; 2) small obstacles on the ground in flat terrain can slow the wind considerably; and 3) wind patterns are greatly affected by hilly and mountainous terrain. This handbook will help identify the sites with the highest wind power potential. This is most important for wind-generators, which take advantage of the high range of winds at a site, for maximum electricity production. The manual will also be of value in choosing sites for waterpumping windmills, which need more protection from high winds and operate in the low range of windspeeds to allow more dependable water pumping.

Most of the information included can be used anywhere in the world. The

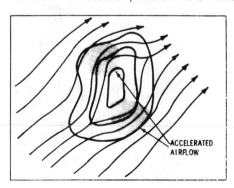

Airflow around an isolated hill (top view)

ACCELERATED AIRFLOW

core of this book is a well-illustrated presentation on the effects of trees (including wind-breaks), shrubs, and buildings in flat terrain, and the effects of ridges, passes, valleys and other features in mountainous or hilly terrain. Groups in other countries could substitute their own data for the section on special weather hazards of the United States (snow, hail, icing, tornadoes, thunderstorms, high winds and dust storms), with maps that identify affected areas. Most Third World countries do not have as firm a data base for these country maps, but some of the problems are avoided also.

"To understand and apply the siting principles discussed, the user needs no technical background in meteorology or engineering; he needs only a knowledge of basic arithmetic and the ability to understand simple graphs and tables."

"According to manufacturers...the greatest cause of dissatisfaction among owners has been improper siting...This handbook incorporates half a century of siting experience...as well as recently developed siting techniques."

Vegetation as an Indicator of High Wind Velocity, and **Trees as an Indicator of Wind Power Potential**, papers, 24 pages plus bibliography and 10 pages, by J. Wade, E. Hewson, and R. Baker, $2.00 and $1.50 respectively, from Dept. of Atmospheric Sciences, Oregon State University, Corvallis, Oregon 97331, USA.

These papers describe the development of a technique for using trees as indicators of the long-term average winds in a particular place. "Plants provide a quick, at a glance, indication of strong winds and when calibrated by the degree of wind shaping provide a rough, first-cut assessment of wind power potential...This technique could appropriately be used as a first stage in a wind survey prior to instrumentation with anemometers."

A widespread obstacle to the use of windgenerators is that the energy available—and therefore the economic feasibility—varies dramatically from site to site. The approach described here is intended to aid in the selection of sites for windgenerators, which require relatively high average windspeeds if they are to be economically feasible. The basic approach could also be used in identifying sites for waterpumping windmills, but they do not use—and in fact need protection from—the higher winds. New calibrators would be required for species of trees common to other areas, and a substantial amount of long-term wind speed data is needed in order to do such calibrations. Exposure and slope also affect the data. A bibliography is included in the first paper.

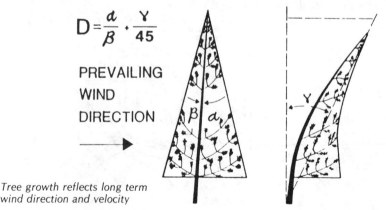

$$D = \frac{a}{\beta} + \frac{\gamma}{45}$$

PREVAILING
WIND
DIRECTION

Tree growth reflects long term wind direction and velocity

Low Cost Wind Speed Indicator, publication No. T-113, single page of blueprints, 1979, $2.50 from Brace Research Institute, MacDonald College of McGill University, Ste. Anne de Bellevue, Quebec, Canada H9X 1CO.

Plans for a simple tilting pointer wind speed indicator. Requires plastic tubing, aluminum sheet and aluminum rod, steel tubing, and a piece of wood.

Energy from the Wind: Annotated Bibliography (July 1975), First and Second Supplements (August 1977, December 1979), compiled by Barbara Burke, all three together are available for $25 in the U.S. and $27 overseas, form Publications, Engineering Research Center, Foothills Campus, Colorado State University, Fort Collins, Colorado 80523, USA.

See our review of the July 1975 bibliography (page 151). Two supplements are now available, for the total of 3800 references covered by the three-volume set. About half of these are from the period 1975-1979. Very few of the documents listed contain any practical construction information, and the index does not identify them for the reader. No addresses are provided for the documents. For people with special topic interests and access to a university library, this bibliography set will, however, provide very helpful access to a wide literature on the economic, policy, and theoretical design aspects of wind power.

ADDITIONAL REFERENCES ON WIND POWER

Waterpumping windmills for India are discussed in articles in **Renewable Energy Resources and Rural Applications in the Developing World** (page 563) and **"Technology for the Masses"** (page 369).

Locally built waterpumping windmills in Peru are pictured in **MINKA**, reviewed on page 376.

Several articles discuss low cost waterpumping sail windmills and other wind technologies in **Appropriate Technologies for Semi-Arid Areas: Wind and Solar Energy for Water Supply**, reviewed on page 566.

Commercially available windgenerators and windpumps are listed in **The Power Guide**, page 568.

Gemini Synchronous Inverter Systems describes an electronic device that allows a windgenerator to be linked directly to the electric grid, thereby eliminating the need for batteries (the system described in **Windmill Power for City People** uses such an inverter); see page 578.

ENERGY: WATER

Energy: Water

For 2000 years, waterpower has been harnessed to do useful work. Waterwheels played a vital role in early industrialization in Europe and North America, powering a wide variety of decentralized manufacturing and processing enterprises. The steel water turbine provided more power at a given site than the waterwheel, and in the U.S. many waterwheel-powered mills were converted to water turbines in the late 19th and early 20th centuries. Blacksmiths and foundrymen produced the turbines and modified the designs during this period of great innovation and profitable production. Water powered mills produced

''...such household products as cutlery and edge tools, brooms and brushes... furniture, paper...pencil lead...needles and pins...watches and clocks, and even washing machines...For the farm they turned out fertilizers, gunpowder, axles, agricultural implements, barrels, ax handles, wheels, carriages. There were woolen, cotton, flax and linen mills...tannery, boot and shoe mills...and mills turning out surgical appliances...and scientific instruments.''

— **Renewable Energy Resources and Rural Applications
in the Developing World** (reviewed on page 563)

The use of waterpower in the People's Republic of China has reflected the same pattern, with first waterwheels and then turbines being built in great numbers as power demands increased along with technical production capabilities (see chart on page 564). By 1976 an estimated 60,000 small hydroelectric turbines were in operation in south China alone, contributing a major share of the electricity used by rural communes for lighting, small industrial production, and water pumping.

With the rising costs of energy in the United States today, small hydroelectric units are returning in large numbers. Generating stations along New England rivers are being rehabilitated and put back into operation. The number of companies making small waterpower units has jumped. The U.S. Department of Energy has estimated that 50,000 existing agricultural, recreational, and municipal water supply reservoirs could be economically equipped with hydroelectric generating facilities.

In Third World countries, the potential for small hydropower installations has never been carefully measured. Past surveys of hydropower potential have focused on possible sites for large dams, as small hydroelectric units were considered uneconomical or ill-suited to the goals of providing large blocks of electric power for cities, industrial estates, or aluminum production. With the rapidly increasing costs of energy, however, the economics are now very favorable for small hydroelectric units, which are also well-suited to the needs of rural small communities, and do not bring the degree of environmental disruption associated with large reservoirs. Many small units do not require reservoirs at all; where needed, safe, small, low cost dams made of earth or soil-cement bags could be built in many cases by the local population.

The success of small water power installations can be greatly affected by forest conservation practices in the watershed above. Rapid deforestation brings high rates of soil erosion and subsequent rapid silt filling of reservoirs behind dams. At the same time, greater rain runoff causes increasingly violent floods that threaten hydropower installations. During the months following the floods, low water water flows are likely to reduce generating capacity. Promotion of efficient cooking stoves, prevention of clear-cutting logging practices, establishment of village woodlots, implementation of land reforms to eliminate shifting cultivation on quickly eroded hillsides, and support of reforestation programs may be needed to protect small waterpower installations.

The first requirement in estimating the potential for a small water power site is to measure the water flow in a stream during medium and low flow periods, and determine the high water level during flooding. **Low Cost Development of Small Water Power Sites** (reviewed on page 155) is one of several publications that give good stream flow measurement instructions.

For the construction of very small earth dams, the useful but out-of-print publication **Small Earth Dams** may be copied at an appropriate technology documentation center. Also of interest is the soil-cement sandbag technique for low dam construction employed by the Las Gaviotas appropriate technology group in Colombia.

Waterwheels still have certain advantages for rural Third World communities. They can be constructed of locally available materials (e.g. wood and bamboo) by village craftsmen, and they can lift water and perform a variety of important crop processing tasks. They are well suited to the small crop production of small farmers. Existing irrigation channels and small streams offer many potential sites at which the civil engineering works expenses can be minimized, because only short diversion ditches need to be dug and lined with cement.

Industrial Archaeology of Watermills and Waterpower is one of several publications that offer valuable insights into the design evolution of waterwheels. Several pages in **Other Homes and Garbage** (reviewed on page 42) provide the standard formulas used in the design of overshot waterwheels in the United States in the last century. **Overshot and Current Waterwheels** and **Water Power for the Farm** are useful references that were used in rural extension efforts in the 1930s and 1940s and have been recently reprinted. **Watermills with Horizontal Wheels** and **Waterpower in Central Crete** describe in detail the vertical-axis, horizontal wheel, stone flour and corn mills once widely used in Europe and Asia, and still used in large numbers in moutainous countries such as Nepal.

The reader interested in water turbines will find **Harnessing Water Power for Home Energy** a good introduction to the subject, although the case studies are all from single homes in rich countries. **A Pelton Micro Hydro Prototype Design** contains an excellent example of a Pelton wheel installation in a village in Papua New Guinea. The Pelton wheel requires a small flow of water that drops 50 feet or more; installations can be built with plastic pipe and a small stream diversion ditch, for minimum environmental impact. Several reports and sets of engineering drawings are included on the cross-flow Banki turbine design adapted by Balaju Yantra Shala (BYS) in Nepal. These turbines are made of a standard diameter, with blades cut from steel pipe, and they can

accommodate a range of flow and head conditions. *BYS is now reportedly
installing 10 or more units a year for use in generating electricity, pumping
water, and directly driving equipment used in small industries. Those involved
in low head turbine design will want to find out about the pioneering work of
Las Gaviotas (Colombia) in the design of low head turbines.*

In Volume I:

Industrial Archaeology of Watermills and Waterpower, book, 100 pages,
Schools Council, 1975, £3.25 from Heinemann Educational Books Ltd., 48
Charles St., London W1X 8AH, England; or $9.50 from Heinemann Educa-
tional Books, 4 Front St., Exeter, New Hampshire 03833, USA.

This well-illustrated book provides a very good summary of the history of
waterwheel development. The reason for design improvements are discussed.
This will allow the reader to better judge what materials and designs are
needed for a particular application. For example, in the early nineteenth
century, a very large slow-running waterwheel would develop high torque on
the wooden main shaft, which might cause it to break. This was solved in two
ways: by using iron shafts, and by using power drives off of the rim of the
waterwheel (the main shaft then had only to strong enough to support the
wheel). Smaller waterwheels, for example, do not necessarily need these more
expensive design elements.

By 1850, the British had built a number of very large industrial scale
waterwheels, producing 65kw to 190 kw of power, and ranging from 7m to 21m
in diameter. Some of these waterwheels were kept in operation for 100 years.
Such waterwheels would likely be very expensive to build today; but smaller
wheels in the range of 10 kw built locally in the rural Third World may be
economically viable in many places.

Options for different installations are discussed: water course layout, types
of wheel construction, bucket design, mechanisms for flow control, and gearing
systems (belts, wooden teeth, iron gears) to take the power off the wheel and
run it to the equipment.

The second half of the book is a guide for teaching students about waterwheels. This shows how to measure water flow in a stream, figuring torque on the waterwheel shaft assuming a certain wheel rpm, and calculating horsepower. Several working models can be used to illustrate principles. A brief but informative section on water turbines is included.

Those who want to design waterwheel installations will find this book helpful with its background information and 200 drawings and photos. However, the design formulas (for number of buckets, bucket depth, wheel width and diameter, etc.) that were well-developed by 1850 are not included, and will have to be found elsewhere (for example, in **Other Homes and Garbage**—see review on page 42).

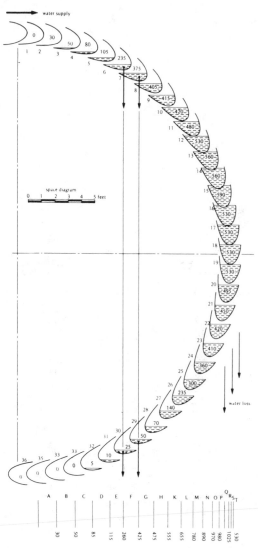

Diagram of waterwheel, showing approximate
amounts of water in each bucket

Overshot and Current Water Wheels, booklet, 30 pages, by O. Monson and A. Hill, reprinted Sept. 1975, Bulletin 398, single copies free from Montana Agricultural Experiment Station, Montana State University, Bozeman, Montana 59717, USA.

This is a valuable booklet to supplement others, such as the **Design Manual for Waterwheels** and **Low Cost Development of Small Water Power Sites**. It gives details of bucket construction and mounting, hubs, and bracing for wide wheels. Useful hints are provided on bearings, wheel mountings, and assembly and balancing of the wheel. A chart compares steel shaft diameters to wooden shaft diameters for equal strength in twisting (shear). There is some discussion of the special problems presented by current (undershot) wheels.

This does not include the design formulas needed to design your own waterwheel if you know the available flow and head (height). For these design formulas you will have to refer to **Other Homes and Garbage** (see review on page 42).

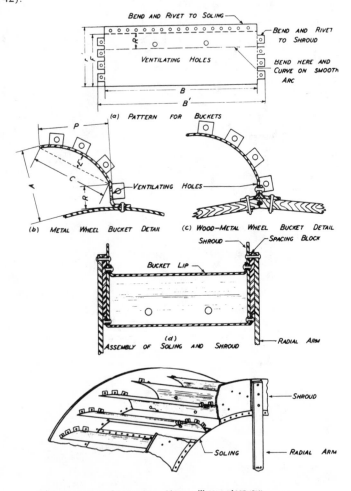

(see review on page 42).

(a) PATTERN FOR BUCKETS

(b) METAL WHEEL BUCKET DETAIL

(c) WOOD–METAL WHEEL BUCKET DETAIL

(d) ASSEMBLY OF SOLING AND SHROUD

(e) CUT-AWAY SECTION OF METAL WHEEL ASSEMBLY

FIG. 6. CONSTRUCTION DETAILS FOR METAL AND WOOD-METAL WHEELS.

Water Power for the Farm, booklet, 39 pages, by O. Monson and A. Hill, Montana Agricultural Extension Service, Bulletin No. 197, 1941, $4.15 from Interlibrary Loan Service, Roland R. Renne Library, Montana State University, Bozeman, Montana 59717, USA.

"Information given here is intended to be helpful in the proper selection and installation of small water power equipment. Directions and specifications are included for the construction of some types of dams and water controls which are practical for use with water power plants." Discusses water wheels and water turbines, how to estimate the power available in a stream, cost considerations, and methods of power transmission.

The water measurement section is inferior to the VITA booklet (see page 155) and the Popular Science article (see page 156). But this booklet provides more than the others on electrical systems. There is brief information on switches and wiring, for people not familiar with electricity. A wiring chart gives sizes of wires needed for different transmission distances, voltages, and total capacity. However, it does refer to equipment that is out of date (32 volts) in the U.S.

Another useful chart shows power output at the wheel shaft and from a generator, given different heads and different waterwheels and turbines.

Mill Drawings, 30 large sheets, measurements and details by W. Foreman, 1974, price unknown, from The Society for the Protection of Ancient Buildings, 58 Great Ormond St., London WC1, England.

Each of these 30 sheets contains a set of perspective and scale drawings of a different watermill installation. These are actual sites, and the drawings

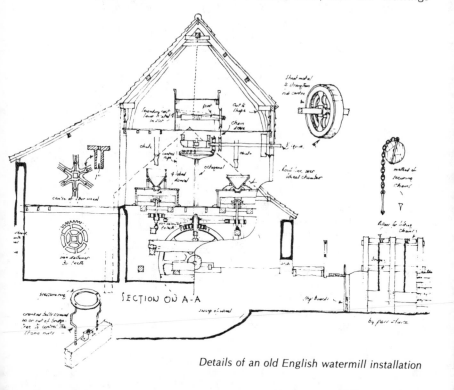

Details of an old English watermill installation

include lots of details of the wheels, machinery, and layout of each mill. Some of them are very unusual. There are overshot, undershot, and breast water-wheels shown.

A great source of ideas.

Watermills with Horizontal Wheels, booklet, 22 pages, by Paul Wilson, 1960, 65 pence including postage, from The Society for the Protection of Ancient Buildings, 58 Great Ormond St., London WC1, England.

This is a survey of the vertical-axis, small ''horizontal stone'' watermills widely used around the world for 1500 to 2000 years, and still in use today in isolated areas of countries like Nepal. A number of different installations around the world and six different mill wheel designs are shown. These machines are ''not as efficient as an overshot or breast wheel, but they have the

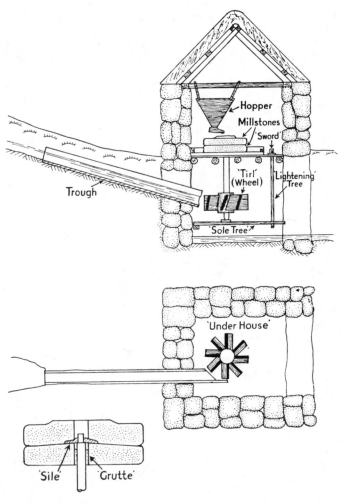

Vertical-axis stone watermill

virtue of simplicity due to the absence of gearing," and they are very cheap to build. Another book notes that this mill is able to produce 40 to 50 pounds of cornmeal per hour.

Most of these watermills had a wooden trough bringing fast-moving water to strike the blades of the wheel, with a head of 4-10 feet. Two types of pressurized systems also evolved, one using wooden channels with nozzles, and the other using stone towers with nozzles. "The Aruba Penstock (water tower) was introduced (in Israel), giving much greater efficiency and enabling power to be obtained from quite small flows of water using heads up to 25 or even 30 feet."

Waterpower in Central Crete, booklet, 8 pages, by N. Calvert, 1973, 40 pence including postage, from The Society for the Protection of Ancient Buildings, 58 Great Ormond St., London WC1, England.

This is a look at two kinds of water wheels in Crete. The most interesting one is a traditional vertical axis watermill, that has a stone tower and a pressurized jet with a deflector, (almost like a modern high speed Pelton wheel). "The constructional materials are of the simplest and most local description. With one important exception (the millstone) stones are small and unworked, timber is of small dimensions...clay is used and a very little iron. An effective and sweetly running machine is built from what is literally little more than a supply of sticks and stones."

The author notes that the basic layout of these wheels is so technically sound that modern improvements (in nozzle and blade design) could improve efficiency by only about 20%.

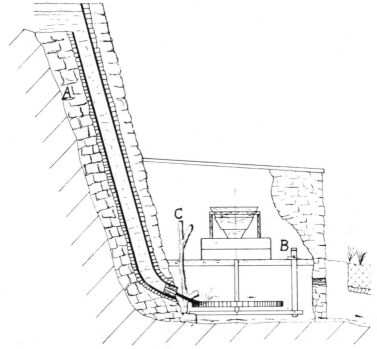

Stone watermill in Crete with pressurizing water tower

Woodbridge Tidemill, set of drawings, by J. Kenneth Major, 1961, 65 pence including postage, from The Society for the Protection of Ancient Buildings, 58 Great Ormond St., London WC1, England.

This is a set of 5 large sheets of detailed line drawings showing plans and sections of a tidemill used for grain milling in England. It operated with a mill pond that filled during high tide, and released its water during low tide, turning the mill waterwheel. The mill was located just above the point where the river Deben enters the North Sea, and thus was protected from the violence of waves during storms. The drawings show the location of the mill on the river, the waterwheel design, the mill layout, and the tidal flows. There is no accompanying text.

Harnessing Water Power for Home Energy, book, 112 pages, by Dermot McGuigan, 1978, $4.95 from Garden Way Publishing Co., Charlotte, Vermont 05445, USA.

This is a good book for someone who wants to learn about the different small-scale water turbines that can be used to generate electricity. The Pelton wheel, Turgo impulse wheel, Banki (Ossberger) cross-flow turbine, and Francis turbine are shown in a total of 6 actual installations in England and the United States. Costs are provided for many of these examples. Only the Pelton wheel and Banki turbine are really suitable for construction in a small workshop. Manufacturers of turbines and whole systems are listed from around the world.

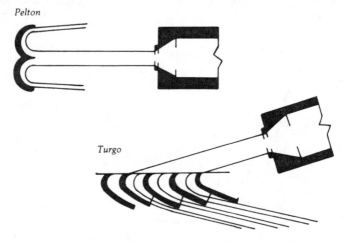

Fig. 12. Turgo and Pelton turbines contrasted. The jet on the Turgo strikes three buckets continuously, whereas on the Pelton it strikes only one. A similar speed increasing effect can be had on the Pelton by adding another jet or two.

Useful notes are included on alternators, transmission drives, dams, and the electronic governor (a device which switches part of the electric current away from the main line—to heat water, for example—when the electric demand falls; this eliminates the need for an expensive mechanical governor which regulates the amount of water flowing through the turbine).

There are many drawings and photos, but these are poorly explained. Electrical circuitry is not shown, and mechanical governors are not explained. Waterwheels are only briefly covered in a few pages. The examples are all single homes in rich countries, using large amounts of electricity. The language is relatively easy to understand, although a number of water-power engineering terms are used without explanation.

You will not be able to build anything from the information contained in this book, but you can get a better idea of what would be required to install a small water-powered electric system, on a useful scale for village electrification.

Hydroelectric Power, promotional leaflet, 12 pages, $2.00 from Independent Power Developers (IPD), Route 3, Box 174 H, Sandpoint, Idaho 83864, USA.

This promotional leaflet describes the low head (propeller) and high head (Pelton wheel) hydroelectric systems offered by IPD. Presents basic information for determining head and flow available and which system to choose. Continuous power production ranges from 300 watts to 8.5 kw with these units; peak power available is 3-12 kw.

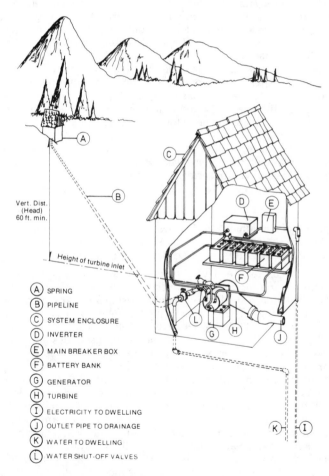

Vert. Dist. (Head) 60 ft. min.

Height of turbine inlet

(A) SPRING
(B) PIPELINE
(C) SYSTEM ENCLOSURE
(D) INVERTER
(E) MAIN BREAKER BOX
(F) BATTERY BANK
(G) GENERATOR
(H) TURBINE
(I) ELECTRICITY TO DWELLING
(J) OUTLET PIPE TO DRAINAGE
(K) WATER TO DWELLING
(L) WATER SHUT-OFF VALVES

Small Pelton wheel hydroelectric system from IPD

A Pelton Micro-Hydro Prototype Design, report, 41 pages, by Allen R. Inversin, June 1980, exchange your publication with Appropriate Technology Development Unit, P.O. Box 793, Lae, Papua New Guinea.

"Is it possible that introducing electricity into rural villages could be one factor towards rejuvenating life in these villages? Can a technically and socially appropriate system with active villager participation in the planning, installation, management, and evolution of their own scheme have beneficial effects?"

"This report describes work to date on a modular design for a Pelton micro-hydro generating set with an electrical output up to...5 kVA and with a 'typical' installation cost of about K300/kVA ($400/kw) including penstock costs...This is less expensive than diesel generating sets, and, when recurring costs are included, less costly than both diesel and petrol. Also covered briefly are ideas on governing, bucket design and prototype performance, and cost/kw of PVC penstock pipe for different site configurations and pipe diameters."

Pelton wheels requiring a head of 50 feet or more were chosen, due to the mountainous terrain and small water flow required by these units. General design guidelines were to develop a low cost but rugged design, which could be locally fabricated with a minimum of special skills, and which could be easily installed with little site preparation.

Notable design simplifications include: 1) a low cost easily made iron pipe cover for the main shaft, which prevents water from entering the bearings; 2) use of holes in steel plate to replace nozzles; 3) pulley substitution to adjust for actual head at the site; 4) bolted assembly. The author also discusses a variety of ways to eliminate the need for expensive mechanical flow governing systems.

This well-illustrated report is a valuable description of the state of the art of low cost micro-hydroelectric systems using Pelton wheels. It incorporates ideas and suggestions based on pioneering work in Colombia at Universidad de los Andes. The technology is widely relevant in mountainous areas of developing countries.

Development of Equipment for Harnessing Hydro Power on a Small Scale, booklet, 20 pages, by Ueli Meier, 1978, $3.00 (including postage) from Balaju Yantra Shala Pvt. Ltd., P.O. Box 209, Balaju, Kathmandu, Nepal.

The author discusses the development of a simplified water turbine for mechanical power and for generating electricity in Nepal. The turbine can be fabricated from standard steel sheet and pipes using familiar general workshop techniques (no casting). There are a limited number of sizes which match a flow of 50 liters/sec to 500 liters/sec, over a head of 4 to 40 meters, with an output of 2 to 50 kw. The turbine is a modified Banki design, 400 mm in diameter, welded construction, with an estimated lifetime of 10 years. Several governing mechanisms being tested are described.

"The selling price of the entire range of turbines produced is in the range of US$1200 to 1500." Total costs vary considerably depending on local topography and the need for dams, ponds, and channels (civil works). For generation of electricity in the range of 20 to 50 kw, total investment costs of all generating equipment, penstock, and civil works (but excluding transmission and distribution) are estimated at $1000 to $2000 per kw.

"Another important application is in lift irrigation where turbines under relatively low head are used to run water pumps mechanically, for lifting water to the generally much higher plateau above the river...two turbines of 30 kw each shall be used to irrigate 50 hectares."

In the hilly areas of Nepal, a wooden vertical-axis type of waterwheel for grinding flour has been in use for hundreds of years. In recent years a number of horizontal axis waterwheels have also been built, with a higher power output. "The fact that there exists a lot of local expertise in diverting water from rivers and streams and in building earth canals for irrigation as well as the more sophisticated devices for harnessing water power with active participation of the rural population in its installation."

Water entering the turbine through the inlet(1) flows through the rectangular nozzle(9) radially into and again out of the rotor(2), thus setting the output-shaft(3) attached to the rotor into circular motion.

Rate of water flow and power output may be governed by the regulator wing(4) which is operated

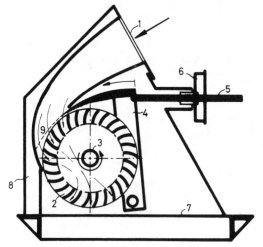

by a hand wheel(6) and push rod(5).
All parts are mounted onto the base frame(7) and in a body or housing(8).

Banki turbine adapted by BYS in Nepal

Harnessing Water Power on a Small Scale: The Example of Development Work by Balaju Yantra Shala (BYS) Nepal, booklet, 43 pages, by U. Meier, 1979, approximately US$10.00 from Swiss Center for Appropriate Technology (SKAT), Varnbuelstrasse 14, CH-9000 St. Gall, Switzerland.

"An introduction to developing small hydro schemes and the manufacture of equipment required." This describes the cross-flow (Banki) turbine systems developed by BYS with Swiss Technical Assistance, many of which are being built and installed in Nepal. These turbines have a power output of 2-80 kw, and can be used either in electricity generation or for direct drive of machinery. Drawings indicate the dimensions and layout of the units, and tables give information on performance. This part is in English.

Part II, "Foto-Documentation of Applications of Turbines Developed by BYS" includes 41 photos of installations in Nepal, with a short descriptive text in German only. An English version will be included in later printings.

Detailed Construction Drawings of a Cross Flow Turbine (Banki) for Heads from 4 to 40 Meters and a Power Output from 5 to 50 kw, booklet, 97 pages, edited by U. Meier, 1980, approximately US$10.00 from Swiss Center for Appropriate Technology (SKAT), Varnbuelstrasse 14, CH-9000 St. Gall, Switzerland.

Large fold-out engineering drawings and parts lists for construction of the BYS cross-flow turbine including housing and hand-operated flow control mechanisms. Steel plate and pipe are the construction materials, and a well-equipped metal working shop is required. The author intends to expand the publication to include "detailed comments and instructions for making the individual parts of the turbine, for the assembly of parts, and also calculation procedures to determine turbine size depending on head available and output power desired." The present version, however, does not include any text. French and Spanish editions are planned for 1981.

Manual for the Design of a Simple Mechanical Water-Hydraulic Speed Governor, booklet, 40 pages, by U. Meier, 1980, approximately US$10.00 from Swiss Center for Appropriate Technology (SKAT), Varnbuelstrasse 14, CH-9000 St. Gall, Switzerland.

This manual describes the mechanical speed governing system developed by Balaju Yantra Shala (BYS) in Nepal to control their cross-flow turbine. Schematic drawings help explain the operation of the unit, and detailed drawings provide information for construction of the key load-regulating valve.

"It is mostly cost that has stood in the way of speedy and large scale development of small hydro power potentials (in developing countries). Imported and sophisticated equipment becomes costlier and costlier and is in most cases not economically feasible. This applies mostly to hydraulic equipment such as water turbines, accessories and governing devices. The case is different for alternators and switchgear which are produced in great number in industrialized countries and are therefore relatively cheap."

"Experience in Nepal shows that it is possible to reduce costs of hydro electricity generation projects vastly by minimizing civil engineering and structural works and by producing hydraulic equipment in local workshops with simple designs and technology."

"Nonavailability of a simple mechanical governor has long been a major obstacle in implementing small hydro projects with acceptable standards of safety. The Swiss Association for Technical Assistance, Helvetas, in Zurich has sponsored a project to develop a simple governor that would be sufficiently accurate and reliable and could be manufactured by local workshops in Nepal... A prototype was built and tested in early 1979. The governor was designed for operating the gate of a cross-flow (Banki) turbine, but may in fact be utilized on the flow regulator of any turbine...this construction manual may enable other organizations and individuals to adapt this governor to their own needs and improve it further as a contribution to the design of simple but reliable hydraulic equipment for small electricity generation units."

Small Michell (Banki) Turbine: A Construction Manual, booklet, 56 pages, by W.R. Breslin, 1979, $4.75 from VITA.

This is basically a reprint of the VITA booklet **Low Cost Development of Small Water Power Sites** (reviewed on page 155), with an expanded (15 page) section on contruction of a Banki turbine.

Specifications are given for only one turbine diameter (30 cm.) but turbine width can be varied to accommodate different volumes of water. The plans require 10 cm diameter steel water pipe for the turbine blades, and steel plate for the sides and nozzle. Welding, cutting and grinding tools are needed. The turbine can be used for direct drive of agricultural equipment or for producing electricity. (You will have to look elsewhere for the information to help you set up a proper electrical installation).

A site with a head (total height water will fall) of 25 feet (7.6 m) and a flow of water of 2.8 cubic feet per second (81 liters per second), would produce about 6.3 hp (4.8 kw) of power at the turbine. Transmission losses or generator losses can be expected to cut this by 1/3 to 1/2.

Design of Small Water Turbines for Farms and Small Communities, book (including working drawings for a selected water turbine), 163 pages, by Mohammad Durali, 1976, $7.00 from Technology Adaptation Program, Room 39-523, Massachusetts Institute of Technology, Cambridge, Massachusetts 02139, USA.

This is a report of a project "to study alternative water turbines producing 5 kw electric power from an available hydraulic head of 10 m and sufficient amount of flow, and to recommend one for manufacture," for use on Colombian coffee farms.

Much of the book presents the sophisticated mathematics and physics of turbine design for optimum performance. This requires some technical training to comprehend. The relationships between the various elements in the design are given in equations, allowing choice for simplicity in particular elements.

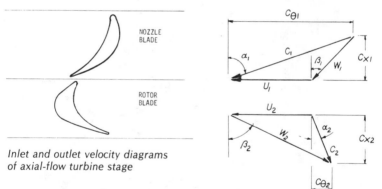

NOZZLE BLADE

ROTOR BLADE

Inlet and outlet velocity diagrams
of axial-flow turbine stage

The design criteria included: simplicity of operation and maintenance (the machine is intended for use by farmers with little technical knowledge); and lower cost of electric power over the life of the machine when compared with transmitted power from the main electric grid. Choice of turbine would be determined in part by which of two production alternatives was selected: 1) use of a simple workshop capable of welding, drilling, and cutting steel parts (local farming area production); or 2) use of more sophisticated production methods like casting and molding with some plastic parts (industrial production at a centralized level).

"The work consisted of the preliminary design of different types of water turbine which could be used for this application. Then one was selected and designed completely. A complete set of working drawings was produced for the selected type."

"Four different types of water turbine were studied: a cross-flow (Banki);

two types of axial-flow turbines; and a radial-flow turbine. Each one has some advantages and some disadvantages (explained in the text). One of the axial-flow turbines...was chosen for detailed design as presenting the optimum combination of simplicity and efficiency.''

The materials range from riveted pieces of thin-wall steel tubing for the blades, wooden bearings, and bicycle sprocket/chain drive (Banki turbine) to molded, extruded, or cast plastic blades (axial-flow turbine).

''A big portion of the price of each of the units is the generator cost. The rest of the construction cost seems likely to be similar for all units for small scale production. For large scale production the cross-flow will be much more costly than the axial-flow types. This is because the material cost for the axial-flow machines is small but initial investments for molds and dyes are required.''

''The design of small water turbine units has not previously been carried to a high degree of sophistication. This might be because of their limited application.'' With the high cost of energy and the need for decentralized power sources for village industry, small water turbines will no longer be perceived as having 'limited application.'

Small Earth Dams, booklet, 23 pages, by Lloyd Brown, 1965, publication No. 2867, U.S. Dept. of Agriculture, out of print 1980, contact an appropriate technology information center for a photocopy.

This introductory booklet has practical suggestions for those who want to build small dams to make ponds for irrigation or watering animals. Although the focus is on U.S. climate and methods, the booklet could be useful as a

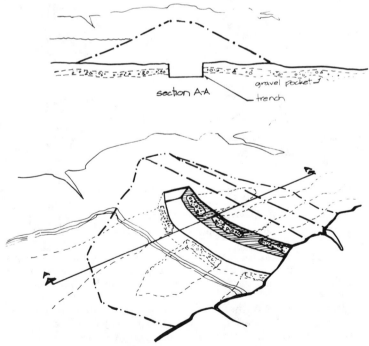

Trench dug and filled to prevent leaks
or failure of small earth dams

starting point for building low-head dams for micro-hydroelectric systems anywhere. The information is only for use in small (6 feet and under) dams and those dams that back up a limited amount of water. There are suggestions for selecting a site, a few hints on construction, maintenance and management practices for the reservoir and spillway (water outlet), and a review of the laws involved (in California).

ADDITIONAL REFERENCES ON WATER POWER

Renewable Energy Resources and Rural Applications in the Developing World describes the evolution of the waterwheel and water turbine in rural industry in Europe, the United States, and present-day China; see review and chart on pages 563-564.

Watermills and small turbines in Nepal are pictured in **Small Scale Renewable Energy Resources in Nepal**, reviewed on page 567.

The Power Guide contains information on commercially available water turbines and a stream flow water pump; see review and drawings on page 568.

Hydraulic ram pumps are the subject of several publications reviewed on pages 541-543.

ENERGY: SOLAR

Energy: Solar

Increasingly, the term "solar technology" is being used to include any renewable energy system that directly or indirectly depends on the sun for energy. This includes water power, wind power, biogas, and wood fuels, for example, which we are covering in separate chapters. This chapter is therefore concerned with direct use of solar energy.

Probably the most significant direct solar technology for the Third World is that of crop drying, which is covered in the chapter CROP DRYING, PRESERVATION, AND STORAGE. Solar distillation for water purification has been covered in the WATER SUPPLY chapter of volume one, on pages 204-206. A solar lumber dryer is described in the FORESTRY chapter.

In this chapter you will therefore find materials on passive solar architecture for house heating (the most significant immediately applicable solar technology for temperate climates), passive solar cooling for the tropics, solar greenhouses, water heaters, cookers, irrigation pumps, refrigerators, and photovoltaic cells. A good general survey of the technologies that may some day have relevance for the Third World can be found in **Technology for Solar Energy Utilization;** most of these, however, are not economically competitive at present.

Several publications on passive solar architecture, now a booming field in the United States, are reviewed. Essentially, passive solar design involves careful choices of building orientation, building layout, location of glass windows, and materials, to best take advantage of natural energy flows. Because they minimize the use of costly primary fuels for space heating and cooling, passive solar buildings will eventually dominate new construction in much of the United States. In North America and other parts of the temperate zones, heating is usually the primary design objective. This is also true in parts of the Himalayas, the Andes, and other mountainous regions, where indigenous structures often reflect certain passive solar principles. In these areas of the Third World there is the potential for new applications of recent advances in the field. By contrast, the cooling of living spaces is the primary object of passive solar design in the tropics. **Elements of Solar Architecture for Tropical Regions** (this section) and **Design for Climate: Guidelines for the Design of Low-Cost Houses for the Climates of Kenya** (see review on page 662) introduce the basic design considerations for passive solar cooling.

Conventional greenhouses consume large quantities of energy to control the climatic conditions inside. Solar greenhouses, on the other hand, are heated primarily by the sun, and are often attached to houses to provide home heating as well. Low cost designs using plastic sheeting and local materials may be of relevance in the mountainous regions of the Third World for supplementary home heating and food production. In the tropics greenhouses are probably mainly of interest because the vegetables grown inside require less water. Two publications reviewed here offer a look at current U.S. solar greenhouses.

Solar cookers have often been identified as a possible alternative in fuel-short and deforested regions. The 1962 publication **Evaluation of Solar Cookers** offers a valuable look at many designs, including most of those still being toyed with today. The high cost, awkward and sometimes dangerous operation, and requirement that cooking be done outside during the sunniest parts of the day have prevented this technology from finding a niche. To our knowledge, despite two decades of scattered attempts, there are no examples of the successful introduction of solar cookers.

The use of solar energy to drive engines for irrigation water pumps has recently received a great deal of attention. In this application, flat plate collectors provide hot water, which is used to heat liquid gas. The gas expands, and drives the engine. The gas then passes through a condenser, where it is cooled by the well water, and the cycle is then repeated. This appears to be one of those solar energy applications that are technically but not economically feasible in developing countries. Though the costs appear to be dropping below $25,000 per installed kw of capacity (the level of a few years ago) they are a long way from being affordable. Locally built waterpumping windmills appear to be a far more cost-effective alternative in areas with even relatively low average windspeeds. In fact, the most thoroughly tested solar pump designs seem likely to remain technological deadends, built in poor countries only through the intervention of rich country aid programs. Some of the completely new concepts in solar pump designs may prove more fruitful.

Solar photovoltaic cells that produce electricity may have a place in waterpumping in the future if their price comes down significantly.* Yet these would present considerable foreign exchange problems, as the very high technology production requirements will prevent domestic production in all but the most technically advanced developing countries. While pumping systems operating on solar cells can be designed without electrical storage equipment, to run when the sun is shining, solar cells seem less likely to be economically attractive for most other rural uses. Whereas the decentralized energy production offered by solar cells is well matched to Third World settlement patterns, electricity is not an energy form well matched to the energy needs of these communities. A whole assortment of electrical equipment would be needed to store and make use of the solar electricity, and very little of this equipment could be produced in the villages.

Direct use of the sun's energy clearly has an important place in energy paths relying on renewable sources. Sunlight is "available" everywhere yet it is also a diffuse or low-grade energy form. This means that while solar energy is an excellent, low-cost means of creating temperature differences of tens of degrees for drying and heating, it is inevitably difficult and expensive to collect and concentrate solar energy to generate electricity or perform mechanical work. For this reason, drying, heating, and cooling are today the most practicable solar applications for most Third World communities.

---

*For an economic analysis of several solar pumping projects, see "Economics of Renewable Energy Systems" by David French, reviewed on page 565.

In Volume I:

Technology for Solar Energy Utilization, 1977 conference report, 155 pages, 1978, in English, French or Spanish, Document No. ID/202, free to local groups in developing countries, $10.00 to others, from United Nations Industrial Development Organization, P.O. Box 300, A-1400, Vienna, Austria.

This is a good overview of some solar energy technologies that may eventually have relevance for developing countries. Most of the solar technologies presented are technically feasible and proven. Because they are unusual they have attracted the attention of scientists and engineers looking for exciting new technologies to work on. Yet most of these technologies are far too expensive for the Third World. Indeed, probably the majority of the technologies presented would at present be reasonable only in an isolated desert region of a rich country. On the other hand, some of these technologies may one day prove economically attractive.

A good state of the art review at the beginning of the book concludes that solar distillation, solar drying, and solar water heating (if needed) are currently attractive in some circumstances. Solar engines, solar water pumps, solar photocells for electricity, and solar refrigeration are labeled not yet attractive. Later, in the contributed articles, favorable evaluation is made of solar timber kilns for decentralized applications, and Tom Lawand provides a thoughtful examination of the potential for solar cookers and solar dryers. The rest of the articles (a solar electric power plant, conversion of solar into mechanical energy, water pumps, flat plate collectors, refrigeration and cooling, active space heating and cooling) review technologies which cost far more than the poor can afford.

UNIDO suggests the local manufacture of low temperature solar devices in the near future and applied research and development efforts for high temperature applications in the more distant future.

Recommended as a reference book for those active in the solar energy field in developing countries.

The Passive Solar Energy Book, 435 pages, by Edward Mazria, 1979, $10.95 (paperback) and $12.95 (hardback) from RODALE.

This is now the best book available on the design of passive solar homes and buildings. ("Passive solar" space heating relies on direct solar energy, the orientation of the structure, and the natural heat storing capabilities of selected floor and wall materials.) The format was chosen to allow the reader to go through the book in about an hour, covering only the most important concepts, and then come back for more detailed technical information on each topic. The excellent illustrations also make this a valuable tool for teaching basic concepts in a classroom.

The author begins with the fundamentals of solar energy and heat theory. He then introduces the major successful design elements and strategies, such as masonry thermal storage, Trombe walls, attached greenhouses and roof ponds. His presentation on building orientation, north side protection, and location of different kinds of living spaces helps illustrate how crucial these factors are to successful passive solar design. The important contributions offered by movable insulation, reflectors, and shading devices, and the concepts behind summer cooling are also discussed.

The author notes that "more energy is consumed in the construction of a building than will be used in many years of operation," and recommends the use of relatively low-energy-consuming materials such as "adobe, soil-cement, brick, stone, concrete, and water in containers; for finish materials use wood, plywood, particle board and gypsum board."

A full third of this book contains the information needed for calculating solar angles, solar radiation falling on tilted and vertical surfaces, shading effects, space heat loss in winter, solar space heat gains, and auxiliary heating required. Data is included on the solar radiation (insolation) received and space heating needs for major U.S. cities and regions.

Highly recommended.

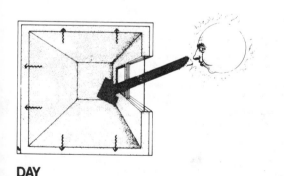 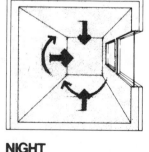

DAY **NIGHT**

Masonry heat storage in a passive solar house

Natural Solar Architecture: A Passive Primer, book, 236 pages, David Wright, 1978, $8.95 from Van Nostrand Reinhold, Litton Educational Publishing, Inc., 135 West 50th St., New York, New York 10020, USA.

In 'passive' solar designs the building itself collects, stores, and circulates the sun's energy without using equipment. In 'active' solar designs solar collector panels, pumps, electricity and various mechanical devices do the things 'passive' can do.

"Nature does not rely on fragile pumps and fans, arrays of gauges and dials, or elaborate tracking collectors to run her vast network. Man's systems are generally complicated, inefficient, expensive and quite fallible...Nature seldom is!"

This book describes, in simple language, the physical processes of natural passive solar systems. Concepts such as the greenhouse effect, heat transfer, heat loss, heat gain, surface-to-volume ratio, and heat sink are fully explained. There are many helpful illustrations.

"In most climates, the natural energies that heat, cool, humidify and dehumidify structures are available throughout the year. The trick is to distribute these energies to the times they are needed for comfort. Since the weather does not adapt to our exact needs, our structures must do the adapting. Buildings can be designed to accept or reject natural energy and store or release it at appropriate times."

North American architectural styles are dominant in this volume, but there are also examples of passive solar buildings that can work in many different climates and are adaptable to other architectural forms.

"Usually, the emphasis placed upon passive solar use is for heating. Cooling by passive means can be effective for controlling excessive heat in most climates." Some of these means are: earth tubes in which air enters the building through pipes buried in the cooler ground; cooling towers in which wind is channeled past wet jars for evaporative cooling as it enters the structure; underground houses to take advantage of the stable, cool temperature of the earth; and shaded, ventilated courtyards.

Readers interested in passive solar design for the tropics should also see **Design for Climate: Guidelines for the Design of Low Cost Houses for the Climates of Kenya** (reviewed on page 662), and **Elements of Solar Architecture for Tropical Regions** (see review on page 635).

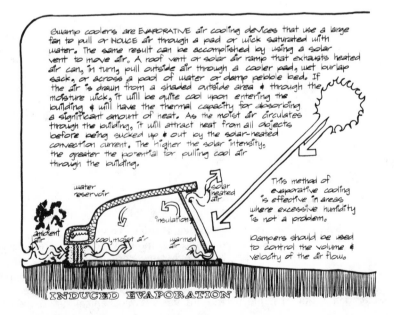

Swamp coolers are EVAPORATIVE air cooling devices that use a large fan to pull or INDUCE air through a pad or wick saturated with water. The same result can be accomplished by using a solar vent to move air. A roof vent or solar air ramp that exhausts heated air can, in turn, pull outside air through a cooler pad, wet burlap sack, or across a pool of water or damp pebble bed. If the air is drawn from a shaded outside area & through the moisture wick, it will be quite cool upon entering the building & will have the thermal capacity for absorbing a significant amount of heat. As the moist air circulates through the building, it will attract heat from all objects before being sucked up & out by the solar-heated convection current. The higher the solar intensity, the greater the potential for pulling cool air through the building.

water reservoir

solar heated air

insulation

ambient air

cool, moist air

warmed air

This method of evaporative cooling is effective in areas where excessive humidity is not a problem.

Dampers should be used to control the volume & velocity of the air flow.

INDUCED EVAPORATION

Basic Principles of Passive Solar Design, paper, 31 pages, Fred Hopman, 1978, available from SATA, P.O. Box 113, Jawalakhel, Kathmandu, Nepal.

SATA has reprinted this paper from the Taos Solar Association of New Mexico, USA. The author presents operating principles and design considerations for passive space heating and cooling systems, including examples of direct gain designs, Trombe walls, roof ponds, attached greenhouses and water circulation systems. Although the examples use Western architectural styles, this excellent introduction to passive solar principles is relevant to building construction in cold climates throughout the world.

Direct gain—note difference between summer and winter sun

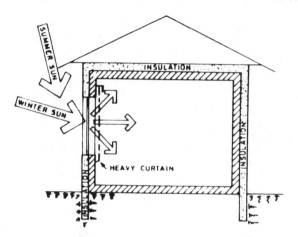

Homegrown Sundwellings, book, 136 pages, by Peter van Dresser, 1977 and 1979, $5.95 from The Lightning Tree, P.O. Box 1837, Santa Fe, New Mexico 87501, USA.

Peter van Dresser, one of the pioneers of solar-heated houses, built his first one in 1958. His book **Homegrown Sundwellings** summarizes a two year program to develop low-cost, owner-built, solar-heated houses. It should be read for its sound observations on sensible solar construction based on local materials, and as an introduction to passive solar home design. More extensive information for designing passive solar homes can be found in **The Passive Solar Energy Book, The Solar Home Book**, and **Natural Solar Architecture** (see reviews in this section).

Although "the Sundwellings concept is firmly rooted in the living construction traditions as well as the socioeconomic circumstances of a natural ecological region—the uplands of northern New Mexico...it reveals principles of universal applicability...To construct using renewable resources is not a sentimental fad in an area without exportable products to pay for imports...In a low cash economy, it is the interactions of human resources with the immediate materials of the land that provide for the richness and fullness of life."

The total solar energy received in winter at sites in Montana, New Mexico and Arizona is greater than the requirement for home heating. The challenge is to store this energy effectively. "The basic strategy is to design the house so that its own masses—mainly walls and floors—are so placed, proportioned, and surfaced that they will receive and store a large measure of incoming solar energy during the daylight hours and will gently release this stored heat to the house interior during the succeeding night hours or cloudy days...A traditional New Mexican floor—either of treated and filled adobe clay or of brick or flagstone laid over sand—is very well suited...its sheer mass gives it great capacity to store this heat with a very slight rise in temperature. If we visualize such a floor 12 inches deep in a room 16 feet square with one exterior wall and an average window, warmed to a mere 72°F (22°C)...it will store 40,000 BTUs of heat which will be released into the room as it cools down to say, 65°F (18°C). This is sufficient heat to take care of a well-insulated room for 26 hours, with an outdoor temperature of, say, 20°F (-7°C).''

Solar Dwelling Design Concepts, book, 146 pages, by American Institute of Architects Research Corporation, $2.30 (order Stock #023-000-00334-1) from US Government Printing Office, Washington DC 20402, USA.

This book is similar in purpose to the **Solar Home Book** (see review in **Sourcebook** Volume I, page 174). It presents principles in easy-to-understand terms for both passive and active solar heating and cooling of homes. Intended for architects, the emphasis is on the integration of solar concepts with traditional western home designs. Factors influencing design are also covered, such as climate, comfort and choice of building site.

32 solar home designs are described, with architectural drawings, to show a variety of passive and active building concepts already in use. Although these designs are from the US, the concepts could be adapted by building designers in other temperate climates.

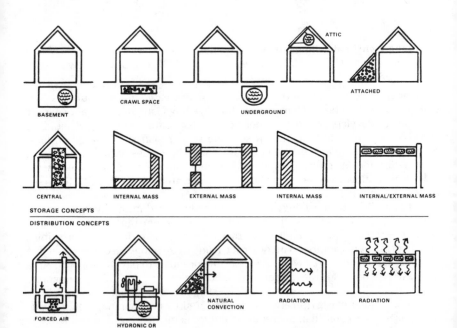

Common solar energy storage and distribution concepts for home heating

Elements of Solar Architecture for Tropical Regions, booklet, 23 pages, by Roland Stulz, 1980, $5.00 from Swiss Association for Appropriate Technology (SKAT), Varnbuelstrasse 14, CH-9000 St. Gall, Switzerland.

One of only a few publications on the design of solar buildings in tropical regions, where cooling and protection from heat are the major objectives. (See also **Design for Climate: Guidelines for the Design of Low Cost Houses for the Climates of Kenya**, reviewed on page 662.) This booklet concentrates on proper building orientation; cross ventilation; reflecting, absorbing, and insulating building materials; shading with trees, shutters, roof overhangs and other techniques; and evaporation of water (in arid climates) for cooling. Tables indicate some of the different considerations for buildings in humid vs. arid regions. A good illustrated introduction to the topic; many of these concepts have long been a part of indigenous architecture in different parts of the world, but have begun to disappear in the last few decades.

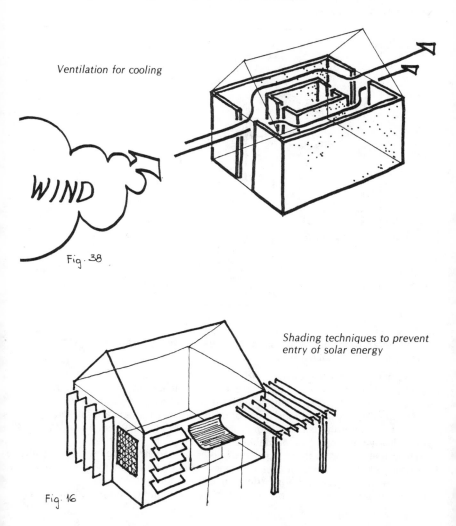

Ventilation for cooling

WIND

Fig. 38

Shading techniques to prevent entry of solar energy

Fig. 16

The Fuel Savers: A Kit of Solar Ideas for Existing Homes, book, 60 pages, Dan Scully, Don Prowler, and Bruce Anderson, 1976, $2.75 plus postage from Total Environmental Action, Church Hill, Harrisville, New Hampshire 03450, USA.

A collection of 20 energy conserving passive and active solar ideas and projects to be used on existing buildings in cold climates. Although intended for North American homes, we have included this book because the projects do not require manufactured solar components, and can be easily completed by the owner.

Included are insulating curtains and shutters, window box air heaters, greenhouses and a thermosiphon (natural circulation) water heater. A drawing and description of each project are given, without exact construction details; the energy savings for each project are discussed.

Useful only in cold climates as examples of simple, do-it-yourself energy conservation measures.

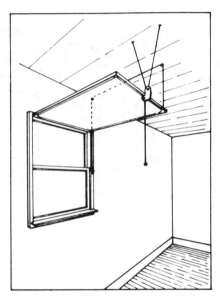

Insulating shutter for cold climates

The Solar Greenhouse Book, 344 pages, edited by James C. McCullagh, 1978, $9.95 from RODALE.

This book covers the design, siting, construction, use and maintenance of sun-heated greenhouses. These are often attached to homes, where they operate as efficient solar heating systems. Such structures can be used from temperate regions to highland areas of the sub-tropics. Plans are presented for a variety of greenhouses for North American readers; others may find the general presentation to be valuable as well.

The effectiveness of these "passive" solar-heated structures is dramatically shown by photos of greenhouses with snow outside and thriving plants inside. The appendix contains information on different types of glass and plastic window materials, suitable plants, available equipment, and a bibliography.

Proceedings of the Conference on Energy-Conserving, Solar-Heated Greenhouses, book, 248 pages, edited by John Hayes and Drew Gillett, 1977, $9.00 ($7.50 to readers in developing countries), from Marlboro College Greenhouse Conference, John Hayes, Marlboro College, Marlboro, Vermont 05344, USA.

This is a collection of papers and reports, with some plans for solar greenhouse construction. The topics covered include the theory, planning, construction and operation of solar greenhouses for food, heat and shelter. The reuse of household water and human wastes is also covered. The language is sometimes difficult for non-native English speakers.

Solar greenhouses are designed to be heated primarily by the sun, even during cold periods. (Traditional greenhouses require large amounts of fossil fuel energy for heating and cooling.) Solar greenhouses are also often designed to heat attached houses, acting as solar collectors.

These greenhouses also work well in hot climates. They can be designed for cooling, by venting hot air up a chimney and drawing cool air in through the house and into the greenhouse. Tree shading can also be used to prevent overheating during warm seasons.

"The greenhouse must also be looked at in light of its water conservation over field crop conditions. Authorities report water usage for greenhouse crops to be 1/10 to 1/30 of the field crop...Tom Rolf of Silver City and the Chavez family of Anton Chico, devised simple systems to trap rain water and snow melt from the roofs of their homes, drain it into tanks in the greenhouse, and gravity feed the water to their plants."

In the many parts of the Third World where cold weather space heating is required, solar greenhouses made of plastic sheeting and local materials may have a role.

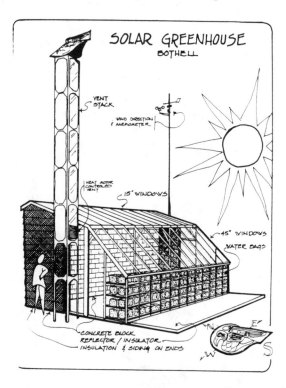

Window Box Solar Collector Design, construction plans with text, large blueprint, 1978, $2.00 from Small Farm Energy Project, Box 763, Hartington, Nebraska 68739, USA.

MATERIALS: wood, insulation, wire mesh (optional), sheet metal, glass or fiberglass sheeting for cover, caulking
PRODUCTION: hand tools

A window box solar collector will heat a room. Air moves by natural circulation, requiring no fan. The six dimensional drawings in this design are somewhat hard to follow, but the design principles are clear. Local materials could be substituted in the design where needed.

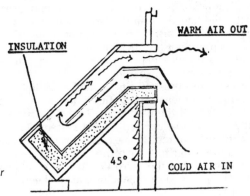

*Window box solar
collector heats room air*

Solar Water Heaters in Nepal, book, 27 pages, Andreas Bachmann, 1977, from SATA, Box 113, Kathmandu, Nepal.

Here is a rare example of a book on solar water heating from a developing country. BYS (Balaju Yantra Shala) Plumbing Division has built systems in Nepal to supply hot water for bathing, washing clothes and cooking. While no detailed drawings are presented, the BYS designs are discussed, component by component. Specifications for the collector and storage tank are given, along with qualitative descriptions of construction and maintenance procedures.

Two systems are described: 1) a thermosiphon (natural circulation) system with separate collector and storage tank, and 2) a "flat tank" collector, where

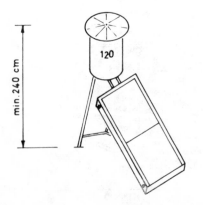

*120 liter solar water
heater in Nepal*

the collector also functions as the storage tank. This is less expensive, but only supplies a small amount of heated water at a time.

This book is intended as a description of Nepal's experiences with solar water heaters, not as a construction manual. It is a good example of a local publication that is increasing the two way flow of information on appropriate technologies between the developing and developed nations.

A Solar Water Heater Workshop Manual, construction manual, 40 pages, 1978, $5.00 plus $0.75 postage from Ecotope Group, 2332 East Madison, Seattle, Washington 98112, USA.

MATERIALS: wood, fiberglass insulation, corrugated galvanized sheet metal, copper tubing, standard pipe fittings

PRODUCTION: soldering equipment, hand tools.

This manual is designed to be used in a teaching situation, with an experienced leader who can provide background knowledge and teach construction techniques. Four pages are devoted to organizing a training workshop.

Ecotope Group and Rain magazine staff have run these workshops in the Northwestern U.S. for several years, usually teaching 30 or more people from a community organization to build a solar water heater in a two-day period. By teaching members of existing groups together, skills are transferred to a naturally supportive network, and more solar water heaters are likely to be eventually constructed. This approach could be used anywhere, with many different technologies.

The manual contains step-by-step instructions, with drawings, for building and installing a solar water heater. This includes siting the system, piping for natural circulation, and various open and closed loop storage alternatives.

Teaching solar water heater construction to a group

Evaluation of Solar Cookers, booklet, 71 pages, by VITA, 1962 (reprinted 1977), $5.95 from VITA.

This 1962 report covers the early solar cooker designs, many of which are still being recommended today. Cost (1962 prices), materials, cooking performance, and problems are presented for each of 12 parabolic reflector and 2 oven designs. The report does not cover any of the post-1962 innovations, some of which are very promising; such an evaluation is needed.

There are some stimulating suggestions that will be of interest to solar cooker experimenters, such as: 1) fixed-in-ground soil-cement reflectors, possible large ones built into the design of a desert home; 2) papier-mache/basketry reflectors made on molds; 3) reflectors made using liquid resin in a revolving pan, catalyzed to harden and retain the parabolic shape; 4) molds made of aluminized plastic balloons. The number of ways of generating a parabolic reflector shape using natural physical properties seems to be very large indeed. (The dangers of parabolic reflectors [burns and eye damage] are not mentioned, however.)

Test methods are described which could be used in evaluating other solar cookers and wood stoves. One test result suggested that the simple shielding of pots from the wind can considerably reduce heat loss for reflector cookers.

There are 27 photos and drawings of ovens and testing equipment.

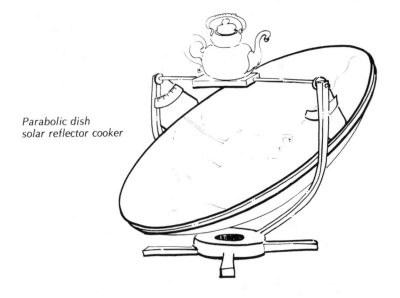

Parabolic dish
solar reflector cooker

The Solar Cookery Book: Everything Under the Sun, book, 122 pages, Beth and Dan Halacy, 1978, $7.95 from Peace Press Inc., 3828 Willat Ave., Culver City, California 90230, USA.

The first third of this book is a construction manual for two types of solar cookers: a solar oven and a solar parabolic reflector hot plate. Most of the rest of the book is devoted to detailed descriptions of solar cooking methods, with many recipes for foods that can be solar cooked. The text is easy to understand.

"Supplies for the solar oven can be bought for as little as 25 (U.S.) dollars. The cardboard and aluminum foil reflector (hot plate) cooker costs less than half that much...both projects can be built by amateurs. Neither of us qualify as experts with tools, but our solar cookers work very well." Materials needed: wood, glass or plastic, aluminum foil, cardboard, some metal and nails. Safety is important when using a parabolic reflector cooker: the focal point can burn or blind a person!

"Don't worry about aiming the solar oven and reflector stove. It's done with shadows and is very easy. In a few days you will be an expert on how fast (or how slow) the sun moves, and how to adjust the oven for best results and desired temperatures." On good days your oven can reach 400°F (205°C).

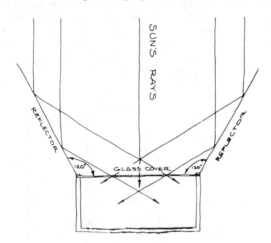

Flat reflectors on a solar oven

A State of the Art Survey of Solar Powered Irrigation Pumps, Solar Cookers, and Wood Burning Stoves for Use in Sub-Saharan Africa, book, 106 pages, by J. Walton, Jr., A. Roy, and S. Bomar, Jr., 1978, limited number of copies available free from Engineering Experiment Station, Georgia Institute of Technology, Atlanta, Georgia 30332, USA.

This is a survey of technologies which might help reduce the serious problems of deforestation and water shortage in the region just south of the Sahara Desert in Africa. Solar-powered irrigation pumps, solar cookers, and wood burning stoves are examined, and research recommendations made. Few construction details are provided.

Unfortunately, what seem to be the most promising technologies for carrying out the functions of lifting water and cooking are not treated in this volume. For example, low-cost, locally made water-pumping windmills (such as can be found Crete and Thailand) appear to be cost competitive in many moderately windy locations and far cheaper than the solar alternative. Improved stoves save more firewood per dollar than solar cookers and do not require imported materials. Highly efficient, low cost designs include the improved chulah stove (India), the FAO-Singer 1961 stove (Indonesia), and particularly the Lorena stove (Guatemala)--yet they are not to be found in this report.

This is a good reference on solar powered irrigation pumps, which, at $25,000 per installed kw of capacity, seem forever doomed to serve as toys of rich country aid programs, and may even divert government funds in a number of countries from more useful pursuits. (By comparison, other renewable energy systems such as micro-hydroelectric turbines cost $600 to $2000 per installed kw of capacity, while locally made water pumping windmills will work at electricity equivalents of roughly $2000 per installed kw of capacity.)

The solar cookers surveyed include some interesting low cost designs, yet the authors note that solar cookers "have never found acceptance at the village level." They tentatively suggest, however, that "lack of acceptance may be due more to insufficient attention to developing an adequate program of village instruction than to technical or social-cultural problems."

The reader is assured of the need for very low cost, locally made technologies, but then shown mainly highly expensive manufactured equipment. Solar pumps are shown to be unrealistically expensive, then described as "accepted as the only solution to providing water by pumping" in Sub-Saharan communities. Contrary to such confusing conclusions, it appears to us that short term new energy resources in the rural Third World can most succesfully be based on proven technologies from the Third World itself.

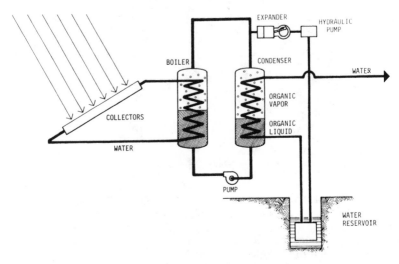

Schematic of a typical SOFRETES solar water pumping system

The Design and Development of a Solar Powered Refrigerator, technical report, 74 pages, R. Exell, S. Kornsakoo, D. Wijeratna, 1976, $8.00 in Thailand, $10.00 in developing countries, $15.00 in developed countries, from Library and Regional Documentation Center, Asian Institute of Technology, P.O. Box 2754, Bangkok, Thailand.

This report describes work on an experimental solar refrigerator designed to be a village-size ice maker or cold storage unit. The experimental version can make 1-2 kg of ice per day in Thailand; larger capacities will be possible in future designs. The cost of the unit is figured to be $750. Ice produced in this unit is calculated to cost about 11 times the wholesale price of ice in Bangkok. This makes it unlikely that such units will be considered "appropriate village technologies" in the near future.

Refrigeration occurs during the night by vaporizing an ammonia solution. During the day, a flat plate collector uses solar energy to pressurize and condense the solution. This type of non-continuous refrigeration, while less efficient, has the advantages of needing no compressor or electricity. Thus it is suited to decentralized applications (operation requires only turning a few valves in the morning and at night. This manual labor substitutes for the electric compressor in other refrigerators).

Most of the report is quite technical. There is also a review of other solar refrigeration work from the last 30 years. Work on these concepts is continuing.

A good example of a decentralized application of sophisticated technology.

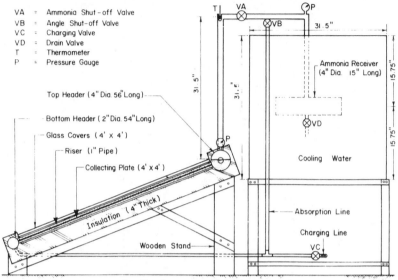

Small solar powered refrigerator

No Heat, No Rent, book, 90 pages, The Energy Task Force, 1977, limited supply of free copies are available, from Community Services Administration, Washington D.C. 20506, USA.

El Movimiento de la Calle Once (Eleventh Street Movement) is a group of tenants who have bought and renovated their old New York city apartment building. Energy conservation measures have been instituted and a domestic solar hot water system and windgenerator installed.

This manual describes the solar heating system, with instructions on how to install and maintain one in similar old urban apartments in the U.S. The authors hope this book can show others involved in tenant cooperatives how to approach self-help design, while recognizing that each housing situation will have different needs: "A careful reading and discussion of this manual can help people make preliminary decisions without the costly advice of an engineer or architect. However, this is a limited tool and licensed technical aid is required for actual design and construction."

Well illustrated chapters cover energy conservation measures such as insulation and storm windows, simple methods of determining if enough sun falls on your roof, determining collector size needed, installation, solar plumbing, and maintenance and repair of solar systems. A glossary and an easy-to-understand economic analysis are included.

An important example of the application of appropriate technologies by a low income group in the United States. The heavy costs involved, mostly borne by government agencies, suggest that in urban high-cost settings it will be very difficult indeed to overcome the problems of housing for the poor, even when using their own labor and appropriate technologies.

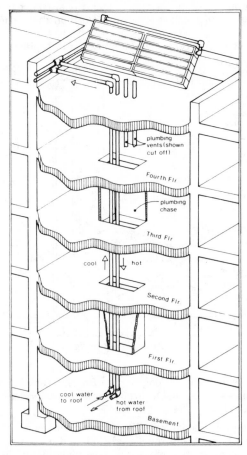

*Eleventh St. Movement
solar water heating system*

Reaching Up, Reaching Out: A Guide to Organizing Local Solar Events, book, 145 pages, 1979, by the Solar Energy Research Institute, $6.00, Stock No. 061-000-00345-2, from Superintendent of Documents, U.S. Government Printing Office, Washington D.C. 20405, USA.

This 'organizing manual' is designed to help groups and individuals organize themselves to achieve awareness of and control over the energy they use. Thirteen events are used as case studies, from solar water heater and solar greenhouse construction workshops to energy fairs and neighborhood energy conservation efforts. Suggestions are made for planning and carrying out these kinds of community events.

Half of the book is a bibliography on small scale solar technologies (all U.S.), general organizing, and a directory of solar groups in the U.S.

The unique value of this book is in its illustration of community action technique. The focus is **entirely** on the U.S., although some of the information might be adaptable to other areas.

Making the collector for a solar water heater

The Solar Survey, booklet, 21 pages, 1979, $1.35 postpaid from Publications Section, National Center for Appropriate Technology (NCAT), Box 3838, Butte, Montana 59701, USA.

A collection of 31 solar designs from various community groups across the US. The designs range from active water and air collectors to passive Trombe wall systems. For each entry, there is a short description and often a drawing (complete designs are **not** given). All were chosen as examples of low-cost, locally-built technologies.

The purpose of the survey is to exchange ideas and designs among community groups in the US who work independently on similar projects. We have included it as a modal for an information exchange approach that could be used in other countries. The specific technologies covered are not very relevant to developing countries.

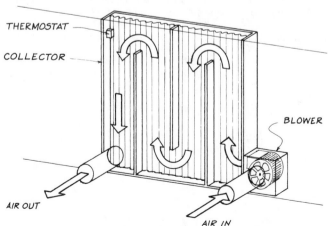

WEST CENTRAL MISSOURI—SYSTEM OPERATION

A Bibliography for the Solar Home Builder, booklet, 38 pages, by Dr. Donald W. Aitken, 1979, $1.00 from California Office of Appropriate Technology, 1530 Tenth Street, Sacramento, California 95814, USA.

"The market is responding to the surging popularity of solar energy with a flood of books and reports...some of these are truly excellent, while others are thinly disguised attempts to sell something...the following bibliography summarizes only the books and reports with which I am personally familiar and that I feel to be the most useful, honest, and worth the cost." This booklet describes 71 publications on solar home design, information for the beginning solar home builder, and advanced solar studies as well as a few general works on solar energy as an alternative for the future. It is especially useful because the annotations are cross-referenced, with notes on which publications contain the most information on particular topics.

Also includes listings of solar energy societies and journals. A good "sourcebook" on solar home building, oriented toward applications for the West Coast of the USA.

Solar Energy Books, bibliography, 118 pages, $4.50 from National Solar Energy Education Campaign, 10762 Tucker St., Beltsville, MD 20705, USA.

An extensive bibliography (almost 500 titles) of practical publications on solar energy. There are sections on solar, wind, and biomass energy, as well as designs for the homeowner, energy conservation, policy and business. Most of the materials covered are relevant primarily in the U.S. and other rich countries.

A short description of each book is provided. All the publications can be ordered from the National Solar Energy Campaign.

ADDITIONAL REFERENCES ON SOLAR ENERGY

Designs for solar dryers can be found on pages 485-488.

Constructing and Operating a Small Solar Heated Lumber Dryer; see review on page 507.

Solar dryers and water heaters are pictured in **Small Scale Renewable Energy Resources in Nepal**, reviewed on page 567.

The Power Guide looks at commercially available solar photovoltaic (electric) cells, water heaters, space heaters, and engines; reviewed on page 568.

Solar pumping, electricity, and distillation are discussed in **Appropriate Technologies for Semi-Arid Areas: Wind and Solar Energy for Water Supply**, reviewed on page 566.

ENERGY: BIOGAS

Energy: Biogas

The term "biogas" is now used throughout the world rather than "methane gas" to describe the fuel produced through anaerobic fermentation of manures and vegetable matter, in devices called digesters. Biogas is generally between 40 and 70 percent methane (CH_4), with the remainder consisting of carbon dioxide, hydrogen sulfide (H_2S) and other trace gases.

While the prospect of generating fuel and fertilizer from organic wastes is an attractive one, significant problems and debate persist about the value of biogas in addressing the energy needs of poor villages in the Third World.

"Biogas technology represents one of a number of village-scale technologies that are currently enjoying a certain vogue among governments and aid agencies and that offer the technical possibility of more decentralized approaches to development. However, the technical and economic evaluation of these technologies has often been rudimentary. Therefore, there is a real danger that attempts are being made at wide-scale introduction of these techniques in the rural areas of the Third World before it is known whether they are in any sense appropriate to the problems of rural peoples."

—Biogas Technology in the Third World, IDRC

Some observers* conclude that the lifetime social and economic benefits of the heavily subsidized Indian family-scale biogas plants do not equal the costs of construction and maintenance. In Pakistan, only moderately prosperous farmers with adequate numbers of animals and significant amounts of capital have been able to afford to build biogas plants. Although the information on community-scale biogas plants is still very scanty, some results in Indian villages are not very promising. It appears that in terms of fuel and fertilizer, biogas may well be a poor proposition without good management, optimal resources, and a suitable social environment. In most villages it may be advisable to invest first in improved wood stoves and village woodlots rather than biogas systems. However, side benefits such as improved village health and increased productivity in associated enterprises (fish farming, livestock, agriculture, etc.) may tip the scales in favor of a biogas project. For example, just one small digester at a rural health clinic can power a refrigerator holding vaccines for thousands of people.

*At the time of the printing of the first volume of the **Appropriate Technology Sourcebook** four years ago, we reported that slow progress was being made in promoting biogas beyond the experimental stage. Since then, there have been spectacular successes reported in the People's Republic of China. Up to 7 million*

---

*See, for example, "The Economics of Renewable Energy Systems", by David French, reviewed on page 565.

family and community-scale biogas plants are reported to be currently in operation there. Many people have talked about or actively tried to duplicate the Chinese successes in their own countries, and a number of new publications have arisen to report these trends and developments around the world.

The Chinese biogas phenomenon is evidence of the PRC's extraordinary level of social and economic organization. In the highly developed Chinese extension system, competent and motivated cadre can be mobilized to promote technical innovations and maintenance of the biogas plants. The large number of pigs and the relatively even distribution of resources are significant factors as well, and manure handling has much higher acceptance than in most other developing nations. It appears that the Chinese designs are resource-conserving, compact, and adaptable to whatever building materials are locally available. Bricks and stones are used with locally produced relatively low cost cement, and in some areas digesters are even carved out of solid rock. Of particular interest are the built-in self-pressurizing mechanisms in the Chinese designs which eliminate the need for costly metal covers.

Recently some observers have questioned the applicability of the Chinese biogas experience. Attempts to replicate the Chinese results outside the PRC have yielded very uneven results. Building materials, such as cement, lime and quarried stones which are produced locally on Chinese communes are unavailable or very expensive in many other countries. Also, the Chinese skill and diligence in construction (particularly for the vaulted dome designs) and maintenance may be difficult to find or develop elsewhere. One observer notes that the Chinese digesters are very similar to septic tanks, and that their gas yields per unit volume may be only a fraction of large scale sewage digesters— meaning the gas production may be significantly lower than commonly assumed. It should also be remembered that virtually all reports on the Chinese successes have come from the Chinese themselves, so that data on construction costs and gas yields need further confirmation.

Until recently, no clear and concise technical reports on the Chinese biogas technology were available outside China. The International Development Research Center (IDRC) and the Intermediate Technology Development Group (ITDG) have produced two fine translations of Chinese biogas manuals: **Compost, Fertilizer and Biogas Production from Human and Farm Wastes in the PRC** (IDRC), and **A Chinese Biogas Manual** (ITDG). The former book covers health and sanitation aspects of biogas fully, while the latter presents more comprehensive information on building materials and construction techniques. IDRC's **Biogas Technology in the Third World: A Multidisciplinary Review** is an excellent review of the social, economic and technical aspects of this technology and the problems encountered in attempting to spread it outside of China. The authors of that publication conclude:

"The viability of a particular biogas plant design depends on the particular environment in which it operates. Therefore, the research problem becomes one of providing a structure in which technologists, economists, and users of the technology can combine to produce both the appropriate hardware for various situations and the infrastructure that is necessary to ensure that the hardware is widely used."

Other Asian experiences, from Nepal, Pakistan, and the Economic and Social Commission for Asia and the Pacific (ESCAP) are also featured in the entries in this chapter. To our knowledge, widespread applications or experiments in the developing world have been concentrated in Asia. Interest and activities in other *parts of the world have lagged behind to date.*

It is hoped that a more realistic and less crusading attitude towards biogas has evolved among appropriate technology advocates. What are needed are carefully planned, well-monitored, locally-built biogas plants that can provide the solid information needed to judge the true potential of this technology option.

In Volume I:

Methane: Proceedings of a One-Day Seminar, by Leo Pyle and Peter Fraenkel, 1975, page 183.

Methane: Planning a Digester, by Peter-John Meynell, 1976, page 184.

Bio-Gas Plant: Generating Methane From Organic Wastes, and Designs with Specifications, by Ram Bux Singh, page 184.

Small-Scale Bio-Gas Plant in India, 1976, page 185.

Gobar Gas Scheme, Khadi and Village Industries Commission, 1975, page 186.

Methane Digesters for Fuel Gas and Fertilizer, by R. Merrill and L. John Fry, 1973, page 186.

Practical Building of Methane Power Plants for Rural Energy Independence, by L. John Fry, 1974, page 187.

Process Feasibility Study: The Anaerobic Digestion of Dairy Cow Manure at the State Reformatory Honor Farm in Monroe, Washington, by ECOTOPE Group, 1976, page 188.

The Anaerobic Digestion of Livestock Wastes to Produce Methane: 1946 to June 1975, A Bibliography with Abstracts, by G. Shadduck and J. Moore, 1975, page 188.

Methane Generation by Anaerobic Fermentation: An Annotated Bibliography, by Freeman and Pyle, 1977, page 188.

Compost, Fertilizer, and Biogas Production from Human and Farm Wastes in the People's Republic of China, book, 93 pages, Michael McGarry and Jill Stainforth, editors, free to local groups in developing countries, $6.00 to others, from IDRC.

"This collection of papers describes the design, construction, maintenance, and operation of Chinese technologies that enable the Chinese to treat human excreta, livestock manure, and farm wastes to produce liquid fertilizer, compost, and methane gas."

From a mere handful of experimental "marsh gas pits" during the Great Leap Forward in 1957, methane plants have proliferated to number 4 million at last report (December 1977). Instead of the often disappointing results reported about methane digesters elsewhere in developing countries, China's experience can only be described as phenomenally successful. It appears that the designs and their implementation were not the result of expert or entrepreneurial intervention, but of mass innovation and spontaneous communal action throughout China.

The lessons represented by this book are profound. The Chinese biogas technology appears promising for other developing countries, but its transfer to other places is probably unlikely without commitment on the part of the people or the support of the government as in the People's Republic of China.

"Since 1964 we have standardized the management and hygienic disposal of excreta and urine, expanded the sources and raised the efficiency of fertilizer, and collected and created a high-quality fertilizer by destroying the bacteria and parasitic eggs that existed in the human and domestic animal excreta and urine. As well, we lowered the morbidity of enteric pathogens, reduced the breeding areas of flies and mosquitoes, improved environmental health, promoted and increased food production, and increased the health standards of all the committee members. Between 1963 and 1971 food production per acre increased by 74%, enteric pathogen morbidity decreased by 80%, and the morbidity of pigs' disease dropped from 5 to 0.3%. Basically, the health profile of the villages was transformed."

The more recent book **A Chinese Biogas Manual** (see review in this section) provides more complete information on biogas plant construction and operation. Both books are recommended.

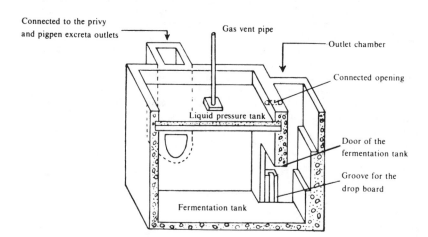

Enclosed three-stage biogas plant

Gobar Gas: An Alternate Way of Handling the Village Fuel Problem, pamphlet, 16 pages, available free to serious groups in developing countries, from Appropriate Technology Development Organization, P.O. Box 1306, Islamabad, Pakistan.

"In China, a simple (biogas) plant has been developed which has gone a long way in relieving Chinese rural areas from fuel shortage. The plant developed by the Chinese is extremely simple, being a brick structure with a dome type roof. This has an added advantage on Indian type plants where a steel gas holder is the common method. The Chinese have been kind to us in sharing their technical know-how and having provided Pakistan with a complete set of working drawings."

In line with the Chinese philosophy of utilizing local material, skills, and labor, the Appropriate Technology Development Organization (ATDO) of Pakistan has built modified versions of the Chinese biogas plant in several locations. These experiments may prove to be very valuable in establishing whether or not a proven technology, like the Chinese biogas phenomenon, can also flourish under very different cultural, economic, and political conditions. Valuable lessons can be learned from these attempts, and we recommend that interested groups and individuals establish contact with Pakistan's ATDO and follow the progress of their work.

The Compleat Biogas Handbook, book, 403 pages, by D. House, 1978, $8.00 ($9.00 outside the USA), from At Home Everywhere, c/o VAHID, Rt. 2, Box 259, Aurora, Oregon 97002, USA.

"This book makes no claim to startling originality or clever breakthroughs. Its usefulnesss comes mainly because here, gathered together in one place, is a great deal of information on biogas generation; what it is, where it comes from and how to make and use it. There are, however, only a few designs for biogas generators given in detail." Readers wanting specific digester designs should look elsewhere.

The author attempts to help the reader understand the complexities of biogas generation. The information covers nearly all problem areas, including safety features, compression ratios for engines, and sizing of effluent algae ponds, in a detailed fashion. There are numerous charts, graphs, and equations to explain the chemical, biological and engineering aspects of biogas generation. The language varies from moderately technical to philosophical. Illustrations are crude, but helpful in understanding the text.

The oildrum digester designs presented in this book are of limited value due to their small size and costly corrosion problems. The rest of the book, however, would be valuable for trained village technology engineers and extension agents in developing nations, or biogas enthusiasts anywhere. A good knowledge of English is required to use this book.

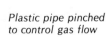

Plastic pipe pinched
to control gas flow

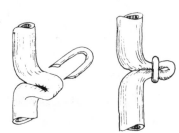

Biogas Technology in the Third World: A Multidisciplinary Review, book, 132 pages, by Andrew Barnett, Leo Pyle, and S.K. Subramanian, 1978, $10.00 (free to serious groups in developing countries), from IDRC.

"In response to the interest in biogas and other rural energy systems shown by a number of Asian researchers, the International Development Research Center (IDRC) commissioned this state-of-the-art review so that it might form a basis of further discussions concerning the direction of future biogas research. This book represents a multidisciplinary approach to the problem and attempts to review existing work rather than to champion particular solutions."

"Our objective is to stress the need to examine a wider range of technical and economic alternatives for meeting the energy and fertilizer needs of rural peoples. It is our hope that this survey contributes to this process by showing what has already been done, by pointing out pitfalls, and by indicating the major gaps that still remain."

The three chapters contain: 1) a broad overview of the energy options facing rural communities in the Third World, detailing what is already known about the technical aspects of biogas production; 2) an approach to social and economic appraisal of rural technologies, particularly successful biogas applications; and 3) a field survey of existing biogas systems and their supporting infrastructure in Asia. The authors are looking for the best uses of the waste material, including options other than biogas production. Estimated gas yields from various crop residues and animal manures are listed. Costs and performance of different digester designs are compared. The Chinese experience is not covered in great depth, due to the lack of information available at the time of publication.

This book can be read by non-technical people, and it deserves wide circulation among development planners, students, and technicians. A strong English vocabulary is required. Not a how-to-build-it book, this is nevertheless extremely valuable to those designing, experimenting, and operating digester schemes. Essential reading.

A Chinese Biogas Manual, book, 135 pages, by the Office of the Leading Group for the Propagation of Marshgas, Szechuan Province, English translation published 1979, Ł3.95 from ITDG.

Some of the most promising developments in biogas technology have occurred in the rural areas of China, where millions of small digesters have been built during the past decade. This construction manual has been used widely since its original Chinese language publication in 1974. It shows how to plan, build, and care for low-cost pit-type digesters. Drawings and text explain the comparative design advantages and construction details of circular pits,

A Chinese biogas plant: bricking a dome without support

rectangular pits and domed covers. Different combinations of stone, lime bricks, traditional cements and mortars, and commercial concrete are also discussed. Simple instructions include notes on why certain designs are suited to certain conditions: "A circular pit made from soft triple concrete with a large volume and a small opening is easy to seal and suitable for the areas where the earth is firm, the underground water level is low, and there is no water seepage. (It is) also quite suitable for plateau regions."

The manual also emphasizes the importance of careful prevention of leaks when the finished pit is filled and pressurized. A chapter on using biogas shows how to make burners for cooking and lighting, out of renewable and recycled materials such as bamboo, iron tubing, and discarded showerheads. An appendix gives an example of how this book has been used by the Shachio Commune of Guangdong Province to spread biogas technology.

A good technical reference, this construction manual is also an example of a tool for sharing skills and experience among rural communities.

Nepal Bio-Gas Newletter, quarterly newsletter, 12 pages, available free to serious groups in developing nations, from Bio-Gas Committee, Energy Research and Development Group, Tribhuvan University, Kathmandu, Nepal.

Reports on biogas technology developments in Nepal and other Asian countries. Issue No. 2 reports on technical, social and economic problems of this technology in Nepal, and includes some articles on locally appropriate burners, lamps, and piping mechanisms. Also shown are some valuable details from the Chinese drumless biogas plants, such as an inexpensive device for pressure regulation.

This newletter is a good model of a low cost problem solving and information sharing channel. Those wishing to obtain this publication should plan to contribute or exchange something for it.

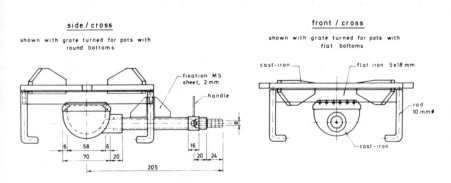

Burner for use with biogas

Methane Generation from Human, Animal, and Agricultural Wastes, book, 126 pages, National Academy of Sciences, 1977, free from Commission on International Relations, JH 215, National Academy of Sciences, 2101 Constitution Avenue, N.W., Washington D.C. 20418, USA.

This book is not for village workers seeking practical instruction. Its highly technical treatment of methane generation and strong scientific flavor is more suited to restricted use in universities or among professional circles. Produced prior to the availability of information from China, this book's sections on developing countries are much poorer than the IDRC's **Compost, Fertilizer, and Biogas Production from Human and Farm Wastes** and **Biogas Technology in the Third World** (see reviews in this section).

ADDITIONAL REFERENCE ON BIOGAS
The Economics of Renewable Energy Systems for Developing Countries examines a hypothetical village biogas plant in India, and concludes that it is not economically viable; see review on page 565.

HOUSING
AND CONSTRUCTION

Housing and Construction

There are housing problems everywhere, in industrialized as well as developing countries. In Jakarta, Manila, Mexico City, and Calcutta millions of squatters camp indefinitely in structures made of cardboard, sheet plastic and flattened cans, on strips of land beside canals and railways, sometimes even in the shadows of high-rise "low-cost" housing. In prosperous Santa Clara County in California, where the median family income is nearly $25,000 per year, the cheapest homes cost more than $100,000, putting them beyond the means of that average family. In these and countless other urbanizing areas, the cost of a place to live is rapidly outstripping the ability of ordinary people to pay. Costs of energy-intensive manufactured building materials, which inevitably rise faster than the other costs of living, are one factor. Inflation of land values triggered by the pathological growth of gigantic urban centers is another. In a vicious circle of unfortunate land use choices, allocation of tracts to expanding industry separates people from their places of work and drives up the cost of remaining land. Sprawl of suburbs and fringe communities eats up yet more space, taking prime agricultural land out of production.

In developing countries, the amount of attention and resources that public works administrations and development assistance agencies devote to housing is probably second only to that devoted to water supply. And the history of housing projects, like that of water supply projects, is largely a history of disappointments worldwide. In **Housing by People** (this section) John Turner notes that

"...it is common for public agencies to build houses or flats to standards which the majority cannot afford, nor can the country possibly subsidize them on a large scale. On top of this, it is not unusual for governments to prohibit private building of the type of housing the vast majority can afford and are satisfied with."

Turner argues that governments should not provide houses built to arbitrary specifications, but should instead reform building codes and provide opportunities for secure access to land. An appropriate housing strategy would rely on a community's initiative, thrift, and ability to organize and turn local resources to advantage to meet the basic human need of shelter. **House Form and Culture, Shelter II, Self-Help Practices in Housing,** and **Shelter After Disaster** (all in this section) provide illustrations and documentation of the power and validity of this approach. Readers interested in locally-based housing strategies for the Third World should also refer to **Architecture for the Poor** (reviewed on pages 217-218).

Most of the remaining entries in this chapter are about techniques for building houses. Wood-framed and stone structures (**The Timber Framing Book** and **The Owner-Builder's Guide to Stone Masonry**) can be built by their owners and are relatively low cost in many areas. Because of its low cost, earth is an important building material, providing housing for what is estimated to be the majority of the world's population. The thermal properties of earth also make it well-matched to passive solar design requirements in many climates. Several books in this section discuss construction of houses made with monolithic earth walls, soil-cement bricks, and adobe bricks. There are two sets of plans for the construction of hand-operated presses that can be used to make soil-cement blocks. Two more entries cover the principles of des.gn of underground buildings; one of these presents an owner-builder approach.

Bamboo has a long history as a flexible, safe, low-cost building material, and is plentiful today in many parts of the tropics and subtropics. Three entries cover some recent innovations and results of research in structural uses for bamboo, as well as reinforcing applications in plaster, cement, and stucco roofs and walls.

Ferrocement—a strong, thin sheet of cement reinforced with wire mesh—is a more recent technology with potential applications wherever durable water-proof walls, roofs, or hulls are required. In addition to the research described in the entries in this section, interested readers can refer to the **'Thailo' Ferrocement Rice Bin** reviewed on page 122.

In Volume I:

Housing by People: Towards Autonomy in Building Environments, book, 169 pages, by John Turner, 1976, $4.00 from META.

With many years of experience working in low cost housing projects in developing as well as developed countries, Turner has written a penetrating analysis of the housing "problem", with broad implications for other kinds of appropriate technology work.

Turner proposes "a radical change of relations between people and government in which government ceases to persist in doing what it does badly or uneconomically—building and managing houses—and concentrates on what it has the authority to do: to ensure equitable access to resources which local communities and people cannot provide for themselves." Throughout the world, so called "low cost housing" projects have repeated the same mistakes by setting a material standard (including building codes) ill-suited and far too expensive for the poor majority. Backed by many case studies, Turner argues that within the constraints of poverty, the poor succeed rather well in providing for their housing needs when they have land tenure and access to materials. "The economy of housing is a matter of personal and local resourcefulness rather than centrally controlled, industrial productivity."

"Personal and local resources are imagination, initiative, commitment and responsibility, skill and muscle-power; the capability for using specific and often irregular areas of land or locally available materials and tools; the ability to organize enterprises and local institutions; constructive competitiveness and the capacity to co-operate." The existence and vitality of "dense local communication and supply networks open to local residents" appears to be a key factor in the "material savings and human benefits of owner-building, rehabilitation, and improvement in the United States."

Government activities in housing often prevent or hamper the use of these resources and networks. Improved income opportunities, guaranteed land tenure, and building codes based on broad function rather than specific requirements would have more effect on housing for the poor than most direct housing projects.

In the Third World subsidized housing has proved a failure, for it usually is occupied by the relatively well-off and the ultimate cost of subsidized housing for all those who need it are far beyond the capability of governments to provide. Whereas, "By far the greatest financial resources are the actual savings of the population from their own earnings, and these are under their direct control. This probably represents between 10 and 15 percent of all personal incomes. It is roughly equivalent to all taxes obtained from incomes and retail sales in an economy such as that of Mexico.

Two non-monetary factors that play a very important role in housing for the urban poor are accessibility (to jobs) and security (of ownership, including the ability to sell so as to recoup the costs of improvements made). By concentrating solely on physical standards for dwellings, without reference to such factors, authorities cannot understand the decision-making context faced by the poor.

The author concludes with "an argument for the redefinition of housing problems as functions of mismatches between people's socio-economic and cultural situations and their housing processes and products; and as functions of the waste, misuse, or non-use of resources available for housing."

House Form and Culture, book, 135 pages, by Amos Rapoport, 1969, $4.95 from Prentice-Hall, Inc., Englewood Cliffs, New Jersey 07632, USA.

A thoughtful look at the way that cultural factors have influenced the form of houses. Although this is not intended to be a book on the practical side of building design, it is full of interesting examples of the ways different peoples have solved a wide variety of problems. Includes about 100 drawings.

"Construction and materials are best regarded as modifying factors...they do not determine form. They merely make possible forms which have been selected on other grounds...Given a certain climate, the availability of certain materials, and the constraints and capabilities of a given level of technology,

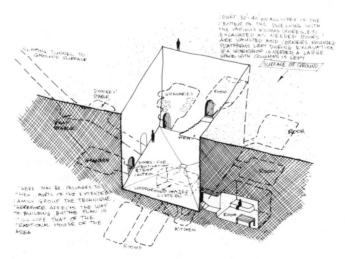

Cutaway view of Matmata dwelling, Sahara

what finally decides the form of a dwelling, and moulds the spaces and their relationships, is the vision that people have of the ideal life."

Recommended reading for those involved in low-cost housing. This book will hopefully dispel any lingering ideas that standardized box-shaped houses built of industrial materials should be imposed upon any people. Rather, if a new technology of construction is advantageous due to lower costs or other reasons, it is best left to the people themselves to apply it in ways that they choose. If housing forms are not allowed to be culturally determined, they will be culturally destructive.

Shelter II, large paperback book, 224 pages, by Lloyd Kahn, 1978, $10.00 postpaid from Shelter Publications, P.O. Box 279, Bolinas, California 94924, USA.

Shelter, the forerunner of this volume, is a handsome, provocative "scrapbook" of building ideas from around the world (see review on pages 219). Citing increasing housing costs everywhere, the editor notes: "The principles outlined in **Shelter** seem even more important today: relearning the still-usable skills of the past, finding a balance between what we can produce for ourselves and what we must buy, and doing more hand work in providing life's necessities."

Hundreds of photographs and drawings in **Shelter II** give an inspirational view of indigenous building styles and techniques, from Nebraskan sod houses to the thatched stick-dwellings of nomadic Kenyan Shepherds. The author discusses the ways structure has related to culture, physical environment, and basic shelter requirements. The book emphasizes innovation and diversity among human dwellings, but the appropriateness of traditional building technologies is also a unifying theme. "Practical builders, wherever they live, work with simple techniques and what is most readily at hand: earth, thatch, stone milled lumber or abandoned city buildings. Weather, purpose, materials govern design. Tradition, experience, practice determine building technique."

Turkmen (Northern Afghanistan)   Kazakh

Some of the most practical shelter alternatives for North America—stud-frame and adobe construction—are explored. An introduction to design of small single-family houses is followed by a guide for pouring the foundations, framing, and roofing a stud-construction house. Also included are sections on interiors, bungalows, yurts, cabins, and dismantling buildings for scrap. The book concludes with pictorial case studies of homeowner rehabilitation in Massachusetts and cooperative homesteading in gutted buildings in New York.

A fascinating book with a broad range of design ideas and useful information. Certain to fire the imagination of all kinds of owner-builders.

Self-Help Practices in Housing: Selected Case Studies, book, 129 pages, 1973, UN Department of Economic and Social Affairs, $5.00 (Sales No. 73.IV.15), from Sales Section, Room A-3315, United Nations, New York, NY 10017, USA.

Adequate housing for the growing poor communites in urban areas of the Third World is the subject of this report. Case studies of selp-help housing projects are included from cities in Colombia, El Salvador, Senegal, Ethiopia, and the Sudan. These projects were undertaken by local and national government agencies, sometimes in cooperation with private organizations. Timetables and size of project varied considerably. The Senegal project involved 90 white-collar and other middle-class family heads in training and construction over a period of four years. The dismantling and rebuilding of houses for 1000 families on new land in Port Sudan was accomplished within one month.

The studies show how certain key factors affect the outcome of any self-help housing scheme. An accurate assessment of loan repayment capability is one important consideration. Continued access to jobs, distribution of manual skills in construction work groups, kind of supervision, and timing of construction work periods are equally important factors that affect the success or failure of a project. The studies also show that communal activities essential to the success of the new neighborhoods, such as maintenance of waste disposal systems, depend on involvement of local leaders and groups from the earliest planning stages. An important reason for the success of the Port Sudan project, for example, was the fact that many people "had already organized themselves into teams which worked in the docks according to the arrival and departure of ships. In this way, the whole team would be free from work two or three times a week" and available to dismantle shacks and rebuild houses together. Other projects, which strictly screened participants according to need and eventual ability to repay, sacrificed these reservoirs of self-help potential by breaking them up in the selection process.

The sometimes difficult language used in this report may present problems for readers with limited English ability.

Shelter After Disaster, book, 127 pages, by Ian Davis, 1978, Ŀ4.75 from Oxford Polytechnic Press, Headington, Oxford OX3 OBP, England.

This fascinating book points out many of the myths about disaster relief that continue to shape aid responses around the world. The author presents the elements of successful shelter rebuilding programs in the light of historical experience over the past 300 years.

Worldwide the frequency and death tolls of disasters are rising, reflecting the increasing vulnerability of the poor primarily in the rapidly growing urban centers of the Third World. This is mostly because they are living in precarious circumstances on hillsides and waterfronts, where damage is likely to be greatest. The author notes that while there is an enormous quantity of post-disaster relief shelter design ideas, most of them are conceived without an understanding of the realities of post-disaster shelter needs. "The vast majority of these concepts mercifully have never left the drawing board or filing cabinet, but this seems no deterrent to the ingenuity and persistence of designers."

Following disasters around the world, local people using their own ingenuity and initiative have accomplished more than 80% of the reconstruction themselves, even in this age of rapid transport and communications. This matches the normal circumstances of the world's poor, where 'development projects' are but a tiny part of local activity. The challenge to national and international agencies is thus quite similar in both cases: to make a genuine contribution by doing something that strengthens and extends what the people are going to do anyway on their own.

"Housing using low technology is more likely to come within the price range of disaster victims, it is probably better suited to local cultural patterns and climate, and it will probably generate local employment." Rubble from collapsed homes should not be cleared, except from roadways, as it is a primary source of building materials. Rebuilding begins almost immediately, and officially provided shelter (particularly oddly shaped houses) will be the least appealing to the people.

Although there are many examples of indigenous housing well-suited to resist the effects of typhoons and earthquakes, for example, these appear to have evolved over an extended time period. Rebuilding following a disaster is usually done in response to everyday needs—not the possibility of a repeat of the disaster in the far distant future. One of the most interesting housing projects in Guatemala is a retraining program that promotes "earthquake-proof construction techniques that use traditional materials and existing (though developed) construction skills. The result is that the traditional character of the houses is retained while the structure is made safe."

An important book, with implications for appropriate technology efforts. Well illustrated.

Shelter After Disaster:
A page from a comic book in use in Guatemala providing guidance on the correct siting of houses

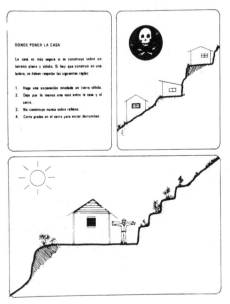

Design for Climate: Guidelines for the Design of Low-Cost Houses for the Climates of Kenya, book, 135 pages, by Charles Hooper, 1975, approximately US$6.00, from Housing Research and Development Unit, University of Nairobi, P.O. Box 30197, Nairobi, Kenya.

"This manual is directed towards house designers and those concerned with the climatic aspect of building design...Section 1 explains the choice of climatic zones, describes the factors affecting human comfort, and considers the presentation of climatic data. Section 2 investigates the impact that the principal climatic elements, namely temperature, solar radiation, humidity, rainfall, and wind, have on building design. Sections 3 to 8 contain the climatic design guidelines for Kenya's six climatic zones."

This is actually a manual on passive solar design. We include it here because, unlike other passive solar references, this one is relevant to housing design and policy in Africa. The analysis of design alternatives takes local customs, materials, and skills into account. "A well-developed, climatically appropriate, system of construction that fully utilizes local materials and skills, exists at the coast. Where regularly maintained this construction system has proven long-lasting and should not be scorned for use in publicly supported housing projects. The walls which consist of mangrove poles, mud, small stones, plaster and whitewash are relatively thin and have a reflective external surface. The makuti (palm thatch) roofs are thermally ideal"; they insulate well and have a low heat storage capacity.

Too many other plans for low-cost housing in the Third World have called for countrywide construction of one particular house design, without regard for variations in climate, culture, and other factors.

Guidelines for each of Kenya's six climatic zones are divided into subtopics: human comfort, site planning, house plan, structure and materials, and openings. Tables indicate month-by-month variation of wind speeds and directions, rainfall, sunshine, and temperature.

The Timber Framing Book, 178 pages, by Stewart Elliott and Eugene Wallas, 1977, $10.00 from META.

Timber-framing is a method of housing construction using interlocking notches and grooves combined with wooden pegs to connect the major wooden beams. The method is labor intensive and requires tools not common to modern Western carpentry but still common and inexpensive in many Third

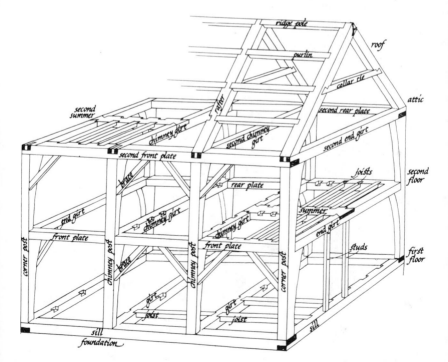

World countries: the adze, auger, draw knife, chisel, and axe.

"Timber-frame house built in Europe as early as the fourteenth century stand proud and sturdy to this day. Compared to conventional construction, timber-frame structures can be 20 to 30 percent less expensive to build. Less energy is expended in both the milling and the construction of the frame...If you have some basic carpentry skills you and some helpers can frame a house using the information in this book. If you do not, and have a carpenter in mind who has not previously used timber framing, he can use this book to teach himself how."

Where wood is plentiful the timber-framing methods of housing construction can be used to build houses of great durability (but you will have to look elsewhere for preservation information). Where wood is scarce, soil-cement, adobe and other wood-conserving construction materials would be more appropriate.

There are illustrations and pictures on almost every page, and a thorough glossary.

Building to Resist the Effect of Wind: Volume 3, A Guide for Improved Masonry and Timber Connections in Buildings, booklet, 48 pages, by S. Fattal, G. Sherwood, and T. Wilkinson, 1977, $2.00, stock number 003-003-01719-1, from Superintendent of Documents, U.S. Government Printing Office, Washington D.C. 20402, USA.

This report discusses the use of connectors in houses and other low-rise buildings to improve their strength under extreme wind conditions. Well illustrated and clearly presented. One half of the report is devoted to detailed discussion of connectors in masonry wall construction. The other half illustrates fasteners used in timber wall construction.

Many of the solutions shown involve the use of manufactured metal parts (such as truss plates and sheet metal fasteners for timber construction or tiebars for masonry construction). These examples may provide ideas and patterns so that locally-produced fasteners could be used to strengthen buildings.

The report is particularly useful in that it identifies the parts of masonry and timber houses in need of greater strength in high wind areas. It is particularly in the rapidly growing urban slums, where people live in makeshift housing often in precarious locations, that damage from hurricanes and typhoons is increasing. This is of great concern in the Philippines and the Caribbean nations. The techniques described in this report could be part of a low cost strategy to minimize that damage.

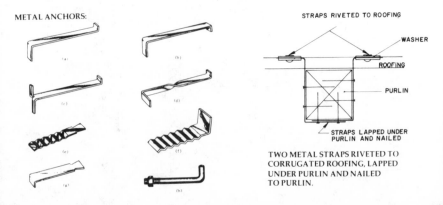

METAL ANCHORS:

STRAPS RIVETED TO ROOFING

WASHER

ROOFING

PURLIN

STRAPS LAPPED UNDER PURLIN AND NAILED

TWO METAL STRAPS RIVETED TO CORRUGATED ROOFING, LAPPED UNDER PURLIN AND NAILED TO PURLIN.

The Owner-Builder's Guide to Stone Masonry, book, 192 pages, by Ken Kern, Steve Magers, and Lou Penfield, 1976, $6.00 (two or more copies $5.00 each to readers in developing countries), from Owner-Builder Publications, P.O. Box 817, North Fork, California 93643, USA.

"The purpose of this book is threefold: 1) We show the inexperienced builder how to 'lay up' stone for various walls, how to 'face' building framework and how to 'cast' stone in a wall with a movable form; 2) ...(we) acquaint readers with the native properties and the availability of useable building stone. Next to earth, there is no more universal nor less appreciated building resource than stone; 3) ...(we) express the aesthetic satisfaction we three authors have experienced building with stone."

The authors carry out the purposes well: teaching the basics of building with stone. The only significant lack of information is on the coverage of earthquake problems. The book covers building with or without concrete, and there is a glossary of masonry technical terms.

"When you trowel mortar use only as much as necessary to provide the bed with sufficient covering. Too much mortar will only squish out and cover the stone face. Do not trowel smooth the mortar; let the stone mash it down. In this way gaps will more certainly be filled. Once a stone is in place try not to move it. Any movement will weaken the bond between stone and mortar."

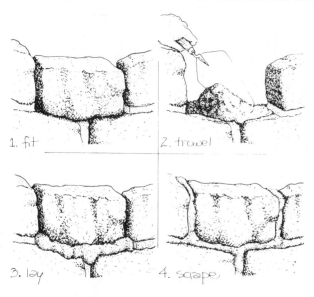

1. fit
2. trowel
3. lay
4. scrape

Fireplaces, book, 192 pages, by Ken Kern and Steve Magers, 1978, $7.95 (two or more copies $5.00 each to readers in developing countries), from Owner-Builder Publications, Box 817, North Fork, California 93643, USA.

"The traditional fireplace not only sends some 80 percent of the fire's heat up the chimney but a goodly portion of the room's heat as well." This is a practical book on good fireplace design to overcome this basic disadvantage of fireplaces in home heating. Step by step construction techniques are presented. The authors discuss the qualities of different building materials: "Most stone cannot withstand intense heat; in a firebox it soon fragments...

due to rapid surface expansion.''

These specific skills and materials will be of most interest to readers in rich countries with cold climates. However, the general theory and principles presented are relevant in any setting. Another high quality Owner Builder book.

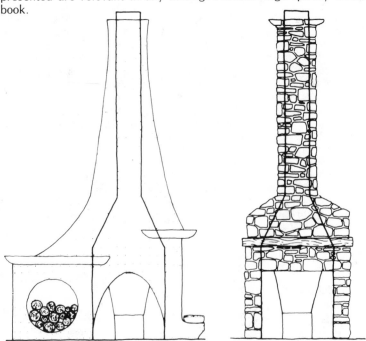

Construire en Terre, 265 pages, by P. Doat et al, 1979, 60F from Groupe de Recherche sur les Techniques Rurales (GRET), 34, rue Dumont d'Urville, 75116-Paris, France.

Produced by a group of architects, this French language manual is one of the best books available on earth building construction. Rammed earth (pise),

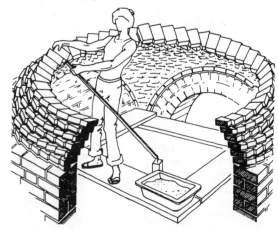

*Mud brick
dome construction*

adobe, compressed blocks, soil analysis, soil stabilization, and earth roofs are the major topics.

This exceptional book contains hundreds of drawings and photos documenting a wide range of indigenous earth construction techniques from Sub-Saharan Africa, the Middle East, China, Latin America, North Africa and elsewhere. Use of local materials and owner or community labor in house construction has obvious advantages in the Third World; in fact more than half the world's population is estimated to live in earth buildings. This volume may contribute to a cross fertilization of ideas and thus better exploitation of the possibilities offered by earth construction in the Third World.

Build Your House of Earth: A Manual of Earth Wall Construction, book, 130 pages, by G.F. Middleton, 1953, revised by Bob Young, 1975, $9.00 from Second Back Row Press, P.O. Box 197, North Sydney, NSW Australia 2060.

"The establishment of a well-tried technique for the identification of suitable earths, and a standard of practice for the methods of construction, should place earth wall construction, which has so much to commend it, high among the accepted building methods. Earth walls have adequate strength and durability to be practicable for building homes and other structures of any size. They can be attractive, hygienic, fire-resistant, dry and soundproof, and can provide good insulation against heat and cold."

Written from experience in Australia, the Middle East, Southeast Asia and elsewhere, this is an excellent introduction to the methods of earth wall construction. The author presents the techniques of pise (rammed earth) construction. This is both an idea book (with many photos, plans, and sketches fo buildings and tools) and a well-organized step-by-step guide to building an earth wall dwelling. Site selection, design, soil requirements, estimation of cost and volume of materials, and building codes are also discussed. The author pays particular attention to the tools and forms needed for earth wall construction; many of these are to be made by the builder.

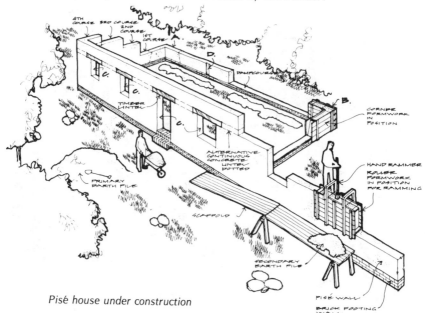

Pisé house under construction

Manual for Building a Rammed Earth Wall, large illustrated booklet, 28 pages plus appendices, $5.00 from Rammed Earth Homes, Lydia A. and David J. Miller, 2319 21st Avenue, Greeley, Colorado 80631, USA.

A concise book providing complete instructions for laying foundations, building and assembling form panels, and the earth tamping (ramming) process itself. Although the authors describe walls built for large homes in the U.S., the method has applications in many areas. Soil testing and stabilizing, making test blocks, and tamping tools are also covered.

"It is our experience that no concrete cap is needed on the wall. We recommend that you insert an eye bolt 12 inches long with a 12-inch piece of reinforcing rod through the eye of the bolt...We have not provided the specifications and plans (for a house). That is beyond our abilities. We urge you to consult one of the many good books on house construction."

Adobe as a Socially Appropriate Technology for the Southwest: Solar-Adobe Sundwellings, report, 45 pages, by John Timothy Mackey, 1980, $5.00 from Center for Village Community Development, 220 Redwood Highway, Mill Valley, California 94941, USA.

Adobe construction (using sun-dried earthen bricks) has been an ecologically sound, low cost building technique in many parts of the world for thousands of years. This paper examines the historical and current use of adobe in the southwestern U.S. Economic, social, and environmental considerations indicate that in this region, adobe is a truly "appropriate" technology: it is long-lasting, conserves energy, uses local building materials, creates jobs, requires little capital, and "fits" culturally. This last factor "also has a 'cost' component in the long term. For example, Indian reservations around the United States have many examples of run-down, poorly maintained modern woodframe constructed homes. These buildings last only a few years because they are not adapted to their environment and their Indian inhabitants refuse to maintain them."

Adobe brickmaking and basic construction techniques are discussed, along

Steps in the adobe building process

with the thermal properties of adobe which have made it ideally adapted to passive solar construction in the southwestern U.S. Mesa Verde, Colorado is an ancient Native American city where '' 'massive stone buildings are clustered under a cliff which protects them from the heat of the summer sun,...at Chaco Canyon (another ancient community in New Mexico)...the buildings were terraced and the roofs of each succeeding unit provided a space outdoors to live and work in contact with nature. All day the sun's heat was buried in these massive walls, and in the great cliff to the north, which also protected them from winter winds.' ''

Because it is low cost and labor intensive, adobe could be an important "self-help technology" today in some areas where families could not otherwise afford their own home. With recent developments in construction techniques, adobe dwellings may soon meet building codes even in earthquake zones.

While it does not provide much technical detail, this report is a convincing illustration of the potential of adobe as an appropriate building technology in the United States.

Assembly Manual for the Tek-Block Press, booklet of plans, 26 pages, by John Dye, $5.00 from Department of Housing and Planning Research, Faculty of Architecture, University of Science and Technology, Kumasi, Ghana.

MATERIALS: steel plate, angle and bar; wood

PRODUCTION: cutting torch, welding machine, milling machine, shaping machine, planning machine, lathe, and metal-working hand tools

This is a complete set of drawings for the production and assembly of a hand-operated press for making soil-cement building blocks. The blocks are about 4% cement and 96% soil. "These soil-cement blocks are nearly as strong and water resistant as sandcrete blocks, while containing about one-third as much cement." The blocks are made at the building site, greatly reducing the amount of materials that must be transported. The press can also be used to make sundried blocks (no cement added) for very low cost construction.

These plans are for a simplified, strengthened version of the machine, which has been widely used in Ghana for more than 5 years. The basic concept is similar to the CINVA-Ram.

"Although a shaping machine, milling machine, and planing machine are all specified, it is possible to fabricate the machine if only one of these is available." The parts are all welded together.

The press can be operated by one person. Up to 10 people can be employed, at which point the machine is being operated continuously while digging and mixing of soil, and stacking of new blocks is going on. Output is 200 to 400 blocks per day with a 3 person crew.

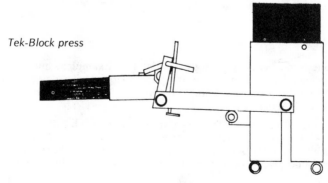

Tek-Block press

La Ceta-Ram, booklet, 14 pages, by Roberto E. Lou Ma, 1977, $2.00 postpaid from the author, Roberto E. Lou Ma, Centro de Experimentacion en Tecnologia Apropiada (CETA), 15 Ave. 14-61, Zona 10, Guatemala City, Guatemala.

This Spanish language booklet (with English summary) provides drawings and photos of a machine for making pressed soil-cement blocks. This machine is unusual in that it makes two holes in each block so that reinforcing rods for earthquake protection can be used.

Only one of the drawings, showing only the block itself, includes dimensions. All other dimensions of the machine will have to be calculated from this. The thickness of steel to be used, and the precise positioning of the pivot points are not provided, which is likely to cause the reader some difficulty.

The construction details provided for the Tek-block press (see review) are much more complete. Readers may wish to combine the two designs, for use in earthquake areas.

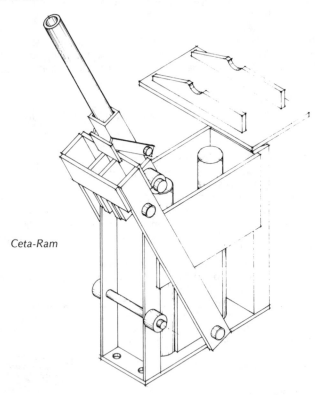

Ceta-Ram

Earth Sheltered Housing Design: Guidelines, Examples, and References, book, 318 pages, by The Underground Space Center, University of Minnesota, 1979, $10.95 from Van Nostrand Reinhold Co., 135 West 50th Street, New York, New York 10020, USA.

"The intent of this study is to present information which will be useful in the architectural design of earth sheltered houses. Part A discusses design guidelines and includes pertinent factors to be considered. Part B gives plans, details and photographs of existing examples of earth sheltered houses from

SEMI - RECESSED (BERMED)

FULLY RECESSED

LIMITED VIEW

RECESSED INTO HILLSIDE

CLEAR VIEW

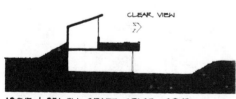

ABOVE & BELOW GRADE SPACE COMBINED

CLEAR VIEW

around the country. These serve to show a number of different ways in which the design constraints discussed in Part A have been dealt with in individual designs. Part C is intended to ease access to further detailed information and includes an annotated bibliography." This summary of design considerations for underground housing, compiled by North American building professionals, emphasizes conventional materials and approaches (e.g. reinforced concrete and planning for mass production). The authors plainly are not advocating independent design by owner-builders: "The provision of this design information should not be construed to mean that no outside assistance with design is necessary. In particular, the structural design for earth sheltered houses should not be treated lightly and professional assistance in this aspect should normally be sought." In fact, the kinds of earth-sheltered homes presented involve so much special architectural and construction expertise that they would be far too expensive for most families in rich countries.

Nevertheless, this book presents the best summary we've seen of factors of influencing earth-sheltered housing design and siting. Sections discussing configuration and thickness of earth "blankets" covering wall and roof surfaces, and the cost vs. energy savings implied by these blankets are especially good. Also covered are basic strategies for heating, cooling, ventilation, drainage and waterproofing, and the fundamentals of passive solar design. Appendices discuss building codes and compare energy use in earth-sheltered vs. above-ground houses.

A useful, well-illustrated reference, even though it was not written by people interested in "appropriate" housing.

Recommended.

The $50 and Up Underground House Book, large paperback, 112 pages, by Mike Oehler, 1978, $6.00 plus postage from Mole Publishing Co., Rt. 1, Box 618, Bonners Ferry, Idaho 83805, USA.

Low-cost underground dwellings are characteristically damp and somewhat dark. The houses described in this manual are designed such that the pitch of the roof coincides with the slope of a hillside, so that rainwater drains off and

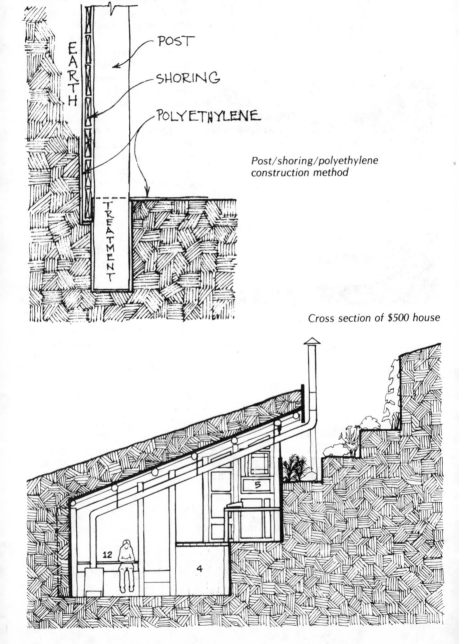

EARTH

POST

SHORING

POLYETHYLENE

Post/shoring/polyethylene construction method

TREATMENT

Cross section of $500 house

12

5

4

away. The author's Post/Shoring/Polyethylene (PSP) construction method should result in a sealed, durable living space. ".In the PSP system treated posts are set into the ground after excavation has been made. Beams for the roof are notched into these. Then a sheet of polyethylene is stretched around the outside of the wall. Shoring is placed between the posts and the polyethylene, one board at a time. The polyethylene is stretched snug, and earth is backfilled behind, pressing the polyethylene against the shoring and the shoring against the posts."

The author has lived in his PSP home for several years and made some adaptations—an uphill patio, a foyer, side-facing windows—which enhance its appeal. Photos and clear sketches show these and other possible modifications.

Underground housing has been used in many parts of the world for thousands of years. It offers, in particular, protection from extreme weather conditions. This book may calm some of those who accuse appropriate technologists of returning to the age of the caveman, with a nice look at the owner-built low technology end of the underground housing spectrum.

Bambu—Su Cultivo y Aplicaciones, book, 318 pages, by Oscar Hidalgo, 1974, $27.00 from Estudios Tecnicos Colombianos Ltda., P.O. Box 50085, Bogota, Colombia.

Available in Spanish only, this book contains a wealth of information about the cultivation and applications of bamboo in construction, engineering, paper processing, and handicrafts. Extensive illustrations and detailed graphs help present the biology and technology of this "wonderful weed" from throughout the world. Some of the more unusual examples include: a 225-meter bamboo bridge (with five supports) spanning the Min River in Szechuan Province, China; an experimental single engine airplane from the Philippines with wings and fuselage of bamboo; and a bamboo geodesic dome seating 2,000 people built in Honolulu, Hawai.

Many variations of bamboo construction joinery, appropriate hand tools, low cost housing, bridge building, preservation techniques, and bamboo-reinforced concrete forms and formulas are described in this fascinating book. Even those unable to read Spanish will find many ideas and much inspiration through the illustrations alone.

Highly recommended.

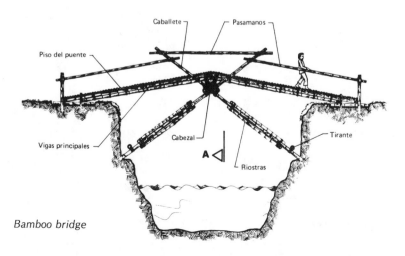

Bamboo bridge

Nuevas Tecnicas de Construccion con Bambu, book, 137 pages, by Oscar Hidalgo Lopez, 1978, in Spanish, $9.00 airmail from Estudios Tecnicos Colombianos Ltda., P.O. Box 50085, Bogota, Colombia.

A companion to Oscar Hidalgo's epic **Bambu — Su Cultivo y Aplicaciones**, this volume provides information on cultivation of a valuable Colombian bamboo species, and several specific applications of bamboo in construction. Featured are A-frame structures for coffee processing and low cost housing, and a soil-cement plaster on a split bamboo base as an innovative roofing material. Another application with a lot of potential is bamboo reinforcing of cement and concrete; here it is used in water containers, flat panels, and concrete beams (with technical information on strength).

The Use of Bamboo and Reeds in Building Construction, book, 95 pages, by UN Dept. of Economic and Social Affairs, 1972, $3.00 from META; or from Sales Section, Room A-3315, United Nations, New York, NY 10017, USA (ask for Sales No. E.72.IV.3).

"Bamboos and reeds are the oldest and chief building materials in rural areas and villages throughout the world's tropical and subtropical regions... more people live in bamboo and reed buildings than in houses of any other material. Bamboo and reed construction is popular for good reasons: the material is plentiful and cheap, the villager can build his own house with simple tools, and there is a living tradition of skills and methods required for construction. This tradition has been augmented in recent years by experiments carried out principally in India, Indonesia, the Philippines and Colombia. The bamboo and reed housing is easily built, easily repaired, well-ventilated, sturdy and earthquake resistant."

"Deterioration by insects, rot fungi and fire is the chief drawback of bamboo and reeds as building materials."

This study was produced to inform government planners, extension officers, contractors and villagers of new or less well-known techniques of construction, and to stimulate additional research to improve the material properties and

Sprung strip construction

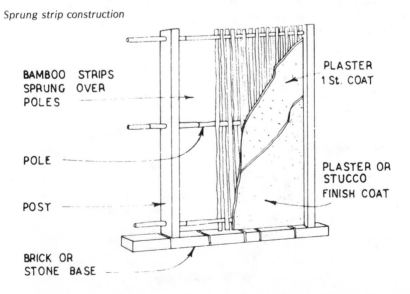

BAMBOO STRIPS
SPRUNG OVER
POLES

POLE

POST

BRICK OR
STONE BASE

PLASTER
1 St. COAT

PLASTER OR
STUCCO
FINISH COAT

techniques of building construction with bamboo and reeds.

Included are descriptions with photos of common uses of bamboos and reeds, drawings of a wide variety of joints used in building with bamboo, a summary of research (now 25 years old) on concrete with bamboo and reed reinforcing, strength data on selected bamboo species used in construction, tools and species lists, and preservatives for different bamboo end uses. Some of the material was taken from **Bamboo as a Building Material** (see review, page 233), to which this book makes a good companion volume.

Ferrocement, a Versatile Construction Material: Its Increasing Use in Asia, book, 108 pages, edited by Ricardo P. Pama, Seng-Lip Lee and Noel D. Vietmeyer, 1976, $2.00 from International Ferrocement Information Center, Asian Institute of Technology, P.O. Box 2754, Bangkok, Thailand.

These are the proceedings of a workshop held in Bangkok in November 1974. It offers a general survey of ferrocement use and research in Asia, including activities in Korea, Fiji, Thailand, India, Sri Lanka, Malaysia, Singapore, Papua New Guinea and Bangladesh. The economics, labor and materials requirements, versatility and durability are explored. Specific construction details are usually not included, although some of the things described—for example, the water jars—could be built using only the instructions in this book.

"In India ferrocement is being introduced for silos in sizes to hold about 1 to 30 tonnes of grain. Methods developed for ferrocement boat building are being applied to these storage structures to obtain a structure of high quality."

Ferrocement products discussed include boats, housing, food and water storage silos and tanks, roofing, biogas plants, road surfaces and tube well casings. Over 60 photos and drawings.

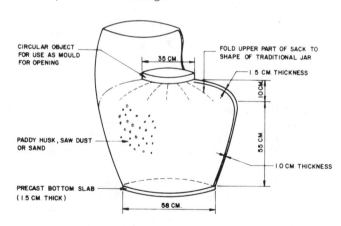

Gunnysack water jar mold

Journal of Ferrocement, quarterly, usually 50-60 pages, edited by the International Ferrocement Information Center, subscriptions from USA, Europe, Canada, Australia, New Zealand and Japan are annually US$25 for individuals and US$50 for institutions; subscriptions from other countries are annually US$15 for individuals and US$30 for institutions, add $8.00 for airmail in North and South America, Africa and Europe, add $6.00 for airmail in Asia and the Pacific, from Journal of Ferrocement, IFIC/AIT, P.O. Box 2754, Bangkok, Thailand.

"The purpose of the journal is to disseminate the latest research findings on ferrocement and other related materials and to encourage their practical applications especially in developing countries. The Journal is divided into three sections: (a) Proceedings (of conferences and workshops); (b) Technical notes (covering specific plans and construction methods); and (c) Annotated and indexed bibliography, current awareness service, news, etc."

Although the articles are thorough and detailed, many are very technical and use complex mathematics. This journal is therefore recommended primarily for serious researchers.

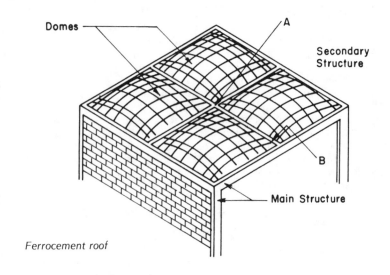

Ferrocement roof

Standard Trail Suspended and Suspension Bridges, 2 volumes, 400 pages, by Ministry of Works and Transport, Roads Department, H.M.G. of Nepal and the Swiss Association for Technical Assistance, 1977, available from Swiss Association for Technical Assistance, P.O. Box 113, Kathmandu, Nepal.

"This manual for construction of suspension bridges will be quite helpful to the engineers who will construct suspension bridges in Nepal. It contains the details of methods of surveying, calculations, and design procedures."

This set of books is specific to steel cable unstiffened suspension and suspended trail bridges. Spans described range from 40 to 170 meters. Includes bridge design, structural analysis, survey of bridge sites, cost

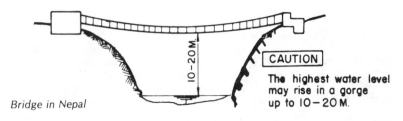

Bridge in Nepal

estimates, construction practices, and maintenance. Most sections have examples of calculations, necessary engineering tables, and ample photos, plans or sketches. The manuals contain a wealth of information, but this is mostly of a type only useful to engineers. Poorly organized, these books may be very confusing to one without previous experience in the subject. The many sections have different formats and no continuous explanatory text.

Simple Bridge Structures, Project Technology Handbook No. 2, book, 28 pages, by Project Technology/Schools Council, 1972, $4.25 from Heineman Educational Books Inc., 4 Front St., Exeter, New Hampshire 03833, USA; or Heineman Educational Books Ltd., 48 Charles St., London W1X 8AH, England.

This British book introduces students to basic bridge designs. Included are class activities to test models made of different kinds of wood. A well-illustrated set of experiments demonstrates the properties and functions of beams, frames, and columns in bridges and other structures. Simple methods for calculating the forces acting on the members of a framework are explained.

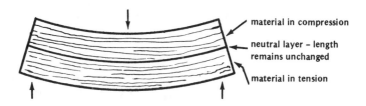

material in compression

neutral layer – length remains unchanged

material in tension

Effect of weight on wooden bridge model

ADDITIONAL REFERENCES ON HOUSING AND CONSTRUCTION

The design of solar heated and cooled houses is the subject of books reviewed on pages 631-638 and 646.

Low cost housing in India is discussed in an article in **"Technology for the Masses,"** reviewed on page 369.

Fichier Encyclopedique du Developpement Rural includes soil cement block making among its leaflet series; see review on page 375.

Small Scale Cement Plants describes the economics of small scale vertical shaft cement kilns in China; see review on page 388.

For water systems in buildings, see **Drinking Water Installations and Drainage Requirements in Buildings in Nepal**, reviewed on page 536.

TRANSPORTATION

Transportation

Readers investigating transportation alternatives in the light of appropriate technology principles will find that this topic area reflects many of the fundamental problems and issues of technology choice and development goals. In the Third World, traditional transport technologies (e.g. backpacks, bicycle-loading baskets and frames, bike trailers, animal packs, carts, pedicabs, push carts, wheelbarrows, and small boats) have been almost completely neglected in engineering design efforts.

Informed observers note that few modern vehicles fit the needs of the developing countries very well; in most cases the designs or the machines themselves are imported. These vehicles—trucks, busses, cars—bring with them high foreign exchange requirements for purchase and fuel costs, problems of maintenance and spare parts, and low durability when operated over rough terrain. Although busses can be an effective means of transporting people (especially in heavily populated areas and for long distance travel), trucks are often ill-matched to the basic transport needs of small farms. Trucks also require heavy investments in high-quality road construction, without which they have even higher maintenance costs. Furthermore, World Bank studies in Kenya give evidence that the concentration of ownership of transport vehicles (in the form of trucks) significantly limits prices received by farmers for their produce.

The technologies of transport include not only the vehicles themselves, but also the roadbed and surface materials, as well as design speeds and road routes. In road-building, the technical choices reflect the goals of the programs themselves. Given the goals of most current road-building programs, once the route, design speed, and road surface and strength have been chosen, the most economical vehicle is likely to be the truck. Roads built with high strength roadbeds and high design speeds provide a hidden subsidy and competitive advantage for heavy-weight, high-speed trucks compared to other vehicles. The range, speed, and capacity of small motorized, animal-drawn, and pedal-powered vehicles would be served just as well by less-expensive roads.

A peculiar characteristic of transportation systems is that the availability of higher speed transport technologies creates first a demand and then a dependence on long distance transport as settlement and industrial production patterns adjust. Increasing the speed of transportation increases the energy consumption at an even greater rate, because both wind resistance and weight of the vehicle increase. Thus one observer of transport in urban and rural Southeast Asia notes: "A shift in mobility from 2500 km to 5000 km in a 500 hour transport budget would increase energy requirements sevenfold through its greater emphasis on faster car-bus-rail-air modes at the expense of walk-cycle-subsidiary motor modes."[1] Ivan Illich has written insightfully on this and the inequitable political effects of high-speed transport technologies in the essay "Energy and Equity"[2] The works of Leopold Kohr[3] on scale and velocity also contain a number of important observations on urban transportation dynamics.

There are also other negative effects that come with smoother, faster transportation links. Road-building is justified by arguments about the increased inputs that will be made available to small farmers, the increased market for farm surplusses that will be created, and the reduced running costs for the trucks that will operate over the better roads. Yet many observers have noted the classic pattern of destruction of crafts and cottage industries that also comes with the opening up of a road. Manufactured tools and household items trucked in from the towns and cities outsell the products of local potters and blacksmiths. The types of jobs available change, and the variety of income-producing activities in the community may be reduced. A whole chain of negative economic multiplier effects may be set in motion. The greater commercialization of agriculture is likely to also mean lower crop diversity and thus greater risks from pest attacks and global market fluctuations. Income in the community may shift significantly to land-owning farmers who sell their surplusses without being affected by higher rents.

These are reasons why national transport strategies dramatically affect the strength of local industry and decentralized development. In China, the development of the substantial rural small industry sector has brought with it a unique blossoming of managerial and technical skills among the rural population. One factor that seems to have greatly aided in this process is the existence of protected markets that result from the commune system and the poor transportation infrastructure. (See for example the discussion of China's 2800 cement plants, in **Small Scale Cement Plants**, reviewed on page 388). It appears that there may be an optimum degree of transportation integration, beyond which damaging centralizing effects are felt.

The Transport Panel of the Intermediate Technology Development Group has done some of the most consistently thoughtful reevaluation of the transport problem and possible avenues for appropriate technology solutions. Four of the first five entries in this section offer their perspectives. Transport Panel members give high priority to improvements in intermediate motorized vehicles, traditional vehicles, animal packs, and bicycle-powered units; these represent a better match of available capital to the transport needs of small farming communities.

The farm family needs to move small quantities of inputs to the farm, harvested crops from field to home (often over rough terrain), and small surplusses to market. Small vehicles travelling at low speed over simple roads and tracks seem to best match this need. In fact, the animal-drawn cart in most developing countries still dominates this activity. In India, it has been calculated that the total national investment in bullock carts exceeds the investment in either the national railroads or the national road network. The number of ton-miles of material moved is also comparable. Recognizing that bullock carts are going to be part of the Indian transport network for many years to come, organizations such as the Indian Institute of Management have initiated work to improve cart designs through better bearings, lighter frames, and wheels less damaging to roadways. They are also developing harnesses that are not injurious to the draft animals. (See **The Management of Animal Energy Resources and the Modernization of the Bullock Cart System**, reviewed on page 566.)

The bicycle has often been cited as the most efficient machine for personal

1. Peter Rimmer, in an article entitled "A Conceptual Framework for Examining Urban and Regional Transport Needs in Southeast Asia," 1978.
2. This essay is included in **Toward a History of Needs**, by Illich.
3. See for example the review of **The Breakdown of Nations** on page 386.

transport ever invented. It is also a major mover of goods in the Third World. Loads can be tied directly to bicycles or placed in special frames and baskets, and bike trailers and 3-wheelers can be used to carry several pounds of goods. **Bicycling Science** offers the reader an excellent summary of the physics of bicycles and the human body as a power producer. **Bicycles and Tricycles: An Elementary Treatise on Their Design and Construction** provides an encyclopedic treatment of successful and unsuccessful design ideas. Two publications reviewed in the ENERGY: GENERAL chapter evaluate the potential of pedal-powered vehicles for transport and agricultural uses in the developing countries. **The Design of Cycle Trailers** details the basic considerations in the design of two-wheeled trailers for hauling goods behind a bicycle, and includes design examples from around the world.

Small engine-driven vehicles, including motorized bicycles, motorcycles, motorcycles with sidecars and other three-wheelers, and two-wheeled tractor-cart combinations have found a niche in many countries. These small vehicles with very low fuel requirements seem to have an almost unlimited number of possible applications. Discussion and examples can be found in **Notes on Simple Transport in Some Developing Countries, Intermediate Transport in Southeast Asian Cities,** and **Three-Wheeled Vehicles in Crete**. (Rough terrain vehicles produced in industrialized countries are catalogued in a World Bank report **Appropriate Technology in Rural Development: Vehicles for On and Off Farm Operations**. These are primarily expensive, relatively high speed vehicles with poor fuel economy, beyond the means of most farmers in developing countries.)

The kinds of roads needed by small vehicles can be built using labor intensive road construction methods. Two of the books in this section discuss the requirements for labor intensive programs that are economically competitive with those using heavy equipment in the construction of conventional roadways. The success of such labor intensive road-building techniques has been demonstrated, but greater savings may be achieved if road standards are kept flexible, and labor intensive programs are not required to always produce roads of equipment intensive standards. Where transport needs are defined as including roadways for lightweight small vehicles to travel at moderate speeds, labor intensive road construction methods are more likely to be the first choice. In any labor intensive road construction program, good quality hand tools and simple equipment are essential to high productivity; see the review of **Better Tools for the Job.**

A key consideration in vehicle choice and transportation strategy is, of course, fuel supply. Alcohol fuels, distilled from grains or crop residues, have been generating great interest as alternatives to gasoline. Brazil has taken the lead in alcohol fuel production, based on cassava and sugar cane. Several books reviewed in the ENERGY: GENERAL chapter offer plans and instructions for small scale alcohol fuel producing units. However, the author of **Food or Fuel: New Competition for the World's Croplands** (also reviewed in the ENERGY: GENERAL chapter) concludes that large national alcohol fuel programs are likely to greatly increase world hunger by diverting vast areas of agricultural land into fuel production for the relatively wealthy few. And the basic futility of such programs is indicated by the fact that the entire U.S. grain crop, if converted to alcohol fuels, would replace only about 30% of the gasoline currently consumed in the United States. Already, with the recent dramatic increases in oil prices, it is the citizens of poor countries who have been hardest hit, who are getting off the busses and walking, who are shutting down small town power plants, who are spending hours collecting scarce firewood as they shift away from kerosene cooking fuel. Despite the public outcry over higher gasoline prices in the United States, consumption has dropped only a few percentage points (approximately 8% from 1979 to 1980), and one out of every 8 barrels of world oil production is burned as gasoline in U.S. automobiles. As long as 6% of the world's population continues to consume about 1/3 of the world's energy supply, the problems of conventional motorized transport in poor countries will intensify.

Given the stiff buyer competition for scarce gasoline and the vast amounts of agricultural materials required for large volume alcohol fuels production, it can be reasonably predicted that there will never be enough fuel for more than a tiny part of the world's population to be driving automobiles. Decades of settlement patterns based on the daily use of the automobile in the United States have locked this country into a high fuel demand pattern that will be broken only through a very difficult and expensive process of great change. In Third World oil-importing countries, the decentralization of industrial production, along with increased emphasis on bicycles, improved carts, and small engine driven vehicles appear to be important elements of a practical transport strategy for the future.

In Volume I:

Carts, ITDG, 1973, page 100.

Dahomey Ox-Cart, page 101.

Single-Axle Power Tiller, page 104.

Power Tiller, page 104.

Oil Soaked Wood Bearings: How to Make Them and How They Perform, page 111.

Pedal Power in Work, Leisure and Transportation, edited by James McCullagh, 1977, page 191.

Pedal Power, by S. Wilson, 1975, page 192.

Transport Bicycle, page 194.

Bicycles: A Case Study of Indian Experience, UNIDO, 1969, page 194.

Pedal Power News Notes, by Stuart Wilson, page 196.

Automotive Operation and Maintenance, by E. Christopher Cone, 1973, page 275.

Rural Roads Manual, by PNG Dept. of Public Works, 1976, page 226.

Appropriate Industrial Technology for Low-Cost Transport for Rural Areas, booklet, 54 pages, UNIDO, 1979, Document No. ID/232/2, available from Editor, UNIDO Newsletter, Industrial Information Section, UNIDO, P.O. Box 300, A-1400 Vienna, Austria.

A 24-page background paper entitled "Appropriate Transport Facilities for the Rural Sector in Developing Countries" by I.J. Barwell and J.D. Howe of the I.T. Transport Panel is the most valuable part of this booklet. It provides an insightful and thorough examination of the elements of appropriate rural transport technology (both equipment and roads), and what actions policy makers and R&D groups can take in support of such technology.

The authors note that crucial on-farm transport technology has been almost totally neglected, along with virtually all of the low-cost technologies of the traditional sector: backpacks, bicycles, hand-carts, wheelbarrows, animal packs, and animal carts. The road networks do not effectively serve the majority of the population, but in fact subsidize the privately-owned imported motor vehicles and can bring real disadvantages to the rural poor (e.g. by destroying local crafts through transport of manufactured goods into the area). "Few vehicles have ever been designed specifically to meet the needs of developing countries. Their use in developing countries indicates not that they best meet transport needs but rather that they are better than anything else currently available." The authors note a variety of existing basic traditional vehicles which should be improved and more widely used.

This background paper is an excellent summary of the issues involved, and the change in orientation that planners will have to make if rural transport technology is to serve the poor.

Proceedings of ITDG Seminar "Simple Vehicles for Developing Countries", Information Paper No. 3, ITDG Transport Panel, report, 66 pages, 1977, for information on how to obtain this paper write to: Intermediate Technology Transport Ltd., 9 King St., London WC2E 9HN, United Kingdom.

The papers included in this report cover "vehicles presently in use and prototypes currently being developed; the role of transport in agriculture; the use of simple vehicles in labor-intensive construction; manufacturing strategies for local production; and the transport needs and economic constraints in the rural areas of developing countries." The need for improved wheels and tires for rural dirt roads is noted as an important research priority. 36 photos are included.

One of the more interesting motorized vehicles discussed is the TRANTOR, a multi-purpose vehicle able to serve as a truck, tractor, and passenger vehicle. The TRANTOR has been designed to be economically produced in developing countries in very low quantities (1000 units a year). Simple machine tools and jigs are used in production, and components are grouped into similar categories to allow the benefits of a large batch production to be achieved without requiring the production of a large number of completed vehicles.

This report provides additional evidence that an appropriate technology approach to rural transport would be considerably different from the prevailing approaches. "It is a technical fact that the design of roads in developing countries is dictated by the characteristics of the private car and the lorry (truck). The 'desired' speed that it is assumed car drivers want dictates the overall horizontal and vertical alignment of the road, whilst the frequency and load carrying capacity of the lorries that will use it decide the strength of the road's structure. It has never been shown that either or both of these vehicles is in any sense necessary, much less optimum, for development to take place. The possibility that other, simpler and probably cheaper vehicles might be more appropriate to needs does not appear to have been given serious consideration...The simpler and thus most probably lighter, the vehicle, the cheaper the cost of having an adequate road."

Contributing to the above problem is the fact that in developed countries "there appears to be a misunderstanding about the nature of movement demands (in the developing countries). For passenger transport the existing buses and various forms of share taxis probably meet demands very well. But for goods transport the available evidence suggests that the fundamental demand is for the movement of small consignments over relatively short distances. Smallholder agriculture, almost by definition, gives rise to limited crop surpluses and farm inputs."

The Manufacture of Low-cost Vehicles in Developing Countries, booklet, 31 pages, UNIDO, 1978, $3.00 (ID/193, Sales No. 78.II.B.8), from Sales Section, Room A-3315, United Nations, New York, New York 10017, USA; free to readers in developing countries from Editor, UNIDO Newsletter, UNIDO, P.O. Box 707, A-1011 Vienna, Austria.

This is a report of a Feb. 1976 meeting that discussed the obstacles to wider use of low-cost vehicles. An interesting variety of motorized vehicles are described, many of them from India and the Philippines. These include motorized bicycles, mopeds, motorcycles (often with a sidecar), three-wheelers (150 to 1200 cc engine) and small trucks (600 to 1600 cc engine).

Particularly for the 3- and 4-wheeled vehicles, the pattern has been first to import, then assemble, then partially produce vehicles locally. This has prevented radical innovation in design that might have led to vehicles more suitable to developing country circumstances.

A number of points are made in the report which suggest that perhaps the widespread use of small motors on existing non-motorized vehicles could provide additional power and speed where needed, while avoiding the need for enormous investment in high-speed, heavy vehicle roads. This might allow improved traditional vehicles to recapture some of the activity now monopolized by the imported high-speed cars and trucks. Transportation activities would thus remain in the hands of a broad section of the society.

In India there have ''been moves to motorize cycle rickshaws, and at least two companies manufacture small two-stroke engines and conversion kits for fitting to rickshaws. One of these...uses a 35-cc, two-stroke general purpose engine developed for agricultural and other use. It is used to provide chain drive to only one of the back wheels, which makes a differential unnecessary.'' Power packs for bicycles are also manufactured in India.

In the Philippines ''about 90 percent of the motorcycle population is fitted with side-cars.'' In the countryside these vehicles perform quite well on rough roads and paths.

''In terms of capital, space and economic volume, the requirements for making simple two-stroke engines are all only a fraction of those for manufac-

Auxiliary motor on a bicycle

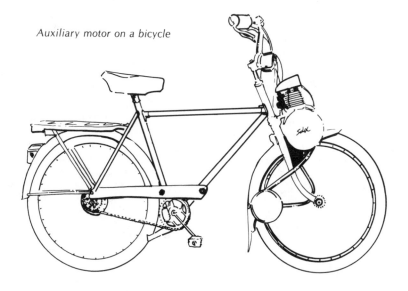

turing conventional four-cylinder engines. The technical requirements, although still exacting, are also much lower."

"In a world where high performance vehicles did not exist, the required road structure and system of traffic regulations would be considerably different from that of either an industrial or pre-industrial economy...Rigorous traffic separation reinforces the advantages of the more powerful vehicles."

Much of this report indicates that the traditional non-motorized transport vehicles have survived because they are better suited to existing needs for many tasks. Imported vehicles (and imported designs) suffer from high cost, difficulty of repair, lack of smooth supply of spare parts, high fuel consumption, and poor durability when operated over rough terrain. Thus there would appear to be an opportunity for low-cost motorized vehicles to replace many of the functions of the more expensive machines. Rather than argue for this, however, the expert group discusses ways to eliminate indigenous non-motorized vehicles: "The possible of designing a motorized competitor for the rural bullock cart was discussed...Indeed, it is the extent that low-cost vehicles replace the more primitive means of transport that will effectively measure their success." The failure to consider vehicles in the light of who makes and owns them is a serious shortcoming here.

A different, more socially appropriate approach would defend the role of existing vehicles such as bullock carts, and try to improve them rather than eliminate them. By ending hidden subsidies (in the form of more costly road construction required for high-speed heavy vehicles), low-cost motorized vehicles could very favorably compete with the larger vehicles in many activities.

Notes on Simple Transport in Some Developing Countries, Information Paper No. 2, ITDG Transport Panel, report, 26 pages, for information on how to obtain this paper write to: Intermediate Technology Transport Ltd., 9 King St., London WC2E 8HN, United Kingdom.

"The report discusses intermediate transport in Papua New Guinea, the Philippines, China and India, and describes the simple vehicles, human-powered, animal-powered and motorized, which are used in those countries." 22 photos of simple vehicles are included.

In Manila, among the unusual vehicles is "a bicycle and sidecar with a 50cc two-stroke engine mounted in the cycle frame. This drives the rear wheel through an additional chain drive, the original pedal chain drive being retained so that the rider can augment the power of the engine when necessary."

In China "many of the minor roads and tracks in the rural areas are narrow and unsealed but are quite satisfactory for the types of vehicle which travel on them" (bicycles, handcarts, wheelbarrows, and two-wheeled tractors).

The Chinese have shown considerable innovation in the design of tricycles for proper gearing and effective braking, while the Indians have not. (This suggests that the subservient role of the tricycle driver in India may have led to neglect from designers.)

A 1974 commission in Papua New Guinea recommended that 1) private cars were inappropriate to PNG conditions, 2) use of bicycles and pedal drive car vehicles should be promoted, 3) a bicycle path system should be built in Port Moresby, 4) the feasibility of using electric vehicles powered by PNG's substantial hydro resources should be investigated, and 5) aerial ropeway systems for mountainous areas should be investigated.

"Simple vehicles can play an important role in the transport systems of all developing countries, yet their use is largely confined to the urban areas of

Asia, and their design is often based on imported technology rather than on local requirements.''

Intermediate Transport in Southeast Asian Cities: Three Case Studies, Information Paper No. 4, ITDG Transport Panel, 30 pages, by A.K. Meier, 1977, for information on how to obtain this paper write to: Intermediate Technology Transport Ltd., 9 King St., London WC2E 8HN, United Kingdom.

This report provides three case studies of intermediate transport technology in Asian cities (Penang, wartime Saigon, and Jakarta). The author ''records the very rapid changes in intermediate forms of transport that have taken place in these cities, gives an account of the factors which have influenced this,'' and compares the three experiences. 12 photographs are included.

Among the examples discussed are: the motorized three-wheelers that were driver-owned in Saigon; the collapse of the Saigon government bus system in 1965 and the subsequent import of nearly one million small motorcycles to soak up the dollars flowing out of the American war effort; the becaks, bemos, opelets, helicaks, super helicaks and pickups of Jakarta; and the freight tricycles of Penang.

A widening zone of Jakarta is becoming off-limits for the becaks (3-wheeled pedi-cabs) to make way for faster vehicles with higher fares. ''The implications of a ban on becaks are serious and serve as an example of the imposition of a technology on a society which is neither economically nor socially prepared. A large number of persons will become un-employed (or underemployed) while a much smaller number of new jobs will be created. Unless there is a substantial lowering of fares on the becak replacements, the ban will deprive the lower classes of their only means of individual transportation.''

Freight tricycles in Penang have a large metal box between two front wheels, cost about $175, and number about 3,000. They ''are ideally suited for Penang's narrow streets. They can efficiently carry loads of 5 to 200 kg. for distances up to four kilometers. A number of tricycles travel much greater distances, leaving the city entirely to gather a crop of coconuts, sugar cane, or bananas, for use in the food stalls.''

''Each city has several unique aspects due to its size, organization, social composition, climate and geography. The direct consequence of this is significantly different intermediate transport networks...There is a general tendency to shift into motorized vehicles, which, from (the owner's) viewpoint are almost always more efficient, especially in cities where a premium is associated with time.'' However, the foreign exchange burden created by fuel imports, the loss of jobs making and operating non-motorized vehicles, and the loss of transport for a segment of the population must be weighed against the gains provided by motorized vehicles.

''There is a wealth of expertise and a great diversity of vehicles available... At the same time, the vehicle cannot be considered alone. Its function, service characteristics, and social context must all be considered in a complicated equation that determines an appropriate technology.''

Manual on the Planning of Labour-Intensive Road Construction, book, 253 pages, by M. Allal, G. Edmonds, and A. Bhalla of the International Labour Office, Geneva, 1977, $22.80 from International Labor Office, Washington Branch, 1750 New York Ave., N.W., Washington D.C. 20006, USA.

This is not a how-to manual for the construction of rural low-cost labor-intensive roads on a self-help basis. (See **Rural Roads Manual** on page 226.) Instead it is intended for use by planners who are responsible for national road programs, including people involved in evaluation and design of road construction projects. "They may fully agree with the notion of appropriate technology, but they must first be presented with viable alternatives to the technology they are using. They also need to be given the means of evaluating, assessing and taking advantage of these alternatives. The present manual constitutes an attempt to meet that need."

The authors note that an opening up of the spectrum of choice is required. Labor-intensive techniques should not be viewed as simply one way to achieve roads of equipment-intensive standards, but rather the design standards themselves should be flexible to allow for the most beneficial selection of technique, total road cost, user costs and maintenance costs.

There is a chapter (24 pages) on the range of labor-intensive techniques (mostly for hauling), including drawings, photos, and comments about relative costs and suitable applications. Included are headbaskets, stretchers, small trucks on rails, spades, pack animals, animal-drawn carts, animal-drawn scrapers, wheelbarrows, trailers, and small aerial ropeways.

"The planner of any labor-intensive scheme must bear in mind that the choice of the right sort of tool is as important as the choice of the right type of machinery in a capital-intensive project: given the right tools a worker's productivity can be enormously increased. A small research unit to consider the appropriate designs of small tools and equipment would be very useful."

"Earth roads are the most suitable for labor intensive construction...Also, earth roads are generally used to transport farm produce to market or to provide access to remote villages. Accordingly, they are of obvious and direct benefit to the local population."

Information is presented on the relationship between design speed and construction costs. Roads designed for vehicle speeds of 40 km/hr have

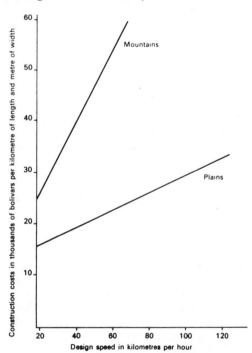

Example of the relationship between design speed and construction cost in different kinds of terrain

substantially lower costs than those designed for vehicle speeds of 70 or 80 km/hr, because there is a larger acceptable range of curves and gradients.

The later chapters discuss cost-benefit analysis, problems of organization and management in labor-intensive works, and action to eliminate capital-intensive biases in policy and attitudes among engineers.

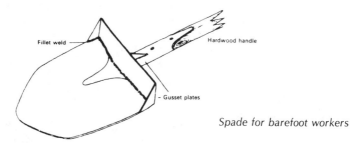

Spade for barefoot workers

Roads and Resources: Appropriate Technology in Road Construction in Developing Countries, book, 200 pages, edited by G. Edmonds and J. Howe, 1980, ₤4.95 from ITDG.

This study prepared for the International Labor Organization is about road construction programs. The authors note that "the use of more labor-intensive techniques can be technically and economically efficient." Part I, "Institutions and Issues of Implementation", is concerned with how to best organize labor-intensive programs and choose intermediate technologies. The special planning and administrative requirements of labor-intensive programs, and other institutional biases favoring equipment-intensive approaches are discussed. One chapter identifies a range of simple tools and equipment that, if given proper design attention, could multiply labor productivity by a factor of 3-6. These include Chinese wheelbarrows and animal-drawn carts on portable light rail systems.

Noting that road systems in many former colonies have been designed to move goods for export, one author urges that the emphasis should be "on providing inputs into the rural areas which will stimulate growth rather than access to ensure the maximum level of exportation."

Case studies in the second half of the book examine labor-intensive road construction programs in Mexico, Afghanistan, India and Iran.

"The use of labor-based methods would seem...to meet all the criteria upon which development planning is based. They serve the mass of the population, their implementation can involve popular participation in the decision-making process, they are an instrument of self-reliance, they can enhance the potential for rural development and they can, by providing income, serve to improve the standard of living of the mass of the population." Despite these claims, the authors have limited themselves to looking at how labor and good small tools and equipment can compete economically with heavy equipment. The goal of this strategy is twofold: income distribution and skill development on a broader scale.

There are a number of crucial questions that are beyond the scope of this book, such as : Who decides where a road will go? Who benefits from the road? Who loses? Who decides the road's design speed and strength (and hence cost and which vehicle owners will benefit)? What road-vehicle combinations go well together? Readers interested in these kinds of issues will find relevant material in other entries in this chapter.

Better Tools for the Job: Specifications for Hand Tools and Equipment, booklet, 43 pages, by William Armstrong, 1980, available from ITDG.

Labor intensive road construction projects need large quantities of good quality hand tools and carrying devices like wheelbarrows. Available tools are often of poor quality, due to the practice of seeking the lowest quoted price without specifying quality standards. A Kenyan technology unit has developed a set of specifications which have been successfully applied to tools for road projects in that country. By including these specifications with price requests, the tools may initially cost 30% more, but are likely to have a 500% longer working life.

Materials, strength and hardness, and construction specifications are provided along with detailed drawings for each tool and piece of equipment (shovel, hoe, wheelbarrow, forked hoe, crowbar, machete, mattock, axe, pickaxe, spreader, and rammer).

"A very large proportion of the problems encountered with hand tools in the field (on road, irrigation and construction projects, for example) arises from the use of handles made from cheap unseasoned softwood...The cost increase for a specified handle (of seasoned hardwood) as compared to a cheap handle is modest, and no other single step can return such high dividends in terms of cost effectiveness and productivity."

Steel strength and hardness depends on chemical composition and heat treatment. The author notes that by pressing a diamond or hardended ball into a small number of samples of the equipment, the buyer can perform a low cost reliable test of hardness to insure that tools have been made to specification.

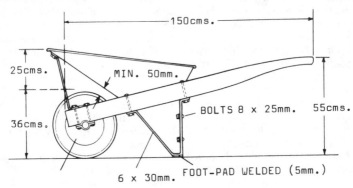

Specifications for a sturdy wheelbarrow

Bicycling Science, book, 243 pages, by Frank R. Whitt and David G. Wilson, 1974, $5.95 plus $0.75 postage from MIT Press, 28 Carleton St., Cambridge, Massachusetts 02142, USA.

This very readable book describes the physics of bicycles and other human powered machines, and the characteristics of the human body as a power generator. Topics include the power needed for movement on land, maximum performance of cyclists, optimum pedaling rates, comparison of human power with internal combustion machines and electric motors, bicycle design for rough roads, braking of bicycles, construction materials (including bamboo frames and plastic frames), water cycles, railway cycles, and possible future enclosed pedaling vehicles. This book will be a useful reference to people designing pedal-powered equipment or bicycle modifications for low-cost transport.

Regarding power, the authors note that tests of human energy production indicate that for prolonged periods (e.g. one hour) an ordinary college student would produce about 0.05 hp (37 watts). Other tests show highest power production for 1 minute of .54 hp (403 watts), and for 60 to 270 minutes, 0.28 to 0.19 hp (208 to 142 watts). Hand-cranking is not as efficient as pedaling. There is some evidence that screw-pedaled boats are more efficient than oar-driven boats. A gasoline engine added to a bicycle to give it the equivalent power of an extra human being would weigh 20 pounds. The same power using an electric motor would require adding the weight of one man, in the form of batteries and electric motor.

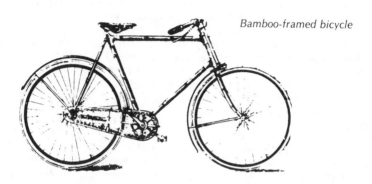

Bamboo-framed bicycle

Some designs might be of particular interest in poor countries. Bamboo framed bicycles were marketed prior to 1900. For rough roads, "some form of sprung wheel or sprung frame can greatly reduce the kinetic energy of momentum losses by reducing the unsprung mass and ensuring that the wheel more nearly maintains contact with the surface." (In developing countries, large diameter wide tires ensure an acceptably comfortable ride without a sprung frame.) Cycles with steel wheels developed for running on steel rails have less friction to overcome than the best bicycles on excellent roads.

The standard bicycle has seen few changes in design since 1890. The authors claim that this is partly because at about that time the automobile began monopolizing the attention of the inventive mechanics and mechanical engineers. There is certainly potential for new concepts and designs.

The Garbien semienclosed bicycle. Springing of both wheels is provided in this design. Power to the rear wheel is through swinging constant-velocity cranks and an infinitely variable gear.

Bicycles and Tricycles: An Elementary Treatise on Their Design and Construction, book, 536 pages, by Archibald Sharp, 1896 (reprinted 1979), $8.95 plus $0.81 postage from MIT Press, 28 Carleton St., Cambridge, Massachusetts 02142, USA.

This is a fascinating book on what is perhaps the most efficient machine ever created. A valuable reference for designers of bicycles, tricycles for hauling goods and people, bicycle trailers, and stationary pedal-power and treadle-power machines.

The first 140 pages contain an introduction to the physics and mechanical engineering of bicycles, essential to the design of successful machines. This would make a good text for design classes.

Part II reviews the history of the development of the bicycle, with the various improvements, when they were made, which ideas were dead ends and which ideas led to further refinements. The basic bicycle design we use today evolved by 1886.

The section on tricycle design may prove useful to those investigating design changes to make the pedicab more efficient and easier to operate. For example, the effects of different tricycle configurations on steering capability are explored. Several special gears for driving tricycle wheels at different speeds when rounding corners are discussed in detail (clutch gear and differential). The author notes that great accuracy is not needed in the production of the bevel gears for a differential; these gears only move slowly even when a tricycle turns a corner at 20 mph.

In the remaining chapters (300 pages) the author discusses motion over uneven surfaces, frames, stresses on frames, different kinds of spoked wheels, hubs, bearings, chains, chain gearing, toothed wheel gearing, lever and crank gearing, tires, pedals, cranks, bottom brackets, seats and brakes.

This is truly a classic and monumental work on bicycle design.

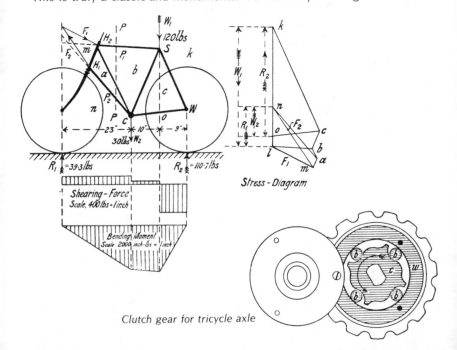

Clutch gear for tricycle axle

The Design of Cycle Trailers, Information Paper No. 1, ITDG Transport Panel, 44 pages, by Ian Barwell, 1977, for information on how to obtain this paper write to: Intermediate Technology Transport Ltd., 9 King St., London WC2E 8HN, United Kingdom.

"The intention of this report is to provide basic design information for those people in developing countries who wish to build bicycle trailers." Detailed recommendations about the critical aspects of trailer design are made, to aid the reader in designing a trailer most suited to local circumstances. These design aspects include size of cargo space, center of gravity of the load, length of tow bar, hitch design, ground clearance, type and mounting of wheels, and chassis design. 15 bicycle trailers are described, along with photos and drawings of each.

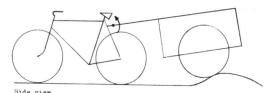

Side view

i) Rotation about lateral horizontal axis for traversing bumps

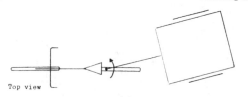

Top view

ii) Rotation about vertical axis for cornering

Movements required in
a bike trailer hitch

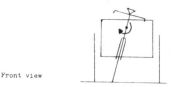

Front view

iii) Rotation about longitudinal horizontal axis for cornering

"The trailer is a convenient way of extending the usefulness of bicycles...It is cheap to produce, suitable for small-scale manufacture by local industries and can, if designed for that purpose, also function as a hand cart. For these reasons the development of trailer designs which meet particular local requirements is to be encouraged."

A bicycle with a trailer can carry an amount of cargo almost as great as a tricycle (100 kg. vs. 150 kg.), yet the trailer can be quickly disconnected to allow use of the bicycle for personal transport. It should be possible to produce the trailers for not much more than 1/2 the cost of the a bicycle.

An easy to read, valuable report.

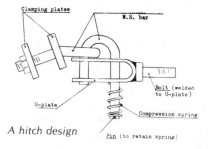

A hitch design

Coaster and 3-Speed Bicycle Repair, book, 132 pages, by Xyzyx Information Corp., 1972, $4.95 from Crown Publishers, One Park Ave., New York, NY, USA.

This book covers all the basic elements of the otherwise rather neglected topic of hub disassembly and repair. For that reason alone it is a valuable resource. However, the author's choice to present only what you absolutely **need** to know (i.e., the technical minimum) makes the book less helpful than it might have been. Furthermore, the use of only new components to illustrate the assorted repairs and operations fails to aid the novice in determining whether his/her dirty scratched parts actually need replacement.

Coaster and 3-speed hub repair is an ambitious project for the novice, the apparent audience. Unfortunately, directions such as "replace worn or damaged parts" mean very little to one who's never seen the insides of a damaged hub.

1. Stop wheel. Locate marks on rim (3).

Spokes (5) on same side of rim (3) as marks must be loosened. Spokes (6) on opposite side of marks must be tightened.

2. Adjust four spokes (5,6) on each side of mark on rim (3) by turning nipple (4) 1/2-turn.

3. Hold felt marker (1) at fork arm (2) approximately 1/8-inch from rim (3).

4. Spin wheel. Check that distance between rim (3) and felt marker (1) does not vary.

If distance varies, repeat Steps 1 through 4.

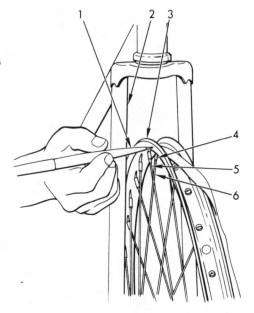

Truing a wheel on a coaster brake bicycle

The Bicycle Builder's Bible, book, 376 pages, by Jack Wiley, 1980, $8.95 from TAB Books, Blue Ridge Summit, Pennsylvania 17214, USA.

This is mainly a buying guide and maintenance manual for bicycles. Variations such as unicycles, tandems, and motocross bikes are also discussed. The author shows how to convert a regular bicycle to a folding bicycle, and how to make some of the other unusual variations from bicycle parts.

Three commercially sold adult tricycles (Schwinn Town and Country, AMF Courier, and Gobby), all with three speeds and rear wheel differentials, are pictured and briefly described. Also shown is an industrial tricycle, The Mover (manufactured by Industrial Cycles, Dayton, Ohio).

There are a few useful ideas included for rigging up pedal-powered equipment. Many of these seem to be taken from **Pedal Power in Work, Leisure and Transportation**, or from the work of S. Wilson in England.

Bicycle Resource Guide, a series of bibliographies averaging 100+ pages, 1000+ entries, $5.00 to $15.00 each, from D.J. Luebbers, Editor, 78 South Jackson, Denver, Colorado 80209, USA.

This is a series of seven bibliographies. The 1976 bibliography, for example, has 106 pages, 1102 entries, an index, and costs $5 in the U.S. and $5.50 overseas. Entries include bikeway studies (both on existing roads and new separate pathways), accident studies, guides for local tours, bike laws, bike repair manuals, conference proceedings, 100 mail order catalogs, 290 newspaper articles, and 300 journal articles on medical, transportation, historical, legal, and industrial topics related to bicycles.

Of special interest to our readers in the 1976 bibliography are: repair and service manuals, case studies of bicycle/bus combinations as low cost solutions to urban transport needs, articles describing new power packs for potential use with mopeds or bicycles, studies of 'low cost' bikeway pavement materials and design, publications presenting methods for using the bicycle in teaching certain principles of physics, and a document on transport planning incorporating bicycles in the city of Nairobi.

Three Wheeled Vehicles in Crete, paper, 10 pages, by Alan Meier, Pub. No. UCID-3968, free from the author, Lawrence Berkeley Laboratories, University of California, Berkeley, California 94720, USA.

A three-wheeled vehicle has evolved in Crete, Greece, and reached widespread use in the rural areas in just 10 years. There are now 20 local factories producing these vehicles. Most have an 8-12 hp rope-started diesel engine. For many of these vehicles the engine can be converted in ½ hour to a rototiller for agricultural use.

"The vehicle appears to have been in part responsible for the economic revival of agriculture in Crete. The three-wheelers borrowed much of their early technology from two-wheeled rototillers but quickly evolved into a unique vehicle." The rapid development and widespread use of these vehicles suggests they fill an important need for rural transport in less developed countries.

"A spectrum of transport alternatives seems to have developed. At the lower end is the cart pulled by the rototiller...The 3-wheeler comes next, providing greater speed and capacity...At the top is a group of light pick-up trucks which offer even higher speeds and slightly more capacity." 6 photos.

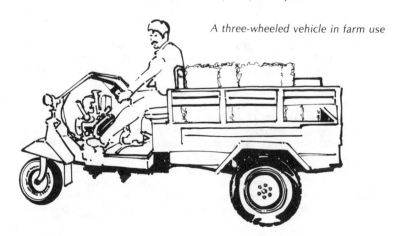

A three-wheeled vehicle in farm use

Appropriate Technology in Rural Development: Vehicles Designed for On and Off Farm Operations, catalog, 150 pages, 1978, free to organizations involved in development work, from the Regional Development Unit, Transportation Department, World Bank, 1818 H St. N.W., Washington D.C. 20433, USA.

A catalog with brief descriptions of vehicles that are produced and sold commercially around the world, that could be used in rural areas of developing countries. The information comes from the manufacturers and has not been verified independently. There are more than 100 photos included.

Good as a summary of the range of rough terrain motorized vehicles designed in the industrialized countries: motorized tricyles and 3-wheeled vehicles, all-terrain vehicles, small tractors, small trucks, and hand tractors. Though a few animal-drawn carts and tricycles are shown, there is unfortunately little information included on the range of transport options affordable to the poor majority in developing countries.

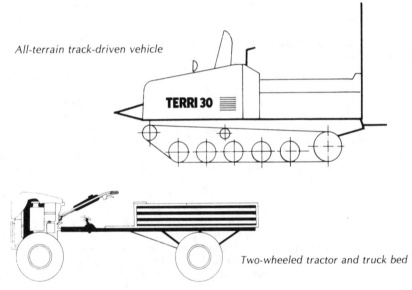

All-terrain track-driven vehicle

TERRI 30

Two-wheeled tractor and truck bed

Electric Vehicles: Design and Build Your Own, book, 210 pages, by Michael Hackleman, 1977, $8.95 within USA and $10.00 to foreign countries, from Peace Press Inc., 3828 Willat Ave., Culver City, California 90230, USA.

This book does not give very many specific construction details, but it does describe the basic principles of the design of small electric vehicles. You will need to know basic electrical theory to use this book.

Electric vehicles are not effective off of hard surfaced roads, and may require frequent battery recharging during a full day of use. The vehicles presented in this book can carry a couple of people but not heavy loads. Unless there are major future technological advances in battery efficiency, battery weight reduction, and solar generation of electricity, it seems clear that electric vehicles will remain too expensive and inefficient for transporting people or things (especially away from surfaced roads). In addition, the materials and production technology required for electric vehicles are much more complex than for internal combustion or steam engines.

A better alternative for low-cost motorized vehicles and power packs for small vehicles and bicycles appears to be small internal combustion engines perhaps using fuels from biological processes (such as alcohol). But if you are interested in designing electric vehicles you should find this book useful. Many illustrations.

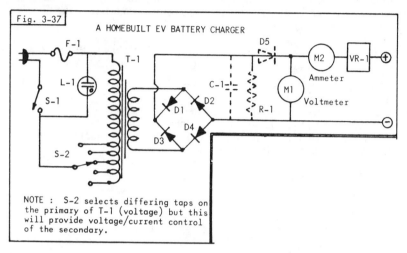

Fig. 3-37

A HOMEBUILT EV BATTERY CHARGER

NOTE : S-2 selects differing taps on the primary of T-1 (voltage) but this will provide voltage/current control of the secondary.

Fishing Boat Designs: 1—Flat Bottom Boats, book, 46 pages, by Arne Fredrik Haug, 1974, $6.00 from UNIPUB.

"The paper contains a selection of designs of flat botton boats suitable for fishing and transport work in lakes, rivers and protected coastal waters. The paper and the designs were prepared to provide detailed technical information to boatbuilders and fishery officers…"

The designs are intended to be built by people having basic carpentry skills and either some boatbuilding experience or a few weeks of training. Building procedures and timber selections are covered.

"The boat designs presented here are suitable where low cost, or ease of

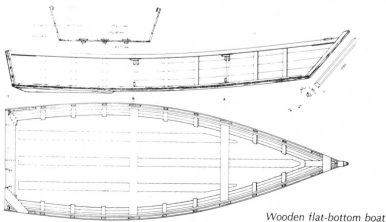

Wooden flat-bottom boat

construction, are all-important factors and where a somewhat reduced sea-worthiness...can be accepted, or where extreme shallow draft requirements are an over-riding consideration."

Materials needed for construction are wood, galvanized nails and screws, caulking compound, wood preservative and sealer, and caulking cotton. The boats could be built with only hand carpentry tools, but a table-saw and hand-held electric drill would be useful. The boats are powered by pole, oar or motor; some of the boats could be modifed for sail power with the addition of side-mounted keel boards.

Boats from Ferrocement, book, 131 pages, by UNIDO, 1972, $2.50 from Sales Section, Room A-3315, United Nations, New York, NY 10017, USA.

"There is no doubt about the urgent need in most developing countries for fishing boats that will help solve their acute food problems and for boats that will facilitate transportation in areas where rivers and channels are the most commonly used communication routes...Ferro-cement boat-building is perfect for developing countries. It requires a minimum of qualified personnel, imported raw materials and capital equipment and the boats produced compare favourably with those made from other materials in terms of price, per-formance, maintenance costs and life span."

This book is a very detailed survey of the equipment, materials and methods used in ferro-cement boatbuilding. There is an enormous number of tables, charts and illustrations. Many sources of further information are mentioned.

"The basic qualities that make ferro-cement ideal for boat construction are the ease with which it can be moulded to any shape, and the unit weight per square foot."

Anyone building ferro-cement boats will find many specific instructions in this book, but there are no complete boat plans. A useful book.

ADDITIONAL REFERENCES ON TRANSPORTATION

"Technology for the Masses" contains an excellent article on the modernization of the bullock cart; see review on page 368.

A Landscape for Humans has some unusual and interesting recommendations for road building; see review on page 387.

Small Scale Cement Plants: A Study in Economics (page 388) and **Rural Small Scale Industry in the People's Republic of China** (page 390) take a look at the relationship between transportation infrastructure and the development of small scale industries in China.

Bearing Design and Fitting should be a useful reference to people working on improved cart designs; see review on page 407.

The Harness Maker's Illustrated Manual contains detailed information on good quality harnesses that could improve the efficiency of animals pulling carts and agricultural implements; see review on page 465.

The Management of Animal Energy Resources and the Modernization of the Bullock Cart System discusses the requirements for improved carts; see review on page 566.

Alcohol fuel production is the topic of the materials reviewed on pages 569-572.

Pedal-powered vehicles for transport are considered in **The Use of Pedal Power for Agriculture and Transport in Developing Countries** (page 572) and **Design of a Pedal Driven Power Unit for Transport and Machine Uses in Developing Countries** (page 573).

Small steam engines and producer gas units can be used on vehicles and in boats; see reviews on pages 572, 574-577.

HEALTH CARE

Health Care

"Each month a civil servant dies in the capital because there is no penicillin. Each day a child in the country recovers from a fatal disease because of a plant growing in the forest."

—Benjamin Owuor, quoted by Aggrey Nyong'o

"We are dedicated to completely eradicating all anti-scientific attitudes and ideas."

—Cuban doctor

These contrasting views are common among people who are dedicated to improving the level of health among the world's poor. There is the romantic who unquestioningly believes in the general effectiveness of traditional remedies, and there is the crusading doctor who sees only superstition in native cures. Both perspectives are partly valid; traditional remedies range from the dramatically effective to the dangerous. The main weakness of traditional medicine has been the failure of its practitioners to question the validity of cures; due to coincidence and the power of suggestion, good and bad remedies are added uncritically to the medical kit of the indigenous healer. Nor has there been sufficient dispassionate review of what is effective, harmless, and dangerous within the drug arsenal of modern medicine. A major challenge in developing appropriate health practices and remedies is to draw together the effective, cheap, and safe treatments in both traditional and modern healing systems.

Equally important is the question of the kind of people and roles that are to be supported in a strategy for the development of a health care system or systems. Much has been written about why modern medical facilities cannot be extended to reach the entire population of most Third World nations. Among the many reasons for this, the great expense of elaborate facilities, the chronic shortage of professionals to work in rural areas, and the high cost of physician training programs are the most frequently cited (see pages 240-241 for a discussion of these and other factors). Because of these problems, health programs are increasingly involving lesser-trained health workers from the communities in which they work. These people have in many ways a more demanding role than the doctor, requiring a broader range of skills and knowledge to successfully offer basic curative care, lead preventive and health education programs, and take part in community organizing. A unique and significant advantage is that as members of their community they know it intimately.* Schemes involving community level health workers are now

---

* Trainers should be careful to avoid alienating village health workers from their communities. Otherwise the effectiveness of health workers may be undermined, and they may even move away from their communities.

operating all over the world. Even in rich countries there is a movement away from expensive hospital care (with its attendant risks of unneeded surgery and problems caused by drug reactions) to informed self-care and the use of midwives for home births.

Unfortunately, those who endorse the use of village health workers frequently pay only lip service to the depth and breadth of indigenous knowledge and skills. In health care, as in other related aspects of community development, outside agencies have often been quick to assert the absolute superiority of their (usually Western-based) methods. And people who have long been oppressed and belittled are sometimes also quick to accept what outside agencies offer, abandoning their own traditions. Making matters worse, the chief medical personnel in programs working with village health workers often have little faith in these people and allow them few responsibilities; the result has been the creation of little more than referral systems that continue to swamp understaffed clinics in towns and cities.

It appears that this situation is changing for the better. In the last five years, a number of manuals for training village health workers (VHWs) have appeared which provide practical medical information while recognizing the validity of traditional health care roles and experience. In this chapter we have reviewed materials for the trainers of VHWs as well as manuals which can be used by both trainers and trainees. Among these books, **Primary Child Care: A Manual for Health Workers** is the most comprehensive. **Where There Is No Doctor**, already adapted and translated into many different languages, is broader in scope and reflects great confidence in the potential capability of the village health worker. **Child Health for Health Extension Officers and Nurses in Papua New Guinea** and **Child Health Care in Rural Areas** (India) are more simplified materials developed for village level health workers in particular countries. The three manuals from India (**Teaching Village Health Workers: A Guide to the Process, A Training Manual for Village Health Workers**, and **Child Health Care in Rural Areas**) are most sensitive to the value of traditional medicine, next to the remarkable compendium **Philippine Medicinal Plants in Common Use**. The bibliography series **Low Cost Rural Health Care and Health Manpower Training** concentrates on materials written for trainers and program leaders. In the future, the Hesperian Foundation (producers of **Where There Is No Doctor**) will publish a lengthy volume on village health worker training that promises to be of great interest. It will be entitled **Methods, Aids and Materials for Helping Health Workers Learn**.

A notable new strategy for health education and community participation in preventive and basic curative health care is described in **Child-to-Child**. Recognizing that small children are often cared for and taught by their older brothers and sisters, the CHILD-to-child strategy is to develop activities in which older children teach younger children simple practices (like the use of homemade toothbrushes) and identify children with hearing, eyesight, and malnutrition problems. These activities make learning an exciting adventure in which children discover for themselves how and why a problem exists, then together take action to do something about it. Where health workers and school teachers have conducted CHILD-to-child activities, the children's eagerness to learn and ability to effectively take more responsibility for the care of younger brothers and sisters have been impressive.

In **Child-to-Child** and the other manuals reviewed here, the words "child" and "children" appear again and again as the focus of rural health care programs. Especially vulnerable to disease below the age of five years, and often malnourished, children constitute the majority of the Third World's sick and dying people. Untreated infant diarrhea leading to severe dehydration in

malnourished children is the leading cause of death in Third World communities. A simple rehydration solution that can be made in any village kitchen
would save these lives. Through formal and nonformal education, including
visits by village health workers, children and adults can be taught how to make
this solution for early treatment at home before the dehydration becomes
serious. The clash between the education/participation/community-based
strategy for health care and the large agency initiative/dissemination strategy
is quite evident in the way the problem of infant diarrhea is being approached.
The World Health Organization is now promoting the dissemination of
pre-packaged powdered mixes in place of simple education about the required
proportions of salt and sugar. This dissemination program is expensive, faces
logistical problems of distribution, and will add unknown risks when supplies
are interrupted in areas that come to rely on the prepackaged mix. It is not
clear which strategy could save more lives in the short term, but the
community-based educational strategy represents at least a step forward in
establishing better health care systems for the future.

In the longer run, hygienic and public health measures, particularly the
provision of a safe supply of water for washing and drinking, are critical steps
in improving basic community health and reducing infant diarrhea and infant
mortality. The control of communicable diseases (four books on this subject are
reviewed in this section) also depends heavily on a safe water supply and
adequate waste disposal systems. Many of the references in the WATER
SUPPLY AND SANITATION are relevant here.

In addition to these factors which visibly influence the health of the
community, land reform and agricultural development have major roles to play
in improving the basic health of rural people. Some health care programs are
now including agricultural development projects and pressure for land reform
as part of a total effort to improve community health.

Where a national commitment to improved health systems exists, the
organizational requirements for successful village health worker programs are
of primary interest. The bibliography **Health Care in the People's Republic of
China** identifies much of the literature on that nation's successful use of
"barefoot doctors" in their decentralized health care system. **Health in the
Third World: Studies From Vietnam** offers detailed insights into the health
care strategy of another nation. It seems paradoxical but true that increased
local self-reliance in health care depends greatly on supportive initiatives taken
by the policy-making centers in government. **Health in the Third World** will be
of interest to readers wishing to examine this issue of political support for
decentralized development.

In Volume I:

Low-Cost Rural Health Care and Health Manpower Training, by Shahid
Akhtar, 1975, page 241.

The Training of Auxiliaries in Health Care, by K. Elliott, 1975, page 242.

Reference Material for Health Auxiliaries and Their Teachers, 1973, page 242.

Health Manpower and the Medical Auxiliary, 1971, page 243.

Medical Care in Developing Countries, edited by Maurice King, 1967,
page 243.

Pediatric Priorities in the Developing World, by David Morley, 1973, page 243.

Nutrition in Developing Countries, by Maurice King et. al., 1972, page 244.

Doctors and Healers, by A. Dorozynski, 1975, page 244.

Health Care and Human Dignity, paper, 25 pages, by David Werner, 1976, $2.00 from Hesperian Foundation, Box 1692, Palo Alto, California 94302, USA.

Written by the author of **Where There Is No Doctor** (see review on page 249), this paper briefly summarizes the major insights gained from a study of nearly forty rural health projects in Central and South America. It is the clearest, most coherent discussion we have seen of the features of "community-supportive" rural health programs and the obstacles to be faced by people wishing to foster these programs on a broader scale. "*Community supportive* programs or functions are those which favorably influence the long-range welfare of the community, that help it stand on its own feet, that genuinely encourage responsibility, initiative, decision making and self-reliance at the community level, that build upon human dignity...the programs which in general we found to be more community supportive were small, private, or at least nongovernment programs, usually operating on a shoestring and with a more or less sub rosa (low-profile, unofficial) status."

Werner goes on to identify key factors tending to limit or slow the growth of community supportive programs: paternalistic attitudes among those in charge of health care delivery programs, overemphasis on medical "safety", bureaucracy (or, the "superstructure overpowering the infrastructure"), commercialization, and government fear of the politically destabilizing potential of increased rural skills and abilities. The paper concludes with a list of steps that might be taken to implement a country-wide approach to community supportive health care. Appendices compare and constrast the objectives, size, financing, and other characteristics of "community supportive" vs. "community oppressive" health programs.

An extremely useful combination of criticism and positive suggestions for future progress, of interest to anyone interested in health as part of community self-reliance. Highly recommended.

The Village Health Worker—Lackey or Liberator?, paper, with charts and drawings, 16 pages, David Werner, 1977, $2.00 from the Hesperian Foundation, P.O. Box 1692, Palo Alto, California 94302, USA.

David Werner elaborates on points made in **Health Care and Human Dignity** (see review above), illustrating how socio-economic context and political objectives of program planners affect rural health programs. Werner and several co-workers visited some 40 health worker programs in Latin America. "In the the majority of cases, we found that external factors, far more than intrinsic factors, proved to be the determinants of what the primary health worker could do...We concluded that *the great variation in range and type of skills performed by village health workers in different programs has less to do with the personal potentials, local conditions or available funding than it has to do with the preconceived attitudes and biases of health program planners, consultants, and instructors.* Inspite of the often repeated eulogies about 'primary decision making by the communities themselves', seldom do the villagers have much, if any, say in what their health worker is taught and told to do. The limitations and potentials of the village health worker—what he is permitted to do and, conversely, what he could do if permitted—can best be understood if we look at his role in its social and political context. In Latin America, as in many other parts of the world, poor nutrition, poor hygiene, low literacy and high fertility help account for the high morbidity and mortality of the impoverished masses. But as we all know, the underlying cause—or more exactly, the primary disease—is inequity: inequity of wealth, of land, of educational opportunity, of political representation and of basic human rights ...As anyone who has broken bread with villagers or slum dwellers knows only too well: *health of the people is far more influenced by politics and power groups, by distribution of land and wealth, than it is by treatment or prevention of disease.*"

Primary Child Care: A Manual for Health Workers, book, 315 pages, by Maurice King, Felicity King, and Subagio Martodipoero, 1978, Ł2.00 from TALC.

Sponsored by the World Health Organization, this remarkable book is the revised, updated version of **The Child in the Health Centre—Book One: A Manual for Health Workers** (see review on page 249). This basic English text is intended to be adapted and translated, for direct use by health workers everywhere. "It contains a selection of the most appropriate technologies for primary child care taken from all over the world."

The step by step approach, from the basics to the needed level of understanding, makes this a valuable book for people with only a limited knowledge of the field. For each major category of illness (e.g. "Coughs"), the authors begin with illustrations and background information on the system or parts of the body affected (e.g. respiratory system). Then they discuss the different combinations of symptoms, diagnosis and treatment. They have included many effective diagrams that explain how infections and diseases spread.

More than 80 pages are devoted to community health problems, supplies and equipment, and procedures for examination, sterilization of equipment, and record-keeping. Dosage information is provided for all drugs mentioned. There is a glossary of 200 key scientific and medical terms, with which "you will probably be able to understand anything written in the rest of the book."

The three authors invested years of hard work in this wonderful book, undoubtedly the most valuable one that WHO has ever sponsored. This is an outstanding resource, which could become the basis for training programs for child health workers at many levels.

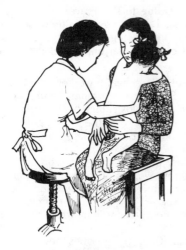

Examining the abdomen

Where There Is No Doctor, revised English translation of **Donde No Hay Doctor**, book, 458 pages, by David Werner, 1977, $7.00 plus postage ($4.00 plus postage to local groups in developing countries) from the Hesperian Foundation, P.O. Box 1692, Palo Alto, California 94302, USA.

In addition to an English translation by the author and associates (see review of the Spanish edition on page 249), there have been a number of other adaptations and translations of this remarkable health care manual in the past few years. Already completed are the following:

1) **Ang Maayong Lawas Maagum**, in the Ilongo language of the Philippines, translated by Teresa S.S. Ludovice, 1980, considerably adapted with all new drawings, much new material, including local herbal remedies, and a reorganization of the chapters. Available from AKAP, 88 J.P. Rizal St., Project 4, Quezon City, or the Rural Missionaries Health Team, Apostolic Center, 2215 Pedro Gil St., Sta. Ana, Manila.

2) **Onde Nao Ha Medico**, in Portuguese, available from Edicoes Paulinas, Caixa Postal 8.107, 01098, Sao Paulo, SP, Brazil.

3) **Mahali Pasipo Na Daktari**, in Kiswahili, available from the Rotary Club of Dar es Salaam, P.O. Box 1533, Dar es Salaam, Tanzania.

4) In the Tzotzil language of the Indians of highland Chiapas, Mexico, from Antropologo Jacinto Arias Perez, Crecencio Rosas #4-A, San Cristobal de las Casas, Chiapas, Mexico.

5) **Where There Is No Doctor**, revised for India by The Voluntary Health Association of India, 55 pages of new material and adapted drawings, July 1980. Available for approximately 29 rupees, from Voluntary Health Association of India, C-14 Community Centre, Safdarjung Development Area, New Delhi—110 016, India. A Hindi edition is in preparation by the same organization.

6) **Where There Is No Doctor**, reprint of the English edition, available to the book trade through the Tropical Community Health Manual Series of MacMillan Publishers (U.K.).

Many organizations around the world are working on additional adaptations and translations. These include the following languages: Khmer, French, Indonesian, Arabic, Bengali, Kannada (India), Thai and Somali.

Teaching Village Health Workers: A Guide to the Process, 2 books, 117 pages total plus several charts and visual aid cards, $5.00 to developing countries, $10.00 to developed countries, plus postage, from Voluntary Health Association of India, C-14, Community Centre, Safdarjung Development Area, New Delhi 110016, India.

Book I vividly illustrates how trainers of village health workers can approach communities in a sensitive manner. Diagrams, cartoons, and text give examples of how knowledge of the community helps village health workers deal with problems more effectively. "Don't be blinded to the social, political and economic forces which will play an important part in the shape and direction of the community health programme."

The first requirement is that the trainer be a "changed person". "Do you really feel that a little-educated, or illiterate woman or man knows more than you do about the village? Are you willing to learn from them and the other 'students' in your class of village health workers?" With this orientation, Book I offers guidelines for curriculum development, teaching methods, and simple communications media.

Book II (Lesson Plans and Curriculum Charts) describes how to teach a limited range of specific treatments and preventative measures to village health workers. Because the health workers in this program were mostly illiterate, the level of sophistication has been limited. Certainly literate health workers in many communities will be able to go far beyond this material, to use of more comprehensive manuals like **Where There is No Doctor** (see review on page 249) or **Primary Child Care** (see review in this section).

A Training Manual for Village Health Workers, book, 110 pages, available from The Project Officer, Village Health Worker Programme, New Paediatrics Block, Government Erskine Hospital, Madurai 625020, India.

Breastfeeding

"This manual has been written primarily as a guide for future trainers of Village Health Workers." Mostly text with a few drawings, it contains some good material for training midwives and mothers who are nearly illiterate, using charts, physical models, and discussions. The emphasis is on child nutrition and the prevention and treatment of common illnesses of young children. Breastfeeding, good nutrition within even a small budget, health charts, and diarrhea treatment are among the topics discussed. Medicines are not identified but are described using color coding; this will be an obstacle to use of this book in other areas.

"Many of the Village Health Workers in the Madurai Project are either the local village midwives or other informal 'health advisers' in their communities. As such their awareness of rural health problems is often quite astute. This experience plus their natural interest in health continually leads to worthwhile exchanges between the 'trainers' and 'trainees'—so that at times these roles become reversed."

Child-to-Child, book, 104 pages, by Audrey Aarons and Hugh Hawes, with Juliet Gayton, 1979, £2.50 from TALC.

"We know a group of community workers who know every inch of the village in which they work, who are accepted by everyone, who want to help their community, who will work hard (for short periods of time) and cheerfully (all the time). Last month the health worker used them to collect information about which children had been vaccinated in the village. Next Tuesday some of them will help to remind the villagers that the baby clinic is coming and they will be at hand to play with the older children when mothers take their babies to see the nurse. Next month they plan to help the school teacher in a village clean-up campaign. These health workers are the boys and girls of the village...This book...calls on us to recognize what children already do towards helping each other and helping us. It suggests ways in which we can support them and in which we can make their contribution more effective, easier, and more fun."

This well-illustrated book was put together with ideas from around the world, from people who believe that development starts with local level action. It provides a selection of possible activities, such as organizing a survey, making a community health map, discovering common accident patterns and preventing them, treating children with diarrhea (including making a special salt and sugar spoon for the water mixture to treat diarrhea), caring for sick brothers and sisters, and finding out what younger children eat and whether it is nutritionally adequate.

Experience thus far suggests that this is an effective approach to health education. These examples of community-based learning and action (in which local resources, skills and problems are identified) are models of the kinds of steps essential for peoples' participation in any type of development effort.

*Older children
often take care
of younger ones*

Child Health for Health Extension Officers and Nurses in Papua New Guinea, book, 209 pages, by John Biddulph, 1976, £4.50 from TALC.

"Child health is an important subject for health extension officers and nurses in Papua New Guinea. This book has been prepared to help them... They will use this book during their three years of training, and so learn to find their way through it...They are not expected to learn the contents; but they are expected to know where to look to find the answer." This is the fourth edition of a training manual for community health workers employed by the Health Department of PNG. It assumes that effective child health care must not rely on treatment alone: "Three functions—Education, Prevention, and Treatment— cannot be separated. The health worker who produces no cures will find it difficult to keep the co-operation of the people which is needed for health promotion and disease prevention. Similarly, the health worker who is so busy treating that there is no time for education and prevention measures, will always remain very busy, and the state of health of the community will not be improved."

Most of the 35 chapters deal with the most common health problems of the children of PNG. Description, diagnosis, causes, and treatment are presented in outline form. Also included are short, straightforward sections on normal growth and development, care of the new-born baby, and accidents. The final chapter lists the important drugs mentioned in the book, and gives dosages for children including instructions for determining dosages based on body weight. Simple English and large-face type make the text easy to understand. Key rules and ideas (like "Breast Feeding Should Continue As Long As Possible") are boxed and highlighted with bold capitals. Excelent line drawings accompanying the text clearly illustrate techniques and symptoms.

This book was written by, about, and for Papua New Guineans. Though it is less polished and comprehensive than **Primary Child Care: A Manual for Health Workers** (see review in this section), much of its contents would be directly applicable in other tropical developing countries. This manual is an important example of a low-cost reference and training resource, published by a national health department responsive to local conditions.

*Examine the ears
of every sick child*

Child Health Care in Rural Areas, book, 364 pages, by Rural Health Research Center, 1974, Ł2.25 from TALC, or $10.00 from Asia Publishing House, Inc., 440 Park Avenue South, New York, New York 10016, USA.

This is a reference and training manual for Auxiliary Nurse Midwives, key paramedical personnel staffing Primary Health Centers and subcenters in rural India. Written in simplified English, the book is intended for broad dissemination to women who work directly with mothers and their very young children. Public health measures, birthing procedures, correct care of newborn babies, and practices to ensure healthy child development are emphasized. Simple treatment of common ailments and administration of drugs are also briefly covered.

The authors stress the importance of establishing effective, mutually respectful working relationships with practitioners of traditional medicine. "The villagers have great faith in vaids and hakims and use their services because: they understand the villagers' ways; their treatment is often cheap and effective; they may be the only doctors close by...The Auxiliary Nurse Midwife who

Vaids and Hakims

works with the village people has to know the vaids and hakims in her area. You should always treat vaids and hakims respectfully, try to understand their work and try to inform them about your work.''

Better Child Care, booklet, 52 pages, 1977, 50 cents plus postage to developing countries, $1.00 plus postage to developed countries, add $1.00 for airmail postage, from Health Publications, Voluntary Health Association of India, C-14, Community Centre, Safdarjung Development Area, New Delhi 110016, India.

Good use of photos and a weatherproof plastic cover make this a model low cost booklet on proper feeding and ensuring normal growth. Excellent color photos will be of great help in identifying anemia, which in 80% of cases is visually evident.

Low Cost Rural Health Care and Health Manpower Training, Volume Two, An annotated bibliography with special emphasis on developing countries, book, 182 pages, by Frances Delaney, 1976, publication no. IDRC-069e, free to local groups in developing countries, $10.00 to others, from IDRC.

This is the second volume in a valuable series (see review of the first volume on page 241). The 700 entries do not cover a specific time period.

Also available in the same series are Volume Three (187 pages, 1977, publication no. IDRC-093e), and Volume Four (1979, 186 pages, publication no. IDRC-125e).

Communicable Diseases: A Manual for Rural Health Workers, book, 349 pages, by Jan Eshuis and Peter Manschot, 1978, $3.50 from African Medical and Research Foundation, P.O. Box 30125, Nairobi, Kenya.

A training and reference manual for Medical Assistants and Rural Medical Aides in Tanzania. ''Most of the common diseases in Africa are environmental diseases due to infection by living organisms—viruses, bacteria, protozoa, or metazoa. These are called communicable diseases because they spread from person to person, or sometimes animals to people. Together with malnutrition they are today the major cause of illness in Africa...For the first time all the essential information on communicable diseases, from both clinical and public

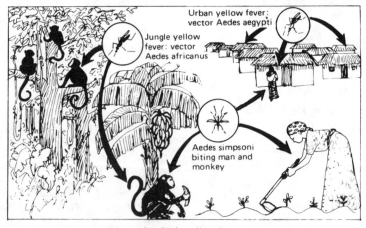

The process through which yellow fever is spread

health aspects, has been collected in one volume, adequate for the training of paramedical staff."

This manual groups diseases by how they spread—by contact, by fecal contamination, by airborne germs, and so on. For each disease, information is provided on where it is found in Tanzania, causes, symptoms and diagnosis, and control. Most of the text is in simple English, although medical terms are also used. Drawings show sources and agents of disease in the African village environment. The authors discuss the kinds of public health measures which interrupt the transmission of diseases and prevent their spread. Historically, adequate supplies of water and safe handling of human waste have been the most important factors in the prevention of communicable diseases.

Like **Child Health for Health Extension Officers and Nurses** (reviewed in this section), this is a book about a particular country written especially for use by village health workers. Much of its content is relevant in other regions, and it is a good example of a low-cost means of sharing important medical information and experience within a developing nation. The African Medical and Research Foundation has published a series of similar books on child health, health education, pharmacology, immunology, and other topics. More information on the series can be obtained from the Foundation at the address above.

Animals Parasitic in Man, book, 320 pages, by Geoffrey LaPage, 1963, $4.00 from Dover Publications, Inc., 180 Varick St., New York, NY, 10014, USA.

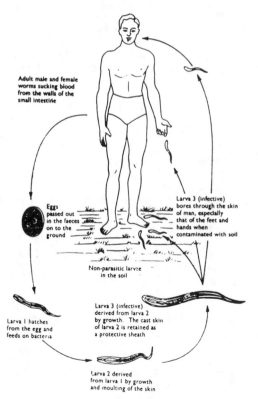

Adult male and female worms sucking blood from the walls of the small intestine

Eggs passed out in the faeces on to the ground

Larva 3 (infective) bores through the skin of man, especially that of the feet and hands when contaminated with soil

Non-parasitic larvae in the soil

Larva 1 hatches from the egg and feeds on bacteria

Larva 3 (infective) derived from larva 2 by growth. The cast skin of larva 2 is retained as a protective sheath

Larva 2 derived from larva 1 by growth and moulting of the skin

The life cycle of the human hookworm

This is a detailed discussion and description of most of the parasites that commonly attack humans. Ths life cycles of the parasites, ways humans are infected, prevention techniques and some medical treatments are discussed in detail.

"In any event each parasitic animal is limited to a certain range of hosts. It is, that is to say, **specific** to these hosts and cannot live in others. This **host-specificity** is an important feature of parasitism and it will be necessary to refer to it throughout this book. It will be evident, for instance, that if a particular host, such as man, is one of the usual hosts of a certain species of parasitic animal, it is necessary, if we wish to prevent the spread of this parasitic species, to know what its other usual hosts are, because all these hosts may be sources from which the parasitic animal may spread. These other hosts are reservoirs of the infection and they are called **reservoir hosts.**"

Parasites are a problem in every part of the world that man inhabits. Many of the parasites are so common in certain areas that it is quite unusual if an individual does not have them. Health campaigns to eliminate parasites must include education of the affected people about parasite hosts and requirements for preventing the spread of parasites. This book could be a useful reference in such an educational effort.

Control of Communicable Diseases in Man, book, 413 pages, twelfth edition, edited by Abram Benenson, 1975, $4.00 from American Public Health Association, 1015 Eighteenth St. N.W., Washington D.C. 20036, USA.

Communicable diseases and malnutrition are the major killers in the Third World. This is a good reference for teachers of health workers, and contains valuable ideas for program leaders who must cope with epidemics. The difficult language means that this book cannot be directly used in explaining the information to health workers. No attempt has been made to deal with the human, social, and cultural factors that must be considered before many of the recommendations can be followed. Some of the recommendations are not "affordable" in the Third World.

Handbook on the Prevention and Treatment of Schistosomiasis, book, 1977, translation of a Chinese publication, DHEW Publication No. (NIH) 77-1290, free to serious groups, from the FIC Publications Office, Bldg. 16A, Rm. 205, National Institutes of Health, Bethesda, Maryland 20205, USA.

This is an English translation of a Chinese handbook, originally published by the Shanghai Municipal Institute for Prevention and Treatment of Schistosomiasis (also known as bilharzia). China is perhaps the only developing country that appears to have been highly successful in controlling schistosomiasis. This book offers some valuable insights into how this can be accomplished.
There are five stages in the life cycle of the schistosome: 1) adult schistosomes, in humans and animals, produce eggs which 2) hatch in water, 3) enter snails and change form, 4) leave the snails, 5) re-enter humans and animals, and develop to adult size. If any of these stages can be eliminated, schistosomiasis can be stopped. Most control efforts concentrate on the destruction of the snails.
The snails in irrigation canals and rivers "are mostly distributed in a line on the water level...Snail elimination can be coordinated with dredging of riverbed soil as fertilizer...Build a dam and drain the water to lower the water level or utilize the dry season to expose the noninfested area of the base of the bank. Dig a ditch...The noninfested soil is piled on the side of the ditch near the center of the riverbed...Pare the soil 3 inches deep. First pare the heavily infested soil near the water level, and dump it into the ditch. Clean up the loose soil, and cover the whole ditch with noninfested soil...A five-inch layer of soil is used which is then pounded and hardened."
Other chapters in this handbook discuss frequently used chemicals for killing snails (some of them very environmentally hazardous), personal protection, diagnostic procedures, treatment of patients, safe treatment of manure, and schistosomiasis in farm animals.

Better Care in Leprosy, booklet, approximately 40 pages, 1978, 50 cents plus postage to developing countries, $1.00 plus postage to developed countries, add $1.00 for airmail postage, from Health Publications, Voluntary Health

Association of India, C-14, Community Centre, Safjarjung Development Area, New Delhi 110016, India.

A simple booklet with good photos and discussion to help distinguish leprosy from other skin problems that are similar in appearance.

Philippine Medicinal Plants in Common Use: Their Phytochemistry & Pharmacology, book, 66 pages, by Michael L. Tan, 1978, revised edition 90-100 pages, to be issued in 1980, expected price approximately $2.00 for overseas orders, from Alay Kapwa Kilusang Pangkalusugan (AKAP), 66 J.P. Rizal, Project 4, Quezon City, Philippines.

Covering the major medicinal plants of the Philippines, this book discusses their cultivation, harvest, storage and medicinal uses. The chemical composition of each plant is provided. Plants are indexed according to Latin names (family, genus, species).

"The present thrust of research into medicinal plants is geared towards the screening of plants for cardiovascular, anti-cancer and anti-fertility drugs. While this type of research has its value, it seems inappropriate in countries where available forms of treatment for widespread diseases such as tuberculosis, malaria, and schistosomiasis continue to be beyond the reach of the majority of the victims. In the Philippines, the situation is even more disturbing, with recent studies revealing that 95% of the materials used to produce 'local' drugs are, in fact, imported."

There are sketches of many of the plants and the text is easy to read. Many of these plants are found or could be grown in other tropical, sub-tropical and mild temperate zones. The book also has a section on weights and measures, a guide to preparation of medicines using the plants mentioned, and a list of sources for further information.

Highly recommended.

Coconut water can be used intravenously for rehydration, but sterile methods should be used. As long as the shell is intact, the water inside is sterile. It is best to transfer the water to another container to filter out some of the sediments that may be in the water. Cut the ends of the nut 1 to 1½ inch into the soft meaty substance. Swab the cut surface with alcohol and allow it to evaporate. Insert a sterile hollow tube into the nut's cavity, and pass the water through a sterile glass funnel packed with sterile gauze, into a sterilized bottle. Use as soon as possible. (noted by Tan, from an article by H.S. Goldsmith, 1962.)

Nutrition Rehabilitation: Its Practical Application, book, 130 pages, by Joan Koppert, 1977, Ł1.00 from TALC.

Nutrition rehabilitation refers to care and dietary supervision for malnourished children and their mothers. It is an attractive alternative to hospital care for undernourishment. "In most developing countries around 1 percent of all children under the age of five years will be suffering from a severe degree of malnutrition at any one time, and in many countries the figure is far higher. In addition, there is a very much larger group of undernourished children. Admitting a tiny minority of the malnourished children to highly expensive hospital wards is almost irrelevant, particularly since studies have shown that a high proportion of such children die either in hospital or in the year subsequent to discharge. A more fundamental and realistic approach to the problem by promoting adequate growth—monitored by a weight chart held by every child—has been developed with the advent of the Under-Fives' Clinics. However, even in the few countries where such services are widely available, some children will develop a more severe malnutrition, and it is for these that nutrition rehabilitation centers are desperately needed."

This book is intended to be "an instruction manual with detailed information on the setting up of a center and its day-to-day running, a place where mothers would learn how to prepare balanced meals for their young, especially weaning children, on returning to their homes. Home economy, household budgeting, home gardening, food values, fathers' cooperation and ways and means of improving the family income have been included. Practical advice is given on the siting and construction of a center along with the financial implications...methods of administration and follow-up care are described."

Also included are helpful sections on community surveys and record-keeping. A useful, low-cost book summarizing a practical approach to an important problem in rural health care.

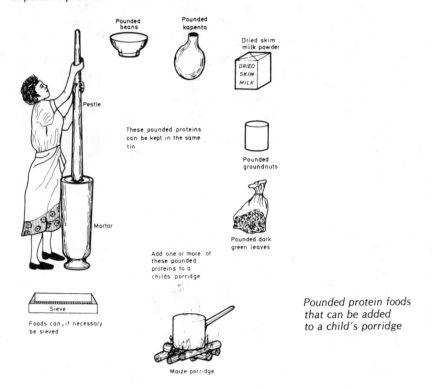

Pounded protein foods that can be added to a child's porridge

Health Care in the People's Republic of China: A Bibliography with Abstracts, book, 182 pages, by Shahid Akhtar, 1975, publication IDRC-038e, free to local groups in developing countries, $15.00 to all others, from IDRC, Box 8500, Ottawa, Canada K1G 3H9.

"The material presented should prove useful to people concerned with providing health service and to those concerned with training auxiliary health workers. The literature concentrates on the now famous 'barefoot doctors' of China, and covers the period 1949-74.''

Major topic areas include health care organization and planning (especially in the rural areas), functions and training of "barefoot doctors", and relationships between health care systems and other community organizations. This bibliography does not cover acupuncture and biomedical research.

There are 560 entries. Complete publisher information is listed for commercially available materials, but no addresses. IDRC will try to supply some of the limited circulation items to requestors from developing countries.

Health in the Third World: Studies from Vietnam, book, 342 pages, edited by Joan McMichael, 1976, U.S. edition 1980, $7.50 postpaid from Alyson Publications, 75 Kneeland St., Room 309, Boston, Massachusetts 02111, USA.

"While outside experts argued—as they continue to do—about the content of primary health care for underdeveloped countries, and how to achieve it, answers were being forged in the Democratic Republic of Vietnam under difficulties well beyond the experience, or even the imagination, of health administrators and others in industrially advanced nations involved in assistance for the less developed countries.'' These edited and translated Vietnamese accounts tell the story of mass mobilization for improvements in sanitary conditions, training of paramedical personnel, and development of indigenous medical and pharmacological research capabilities. The achievements described are truly remarkable among Third World countries: "At present there are on the average, one (double-vault composting privy), one well and one bathroom respectively for every 1.4, 3.3, and 4.7 households... and each of the 5,925 communes in North Vietnam possesses a Health Centre with personnel composed, according to the population density, of three to five individuals of whom the most frequently included are: an assistant doctor, a midwife, and a nurse.''

The text discusses three general topics: the organization and operation of the health care network, the various public health campaigns to control communicable diseases, and the growth of the health care network in response to disasters (particularly U.S. bombing during the Vietnam War). The crucial importance of commitment to community development at every level of political leadership is a recurring theme. Concerning a campaign to build double-vault composting privies for each of 16,000 families in Nam Sach district: "Complete unity of purpose was quickly achieved amongst the leadership of the district. They had then to impress their viewpoint and determination on those who held responsibility in the communes, in the agricultural co-ops, and down to the leaders of the production teams...Every ten days progress was checked, and the communes were rated according to their pace of work and were given a red, blue or white flag according to the results obtained...Everything went so well that in one month, the time set for the campaign, all was complete. In the same period the villagers removed as much earth as in the course of the whole previous year.''

Another central theme of the **Studies** is that decentralization has been the only viable strategy in view of limited resources and the toll of war. With Western medicines unavailable, locally-based research into expanded therapeutic uses for medicinal plants became important. The destruction by bombing of nearly every hospital in the country forced teaching and clinical medicine quite literally "underground", away from the cities.

An intriguing, if somewhat propagandistic, account of appropriate structures and technologies for rural health care.

Manual of Basic Techniques for a Health Laboratory, large paperback book, 487 pages, 1980, Sw. fr. 30, from World Health Organization, 1211 Geneva 27, Switzerland.

"(This) manual is intended for use mainly in medical laboratories in developing countries. It is designed particularly for use in...small or medium-sized laboratories attached to regional hospitals and in dispensaries and rural health centres where the laboratory technician often has to work alone...The manual describes only direct examination procedures that can be carried out with a microscope or other simple apparatus. For example: the examination of stools for parasites; the examination of blood for malaria parasites; the examination of sputum for tubercle bacilli; the examination of urine for bile pigments; the leukocyte type number fraction; the dispatch of stools to specialized laboratories for the detection of cholera vibrios."

This is the second, expanded edition of a book prepared both as a teaching manual and as a reference for health center laboratory technicians. Techniques are explained with text and line drawings in step-by-step fashion, from collecting specimens to recording results. Many photographs and drawings show what parasites and bacteria look like under the microscope. Also included: lists of all reagents (lab chemicals) used, and how to make or obtain them; and a list of all the apparatus needed to equip a laboratory which could carry out all the examinations in the book. An introductory chapter on "general laboratory procedures" gives detailed instructions on using and cleaning a microscope; sterilizing water and glassware; storage and preparation of materials and specimens; and simple plumbing and electrical repairs in a laboratory.

Rattan & Bamboo: Equipment for Physically Handicapped Children, booklet with 13 large sheets of drawings, 1979, by J.K.Hutt, Ł2.00 postpaid from Disabilities Study Unit, Wildhanger, Amberley, Arundel, W. Sussex BN8 9NR, United Kingdom.

Detailed designs of a variety of chairs and walking supports made of rattan and bamboo. The ready availability of the materials, good strength and durability in the tropics, and ease of repair make these designs attractive. All drawings use English measurements. There is no text to explain the particular use of each piece of equipment.

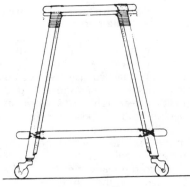

Bamboo walker

Medical Self-Care: Access to Medical Tools, quarterly magazine, average length 60-70 pages, 1-year subscription $10.00 in U.S., $12.00 in other countries, from Medical Self-Care, P.O. Box 717, Inverness, California 94937, USA.

Initiated by an American medical student as a thesis project. "Up until recently I was planning to go through a traditional internship and residency and become a family practitioner. But during my first year on the hospital wards, I realized that over half of my patients were suffering from a partially or totally preventable disease...somehow there was a whole area of health maintenance and preventive medicine that neither the doctor nor the patient was taking responsibility for. I got to feeling like a mechanic working on cars wrecked by people who had never learned to drive. What was needed was not more mechanics, but a little driver's education."

The magazine includes articles teaching basic paramedical skills and how to use medical texts, libraries, and bookstores as well as accounts of the growing self-care/health consumerism movement in the U.S., and the ways American lifestyles affect health. Several pages in each issue are devoted to reviews of "popular" medical books.

Salubritas: Health Information Exchange, newsletter, 8 pages average length, published quarterly in English, Spanish and French, free to serious groups from American Public Health Association, International Health Programs, 1015 Fifteenth Street, N.W., Washington D.C. 20005, USA.

Combines short articles on participatory health and sanitation programs with good short reviews of related books. Also includes a "Readers' Exchange" where subscribers report on needs, successes, and failures of their projects. "Readers are encouraged to share their experiences with their colleagues by submitting articles, photographs and items regarding planning, implementing and evaluating health services in the developing world."

Appropriate Technology for Health, directory, 74 pages, WHO, December 1978, free from the ATH Programme, World Health Organization, 1211 Geneva 27, Switzerland.

A compilation of responses to a questionnaire, listing 382 organizations in 75 countries. The name and address of each organization are followed by a simple list of activites. No details on scale or scope of projects are provided. 1183 listings from developing countries, 199 from industrialized countries.

ADDITIONAL REFERENCE ON HEALTH CARE

Pesticides describes the proper techniques for maximum safety in handling of pesticides, which have become a significant health hazard for farm laborers in poor countries; proper handling of pesticides and eventually change to non-chemical techniques for pest control are badly needed; see review on page 444.

SCIENCE TEACHING: STRENGTHENING LOCAL INNOVATION

Science Teaching: Strengthening Local Innovation

It is a commonplace observation that people with little formal schooling are often more adept at finding practical "hands-on" solutions to the problems of everyday life. In the rich countries, the educated live and consume in an artificial environment of technologies too complex to be understood, much less controlled, by individuals. For most, experience with natural phenomena and basic technical systems is very limited. Thus it is not at all surprising that relatively unschooled villagers in developing countries often demonstrate technical inventiveness and environmental understanding which astonish rich visitors. Farmers, for example, make complex decisions about which crops to plant and when, based on their knowledge of the soil and ecological interactions. Most village-level technical innovation comes from trial and error and careful observation, often over many seasons, unsupported by systematic knowledge of natural science. A result is that the rate of village technology development is much slower than it has to be.

Unfortunately, the science taught in Third World schools does not contribute to useful innovations in village technologies. Science studies should equip young people with an understanding of how to apply the basic principles of physics and biology to address a common problem (whether it be poor grain storage or a broken pump). Related skills—how to systematically control and vary a set of experimental trials, and how to carefully observe and record the results—are equally important. But educational methods and curricula inherited from a colonial past or imported from the industrialized countries mean that science teaching in the Third World actually orients students away from the practical problems they and their families face, by concentrating on phenomena that are beyond their everyday experience.

Science education (and education in general) is, in fact, too often simply part of a sorting process through which a fortunate few may escape the rural areas and qualify for an urban job, usually in government. Science teachers, themselves the products of such a system, do not expect the community to demand that they teach a practical curriculum relevant to local conditions. These teachers are, in any case, ill-equipped to do so; in most cases science is simply one of a number of subjects the teacher is covering each day.

As the science material taught is abstract and has little to do with local conditions, so it is natural that science teaching equipment for demonstrations is composed of expensive and exotic apparatus. The lucky teacher who succeeds in obtaining such apparatus from the education ministry is faced with two alternatives: to use it to demonstrate what are likely to be seen as peculiar and rather magical events, more a property of the equipment than the real world, or to lock it away in a closet to prevent damage to something so valuable. In either case it is rarely (if ever) used by the students themselves, who never get a chance to confirm their abstract understanding through their own experiments.

As the books in this chapter indicate, science teaching can be something quite different. **Towards Scientific Literacy** makes the argument that if the aim of literacy is to enable men and women to better understand the world around them, then relevant science education should be considered a part of literacy, and should be included in nonformal education programs. **The Production of School Science Equipment** and related publications from the Commonwealth Secretariat review the problems and opportunities associated with the local production of science equipment. These educators from around the world recommend the use of objects, devices and tools from the community to illustrate scientific concepts whenever possible. Students are then more likely to see scientific principles at work in phenomena they encounter in their daily lives. Suitable curriculum development and teacher training will have to go with such equipment. The three volume set **Guidebook to Constructing Inexpensive Science Teaching Equipment** presents construction details for a wide variety of easily-made yet sophisticated equipment that can be used to illustrate even the most difficult of scientific principles. **A Method for Cutting Bottles, Light Bulbs, and Fluorescent Tubes** provides instructions for making glassware for science class use. **Adventures with a Hand Lens** takes the science class outside, equipped with a magnifying glass, to study the natural world. **Anti-Pollution Lab** contains projects for students to measure the level and kind of pollutants in their community.

Most of the books reviewed in the other chapters of the **Appropriate Technology Sourcebook** could be profitably used in practical science and technical classes. There are several that deserve a special note. **Minka**, a Spanish language magazine from the Peruvian highlands produced in a popular format, is dedicated to fostering scientific understanding and capabilities among the campesinos, building on a foundation of traditional wisdom. This magazine includes a special project for children in each issue (see page 376). **The Industrial Archaeology of Watermills and Waterpower** examines the design evolution of waterwheels and turbines, and then presents small projects for school children, to allow them to confirm for themselves the operating principles involved. **The UNESCO Science Sourcebook**, available in low cost editions and in a large number of different languages, remains probably the best known general science teaching reference volume.

After decades of neglect, it may be that one of the best ways to improve village technology is through the strengthening of the scientific problem-solving capabilities of millions of rural farmers and craftsmen. Throughout the history of the United States, it has been the farmer-inventor who has most consistently contributed successful innovations in rural technology. Although circumstances in the Third World differ in some important ways, there are good reasons to believe that a similar high rate of technological improvement could be achieved, given the proper relevant science education support.

In Volume I:

Towards Scientific Literacy, book, 96 pages, by Frederick Thomas and Allan Kondo, 1978, available on request from Director, International Institute for Adult Literacy Methods, P.O. Box 1555, Tehran, Iran.

Towards Scientific Literacy "suggests a core curriculum of scientific ideas that should become part of the common human heritage...science can no longer be ignored if the aim of literacy is to enable men and women to understand better the world around them."

"Adults are **not** ignorant in science. They are already raising crops, raising children and managing their lives...Adult education should be designed to use and develop what adults already know. The purpose of teaching science is to enable learners to improve their ability to communicate in science with experts, with peers and with their apprentices."

"Everywhere in the world there are people who look carefully at the things around them and try to understand what they see. Any person who does this is a scientist, even if he or she never studied science from a book."

"People everywhere can do experiments. Anyone who has an idea on how to improve something should do an experiment to test the idea. Perhaps a tailor has an idea for a way to sew stronger seams. If he does, he should try the new way and compare it with the old way to see which is really stronger. If a woman has an idea for a better way to store her family's food without spoiling...she can do an experiment to see if the new idea is really better. She could store some food in the new way and some in the old way, and see for herself which stays fresher longer."

"By doing simple experiments, a farmer can learn for himself what methods are best. He can test different kinds of seed, and he can try different farming methods...All he needs is an idea about how he might improve his crops and a small plot of land on which to test the idea."

"Some people seem to believe that only new ideas are good. A scientist does not care whether an idea is old or new. He wants to see for himself which is better."

The authors discuss a variety of science subjects, using key concepts and key words, and proposing a set of activities for teaching science in non-formal educational settings (which could also be used in a regular classroom). Subjects include soil erosion, plant nutrition, a variety of health topics, energy sources, the internal combustion engine, and electricity. The content of many of these chapters is itself rather conventional, and not directly linked to appropriate technologies. (For example, when discussing herbicides and pesticides, no attention is given to the negative side effects of these and how the farmer might discover such side effects.)

"The learning experience should be related as closely as possible to the learners' life experience." A project method is presented, in which literacy workers help farmers to identify their problems, collect local data, seek out ideas from other sources, devise a procedure for testing of possible solutions, test them, and report the results to others.

Two case studies reveal some of the potential problems and solutions in training teachers. These case studies also "show that the facts of science and procedures based upon science can be taught successfully and usefully to unschooled adults, even though they are illiterate or semi-literate."

This kind of approach could be a very important element in appropriate technology strategy, by increasing the scientific skills of the poor, and thus strengthening their capacity to find their own technological solutions.

This book is intended for use by people preparing literacy materials or training literacy staff—not for direct use by adult learners or most field workers.

Highly recommended.

The Production of School Science Equipment: A Review of Developments, book, 68 pages, by Keith Warren and Norman Lowe, 1975, from Information Division, Commonwealth Secretariat, Marlborough House, Pall Mall, London SW1Y 5HX, United Kingdom; out of print in June of 1980, but an updated edition is planned.

This is a review of the kinds of science teaching equipment that can be made domestically at lower cost than that imported from industrialized countries. Much of the book covers the work of organizations around the world and considerations for the development of appropriate science equipment.

The authors note that much beautifully designed imported equipment has suffered "because the price was totally out of reach of most schools in a developing country" and/or "it is not usable in the situation into which it has been put." Because of over-emphasis on such equipment, "the majority of the children are starved of relevant practical scientific experience."

"It is possible to avoid special manufacture if there is an object already available within the country which can illustrate a concept, or replace a

chemistry beaker, and so on. Indeed, it may be an educational advantage to use an object which the children recognize rather than a foreign-looking object remote from their experience."

If local production is undertaken, there is a need for close collaboration between curriculum designers and apparatus designers. Also, it must be noted that if teachers are to be required to build their own apparatus, this represents a considerable time drain.

The book includes a review of activities in this field in the developing countries, along with 60 photographs of science teaching materials and kits.

Highly recommended.

Publications on the production of low cost science teaching equipment, from Commonwealth Secretariat Publications, Marlborough House, London SW1Y 5HX, England.

With the excellent report **The Production of School Science Equipment** (see review) now out of print, readers will have to look elsewhere to find the ideas and experiences summarized in that report. The Commonwealth Secretariat has issued several recent reports which are of interest:

a) Development and Production of School Science Equipment: Some Alternative Approaches, booklet, 57 pages, by E. Apea and N. Lowe, 1979, ₤1.00.

This report offers a look at national curriculum and equipment development centers in India, Kenya and Turkey, and a regional center in Malaysia. The goals and organizational structure of each are described.

The most interesting of these, from the point of view of practical science education at the primary level, is the Kenya Primary Science Programme, in which a Science Equipment Production Unit has been formed. "The Programme has three basic aims: namely to encourage and assist children to: a) develop the manual and intellectual skills that are necessary to solve problems in a scientific way; b) preserve and acquire the attitudes that are necessary to apply those skills effectively; and c) acquire a deep understanding of the natural phenomena that take place in their environment...Activities in the classroom are made to relate directly to the pupils' environment. This is accomplished by helping children first to acquire problem-solving skills and then to apply the skills in solving problems based on their immediate environment."

"The use of locally available materials is important, not only in the name of economy and feasibility, but to help to prevent young children becoming alienated from their home community and background."

For the upper primary grades, "the topics for investigation will be concerned with applied science and technology, with the expectation that pupils will identify and solve problems of real and practical significance in areas such as agriculture, health and village technology."

The centers in India, Turkey, and Malaysia are primarily concerned with local production of science equipment for conventional science programs.

b) Low Cost Science Teaching Equipment, Report of a Commonwealth Regional Seminar/Workshop, Nassau, Bahamas, Nov. 1976, 98 pages, 1977, ₤1.00.

This report reviews the problems and progress being made in the development and production of low cost science teaching equipment in the Caribbean, and includes recommendations for action by the governments of these nations.

"In the Caribbean, the traditional approach to science teaching—over-emphasis of teacher demonstration and learning by rote—is generally giving way to a new approach which involves inquiry, discovery, and the encouragement of pupil participation...Unfortunately, however, basic equipment needed for this approach to science learning is sparse, or, as in most cases, non-existent. Most primary school teachers have had little or no special training in teaching science at this level. As a result...teachers lack the confidence, knowledge and the skills that are necessary for effective science teaching, and are unable to identify potential sources in their environment that might be used in the classroom for teaching the subject." Teacher training is thus an important aspect of any strategy to develop low cost science teaching equipment.

Experienced teachers are crucial to successful development of relevant equipment. In Kenya, "teachers formed the core of the committee that decided the original objectives; the consultant (a teacher) was a member of that committee...The field trials, the development of the accompanying teacher training programme, the evaluation, and all aspects of production required the involvement of teachers."

c) Low Cost Science Teaching Equipment: 2, Report of a Commonwealth Regional Seminar/Workshop, Dar es Salaam, Tanzania, Sept. 1977, 55 pages, 1978, ₤1.00.

This report examines some of the experiences of African nations in tackling this subject, and includes recommendations for action by African governments.

"We are...presented with a dilemma: on the one hand it is educationally desirable to move towards student experimentation; on the other there are constraints imposed by meager school budgets and a lack of adequate numbers of qualified and experienced science teachers. This dilemma is of fundamental concern when one is considering strategies for the production and use of teaching equipment. Perhaps the best solution is to produce teachers' manuals to accompany the use of apparatus which provide adequate instructions both for work in small groups and for teacher demonstration exercises."

Guidebook to Constructing Inexpensive Science Teaching Equipment, three volumes, 968 pages total, by the Inexpensive Science Teaching Equipment Project, 1972, $10.00 for the set of three books, from Science Teaching Center, University of Maryland, College Park, Maryland 20742, USA.

This is the final product of the Inexpensive Science Teaching Equipment Project at the University of Maryland. The project set out to: "1) identify laboratory equipment considered essential for student investigations in introductory biology, chemistry and physics courses in developing countries; 2) improvise, wherever possible, equivalent inexpensive science teaching equipment."

Simple distillation apparatus

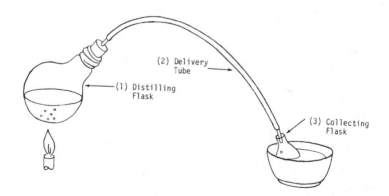

(2) Delivery Tube

(1) Distilling Flask

(3) Collecting Flask

"In designing equipment for production by students and teachers, two factors have been kept in mind. One, project work in apparatus development can be extremely rewarding for students, bringing both students and teachers into close contact with the realities of science, and relating science and technology in the simplest of ways. Two, it is not difficult for cottage (or small scale) industries to adapt these designs to their own requirements."

All the designs have been tested at the University of Maryland, but at the time these books were printed the equipment had not been produced and tested under local conditions in developing countries. A draft edition was circulated for comments from science educators around the world before the current edition was produced. These materials should therefore be considered as ideas to be tried, adapted, and improved when needed.

Only handtools are needed to make this equipment. The drawings and instructions are very clear. Some of the basic materials required will be expensive and/or hard to obtain in some circumstances: plastic lenses, copper wire, glass test tubes, corks, metal tubes and metal sheets. For the most part the equipment is made of simple materials, yet often it can be used to demonstrate rather sophisticated concepts. The emphasis is on qualitative, rather than precise quantitative, measurements.

Notes on the use of the equipment are provided, but the reader will have to refer to other sources to learn how to best use some of it. Each volume has an index for all three volumes, which helps in locating equipment relevant to more than one subject area.

Volume 1: Biology

magnifying lens
microscopes
dissecting tools
tools for collecting: aquatic organisms
                            insects
                            soil organisms
traps for small animals and birds
plant press
aquarium
terrarium
cages

temperature-controlled cages
egg incubator
thermostat
apparatus for growing microorganisms
transfer pipette
kymography device
volumeter
manometer
chromatography apparatus
various uses of syringes

Volume 2: Chemistry

procedures for safely working with
  and cutting glass tubing, glass
  sheets, and glass jars
boring holes in corks
alcohol, charcoal, & gas burners
demonstration thermometer
bi-metal strip
burette
measuring glass
pipette
dropper
specific gravity bottle
supports, stands and clamps
glassware cut from lightbulbs

aspirator
mortar and pestle
sieves
distillation apparatus
electrolysis apparatus
hand drill centrifuge
hand operated centrifuge
gas generators
metalware
heaters and dryers
molecular models
kinetic molecular model
chromatography apparatus
variety of syringe uses

Volume 3: Physics

spring lever balance
rubber band balance
beam balance
extending spring balance
compression spring balance

uses of syringes
slit/aperture light box
reflection apparatus
optical board & refraction apparatus
light filter

pegboard balance
soda straw balance
micro balance
equal arm balance
single pan balance
sundial
water clock
pendulums
ticker tape timer
force and motion carts
ripple tank apparatus
stroboscope

lens apparatus
diffraction apparatus
transformers
rectifiers
simple batteries
circuit board
resistors
electric motors and generators
electromagnets
magnetic field apparatus
galvanometers
ammeter

Adventures with a Hand Lens, book, 220 pages, by Richard Headstrom, 1976, $2.75 from Dover Publications, Inc., 180 Varick St., New York, New York 10014, USA.

This book would be very useful in any educational program such as agriculture, forestry, botany, biology and aquaculture. The book consists of 50 explorations into the natural world using a magnifying glass. Basic natural principles can be taught using the simple experiments and observations. Many of the plants and insect examples are only found in temperate zones. In other regions the book could be a useful model for an approach that examines local plants and insects.

"If we look on cabbage leaves we would likely find conical, pale yellow eggs, and if we viewed them through our lens (magnifying glass) we would see that they are ribbed. The eggs are those of the imported cabbage worm...A little later, when squash leaves have developed, the squash bug, another common insect and rather injurious to squashes and other members of the squash family, appears and lays her eggs on the leaves. They are easy to find, for they are laid in clusters and are oval and pale yellow to brown."

This book can be used by someone lacking in any basic knowledge of botany or biology. It uses the world outside the classroom as the place where learning will take place.

Anti-Pollution Lab: Elementary Research, Experiments and Science Projects on Air, Water and Solid Pollution in Your Community, book, 128 pages, by Elliott H. Blaustein, 1972, $2.25 plus $0.75 shipping from Arco Publishing Company, Inc., 219 Park Avenue South, New York, New York 10003, USA.

Simple methods of testing for pollution are presented in this book. Materials required are lab glassware, chemicals and a few other things such as rubber bands and rubber tubing.

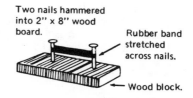

Stretching rubber band to test for ozone

Two nails hammered into 2" x 8" wood board.

Rubber band stretched across nails.

Wood block.

"The great merit of Elliott Blaustein's book is to demonstrate that we can use simple and practical scientific techniques to detect pollutants as well as their effects on the body, and also to develop action programs which will once more render our environment healthy and pleasurable." (From the Preface by Rene Dubos.)

Tests included are: vital capacity measurement (lung breath volume); maximum lung breath pressure; sulfur dioxide (SO_2) air pollution; air dust particles; ozone testing; carbon dioxide (CO_2); air visibility; water turbidity; water particles; algae; detergent in water; thermal pollution; salinity; and fiber decay resistance.

A user of this book should be familiar with simple chemistry. The problem of pollution is one which is increasingly found in all parts of the planet. Industrial manufacturing in Third World countries is increasingly developing and using technologies which are being banned or heavily regulated for pollution reasons in industrialized countries such as Japan and the USA. People in recently industrializing regions need information on effects of pollution and methods of pollution detection and control. This book provides a simple starting point in efforts to detect pollution.

A Method for Cutting Bottles, Light Bulbs, and Fluorescent Tubes, Technical Bulletin No. 36, 6 pages, by Allen Inversin, $1.00 from VITA.

These notes come from the author's efforts to find ways of making science equipment more accessible to the science teacher in developing countries. His initial use of this technique was for "cutting bottles and bulbs to make glassware for use in experiments." He has "cut hundreds of bottles of all sizes and in the process has refined the technique to the point where it should be fairly complete."

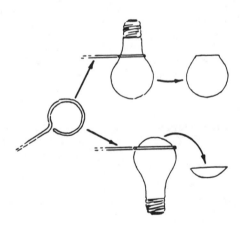

"Occasionally use can be made of cut incandescent light bulbs, as for example, beakers for boiling solutions in chemistry experiments, watch glasses, and glass chimneys for wick lamps." For this, a different technique is presented.

NONFORMAL EDUCATION
AND TRAINING

Nonformal Education and Training

Development workers are increasingly recognizing the inadequacies of formal schooling systems in developing countries. Formal schooling inevitably depends on massive expenditures for schools, teacher training, and centralized administration, in addition to the continuing drain of government revenues to pay teachers' salaries. Typically, the shortage of revenues to devote to education has ensured a chronic shortfall in the number of teachers relative to ever-expanding numbers of pupils at all levels. Inadequately paid teachers cannot afford to devote all of their time to their teaching work, and teacher training based on foreign (often colonial) educational systems means that teachers inherit curricula and methods that have little to do with problems faced by students and their families. For these kinds of reasons, formal education systems are unable to provide relevant educational opportunities for many of the rural poor.*

Given these inadequacies, there is a growing awareness among development workers that the rural poor are often their own best educational resource. Despite lack of formal schooling, they are the greatest source of background and insight on their own recurring problems. They also share, among themselves, a pool of locally-relevant skills and experience for tackling these problems. Recognizing this, many development workers now have two main objectives in their activities: to enable the poor to critically define their own problems and educational goals, and to help them find ways to mobilize the skills and resources to pursue these goals.

Such a strategy implies a belief in the capability of individuals and communities everywhere to define and control their destinies. One of the most powerful voices in this "humanistic" school of thought comes out of the Third World, that of Paulo Freire. His book **Pedagogy of the Oppressed** (this section) tells about how literacy can be a tool for describing and shedding light on people's own lives. This, in turn, is the first step towards taking useful action ("action-reflection: the dialogue with the world"). An important part of Freire's method has been to involve illiterates in discussions about how words and pictures might describe or illustrate the troubling aspects of their lives. This methodology has sparked broad debate and has been adapted worldwide; it has influenced, for example, Latin America's Liberation Theologists, literacy workers in the ghettos of New York City, and field staff in bureaux of adult education from Thailand to Tanzania.

Successes in the application of Freire's methods, which rely on sharing of opinions and ideas in group settings, have triggered increasing interest in how

---

*Science teaching methods, content, and equipment can be made more relevant to rural needs, in both formal and nonformal educational programs. See the SCIENCE TEACHING chapter.

the value of group insights is often greater than the sum of individual contributions. This well-known phenomenon of "synergy" is the focus of **Doing Things Together**, which offers a compelling theoretical illustration of how many individuals, each with different skills and information, pool their knowledge to solve a wide variety of community problems. **Appropriate Technology for Grain Storage** (see review on page 483) provides a concrete example of this effect. Tanzanian villagers—concerned with the drying and storage of their corn—not only knew more than the visiting team of specialists about the situation; they also collectively knew more than any of them had guessed about what the real problems were, and how the problems might be solved.

Key assumptions in this "problem posing/problem solving" approach are a free flow of facts and ideas among group participants, and leadership which is responsive to the group instead of "teaching" it predetermined solutions. "Culture Circles" in Latin America, "Family Life Education Groups" in Asia, and "Study Circles" in Africa are all approaches which rely on the increased creative and productive potential of participatory groups in which leadership is shared and not authoritarian. These Nonformal Education (NFE) programs, supported by a variety of public and private agencies, are efforts to reach and involve the young adults lacking formal schooling who are so numerous in the rural Third World. National NFE projects are often used by governments to channel and disseminate political programs and state ideology. Yet they are increasingly supporting the formation of adult groups based on some mutual interest in language learning, animal husbandry, tailoring, or some other income-generating skill. Examples of this "functional education" are found in a few parts of Indonesia, where members of "learning groups" (ranging in size from 10-25 participants) pool resources to capitalize projects ranging from chicken-raising to silk-screening tee-shirts to installing a locally-built water pump at the community well. Soon the Indonesian NFE directorate plans to make available "learning funds", seed capital for the projects of learning groups.

This chapter includes several publications on NFE approaches and techniques. **Perspectives on Nonformal Adult Learning** is a concise discussion of philosophical bases and practical approaches for NFE field workers and trainers. **Demystifying Evaluation** is a manual on generating useful criticism, or "feedback", about NFE projects. **World Education Reports**, the quarterly journal of a small U.S.-based development education assistance group, offers a continuing series of thoughtful articles on modest-sized NFE projects in Asia, Africa, Latin America and North America.

Nonformal Education's emphasis on local definition of learning needs to be met with local resources is, in effect, a strategy for educational self-reliance. One potential role for outside organizations is to share useful information and problem-identification techniques among various local groups. **Moving Closer to the Rural Poor** and **Participatory Training for Development** both describe training for leaders ("facilitators") and change agents aimed at improving their skills in working with participatory groups. **From the Field** is a compilation of exercises for NFE facilitators and their trainers, emphasizing increased group problem-solving power through broad participation and mutual trust.

Because the success of any training or other educational project depends on a respectful awareness of the specific local situation, community surveys frequently are an initial step in the design of educational programs. Approaches to "needs assessment" vary a great deal. In this section we review two contrasting examples, **Community, Culture and Care** and **Assessing Rural Needs**. Our experience suggests that objective, quantitative needs assessment *techniques may do little more than confirm the survey designer's biases.*

*Practitioners in nonformal, community-based education have always had mixed feelings about "higher" education in colleges, universities, and academies. On the one hand these institutions are a potential source of bright young people with perspectives and communication skills useful in villages. But on the other hand, in practice they often drain the rural areas of the most talented youth, conditioning them for new roles in an affluent urban-based elite; in this way these students are effectively lost as contributors to village progress. Many countries have established study-service schemes which require that college students live and work in a village before they graduate. However, the stay in the village is usually too brief; the student's work role is undefined, and there may be no practical objective which can be reached in a few weeks' or months' time. Some countries are deepening their study-service schemes and setting up volunteer service programs which place college graduates in villages for a year, two years, or more. These programs generate employment opportunities and provide a valuable educational experience for young people previously unfamiliar with the realities of life for their less-privileged countrymen. However, the volunteers almost always return to the city and their impact on village life is seldom a lasting one. FUNDAEC in Colombia has been developing a program that attempts an alternative strategy for linking higher education to village development (see **Papers describing a training program for engineers for rural welfare**). This program draws students from villages and is explicitly geared to dealing with village-level problems; traditional fields of knowledge are combined in a unique "cross-disciplinary" approach. (The experiences of "barefoot doctor" programs have shed some light on the requirements for successful efforts to develop new rural-based professionals. Health workers have proven to be most effective when they come from the communities in which they will work, and when they are genuinely selected by those communities. See reviews of the IDRC bibliographies **Health Care in the People's Republic of China** and **Low Cost Rural Health Care and Health Manpower Training** on page 241 and in the HEALTH chapter of this volume.)*

Both formal and nonformal community education should be based on local situations and challenges. Such an orientation is also crucial to health care and research-and-development programs (see the HEALTH and SCIENCE TEACHING chapters of this volume) which build on local skills instead of eroding them. Communities with these organizational tools for self-reliance will be in a much stronger position to innovate and to seek, adapt, and apply useful technologies from other places.

Pedagogy of the Oppressed, book, 186 pages, by Paulo Freire, 1970, $4.50 from Continuum Publishing Corporation, 815 Second Avenue, New York, New York 10017, USA.

In this pioneering book, Brazilian-born Freire outlines a humanistic theory of education which has become a cornerstone of people-centered development approaches. To be human is to both act and reflect upon the world; "To exist, humanly, is to **name** the world, to change it. Once named, the world in its turn reappears to the namers as a problem and requires of them a new **naming**. Men are not built in silence, but in word, in work, in action-reflection." Education for humanness is, then, an encounter through language—dialogue—with our own real-life situation. In Freire's work with illiterates in Latin America, words and sentences became powerful tools with which peasants could symbolize and define the problems and contradictions in their own lives.

The most important consequence of this approach is that language and any educational content must be rooted in the world of the learner. "It is to the reality which mediates men, and to the perception of that reality held by educators and people, that we must go to find the program content of education...The starting point for organizing the program content of education or political action must be the present, existential, concrete situation, reflecting the aspirations of the people. Utilizing certain basic contradictions, we must pose this existential, concrete present situation to the people as a problem which challenges them and requires a response—not just at the intellectual level, but at the level of action."

In the Third World, Freire's work has helped initiate approaches to education and development which begin with local realities instead of planners' visions. For the "developed" world, **Pedagogy** stands as a warning against homogenized, mass-marketed language and culture which bury the human dialogue with the world in the conformity of products and slogans.

Doing Things Together: Report on an Experience in Communicating Appropriate Technology, book, 108 pages, by Andreas Fuglesang, 1977, $9.00 surface mail or $11.00 airmail from the Dag Hammarskjold Foundation, Ovre Slottsgatan 2, S-752 20, Uppsala, Sweden.

In this report on a workshop on A.T. in village development, the author makes some major contributions towards a new theory of communication compatible with appropriate technology principles. Among the key issues he identifies: Who chooses what is appropriate? What is leadership in a context of people's participation? What is development?

Because individuals have different skills and information "the mass (community) can carry and handle an information burden far beyond any individual's" capability. "The mass (community) is a perfect communication system...It covers all fields of importance to our society's life. It adapts continuously to changes...The communication flow in the mass is controlled by the interests of the individuals." This has major implications for the way in which problems are identified and solutions proposed. Communities collectively know far more than any single individual.

"Leadership is a communication problem. Decisions must be based on information from the mass. Otherwise they are non-responsive to social realities...The ideal leader is an individual in the mass whose perceptions of the need for social change are ahead of the mass, but who recognizes that the ideas originate in the mass itself."

"It has been commonly assumed among information specialists that an information intervention follows a two-step flow, from mass media through opinion leaders to a number of individuals. This idea offers intriguing opportunities for those who have a manipulative outlook, but it is fortunately not borne out by experience. The opinion-leader theory is probably little more than a superimposition of outmoded authoritarianism on modern sociology."

In this book you will also find many of the conflicts and contradictions among appropriate technology principles. The author speaks of Another Development, in which appropriate technology is linked to increasing the innovativeness of people; yet elsewhere he falls back to economic growth and economic surplus as the measure of appropriateness of a technology, and classical diffusionist theories for bringing a pre-determined technology to the people. Cultural sensitivity and respect for traditional technology also sit uncomfortably alongside a clear message that social change must come.

Much of the content is through the eyes of the A.T. workshop participants;

their conflicting motivations are illuminated. This book perhaps raises more questions than it answers, but it carries us many steps further in our thinking about A.T. and how it might be incorporated into programs.

Many photos of the devices and techniques demonstrated during the workshop, with valuable observations about how that approach succeeded and failed. Required reading for those involved in meshing people's participation and A.T. in the Third World.

Perspectives on Nonformal Adult Learning, book, 122 pages, by Lyra Srinivasan, 1977, $5.00 from World Education, 1414 Sixth Avenue, New York, New York 10019, USA.

Nonformal education (NFE), also known an out-of-school education or adult basic education, is a name for programs which offer people the opportunity to learn certain basic skills outside of schools. NFE programs emphasize the specific learning needs of their participants. Special facilities and certified teachers are not usually needed. Thus NFE is inherently flexible, practical, and low cost. It is an important approach in developing countries where formal schooling is expensive and often irrelevant to the lives of the rural majority.

The purpose of this book is to introduce the basic theory behind NFE and to look closely at current NFE learning techniques. In Thailand, for example, "People in nonformal education programs, especially in the rural areas, are not students by profession: they are farmers and fishermen, mothers and market-women. They already have enough problems of their own: the water-pump does not work, the birds are all over the field eating the paddy, the baby is sick. So the approach selected by the Thai nonformal youth and adult education programs focuses on the real and immediate needs of the learners". Thais apply this "problem-centered approach" by using sequences of photographs to illustrate and spark discussion about community problems. Another approach, "self-actualizing education", emphasizes the capacity of individuals to creatively identify their own problems and goals: "...the pace of development will remain restricted if the full creative and visualizing power of rural communities is not turned on...it is not the outsiders as much as the insiders whose imagination holds the key to a major breakthrough in rural development".

Appendices contain exercises to encourage learner participation in groups and a sample of forms used to evaluate a NFE project for rural women in the Philippines.

Nonformal education is an important strategy for self-reliant rural development. This book is an excellent introduction to the field.

Demystifying Evaluation, book, 69 pages, by Noreen Clark and James McCaffery, 1979, $5.00 from World Education, 1414 Sixth Avenue, New York, New York 10019, USA.

An important task for the staff of any village-level development effort is to ask whether or not the project or program is achieving its objectives and addressing villagers' needs. This short book describes a low-cost, flexible approach to program evaluation. It provides a detailed outline of a practical seminar concentrating on "1) helping program administrators and field staff become aware of the need for evaluation to improve decision making and 2) assisting them to ask the right evaluation questions about their projects. The seminar is not designed to produce experts in evaluation; it is intended to assist administrators to identify and initiate evaluation approaches to improve the operation of their organizations."

The seminar includes small-group discussions exploring the reasons for evaluation, developing common-sense evaluation questions, and collecting data. The greatest amount of time in the one-week seminar is devoted to visits to the "case project", in which teams of participants try out and refine their evaluation strategies in the field. Instructions for leading or "facilitating" the sequence of activities are clear and complete. Emphasis is on getting work done in groups, with authority and responsibility shared among members; the group problem-solving techniques included could be adapted for use in a wide variety of cultural situations.

Recommended as a guide to help project staff develop experience in evaluation techniques. Easy to read and well-illustrated.

From the Field: Tested Participatory Activities for Trainers, three-ring binder, 148 pages, compiled by Catherine Crone and Carman St. John Hunter, 1980, $8.00 from World Education, Inc., 1414 Sixth Ave., New York, New York 10017, USA.

Nonformal education practitioners have, in recent years, reached an important conclusion about their "target" groups of rural people without traditional schooling. While their needs for information and skills are many and varied, their own pooled experience is the most important source of knowledge relevant to solving local problems. Thus horizontal or community-wide sharing and exchange of ideas is a crucial key to meeting local needs. Nonformal educators believe that this kind of communication is most likely to occur in a group of people with a mutual interest, in an atmosphere in which all members share authority and submit ideas.

Many teachers and other leaders have never experienced such a "participatory" atmosphere. This collection of group activities is intended to help them learn about this approach. "(Participatory education) emphasizes mutual learning rather than teaching. In this kind of process, the teachers or leaders or trainers take on some roles that may be different from those they are used to. What they are learning is not so much how to teach nutrition or family planning or moral values. What they are learning, rather, is to work with particular groups of people who are affected by their own unique circumstances."

Most of the exercises require less than two hours. Many involve large and small group discussions, demonstrations, role-playing, interviewing, and eliciting ideas with photos and pictures. Each exercise is presented in the context of a particular training session (most of which were held in developing countries): "We introduced this activity to enable the group to identify important facts they should know about the rural people with whom they work, and how they can collect the information they need to develop effective learning experiences." "The trainers wanted to discover whether village-level facilitators could invent and use games in their own educational activities." "The trainer wanted a group of materials developers to become adept at selecting pictures that stimulate active learning, and to discuss and establish criteria for choosing such pictures. He hoped that this three-part mini-course would help to extend the group's use of visual learning materials."

A sequence of exercises on developing and testing learning activities is included. An appendix contains an excellent outline for an introductory planning workshop on simplified PERT (Program Evaluation and Review Technique).

Recognizing the unique value of every individual contribution to a group effort, participatory learning approaches are an important tool in mobilizing local human resources. This manual provides practical material to accompany the introductory book **Perspectives in Nonformal Adult Learning** (see review in this section).

A PERT chart

Moving Closer to the Rural Poor: Shared Experiences of the Mobile Orientation and Training Team, book, 95 pages, 1979, Rs. 10 in India, $1.50 overseas, from Mobile Orientation and Training Team, Indian Social Institute, Lodi Road, New Delhi 110 003, India.

Practical insights for supporting grass-roots-based development are compiled in this report by a mobile training team. The four-member team has spent 2 years in the villages of India, working among the poor and offering training to other voluntary organizations. The group notes that the major task for voluntary organizations is a difficult one: "to shift the emphasis from a predominantly **managing** role in development to a new role of **facilitating** educational processes" and helping the poor create their own organizations.

"The rural poor are 'voiceless' not because they have nothing to say, but...because they have no 'say' in the decision-making structures of society. In this perspective it is legitimate to say that development begins with listening to the people...Unless we begin with an attitude of respect for traditional knowledge we will never to able to make an objective assessment of traditional practices."

The team reports on their goals and methods, the experiences of one member in agriculture, and the training of illiterate women as basic health workers.

"Often we were surprised to see that the rural poor are not even aware of the resources they have...Our agriculturalist...has been able to concretely point out the many possibilities people had in each place we went, to develop their local resources...Who will take the results of significant research on socially appropriate technology to the people who need it most?...There is scope for a mutual give and take. Those involved in research of the type mentioned often are keen to learn from those working at the grass-roots."

In agriculture, training began with and built upon existing agricultural practices. The team always emphasized working with the marginal and small farmer groups who form the poorest half of the population; the others tend to know how to tap available credit and information resources. Special attention was given to low-cost and no-cost ideas that, once introduced, would spread by themselves. One of the goals of the team was to create locally-based teaching material, using the ideas and images of the people themselves.

"Today there is much talk about 'total revolution' and radical transformation of society. But what really matters are the changes taking place in the socio-economic reality of the villages where poverty crushes the poor. In this stark reality of life the rural poor can hardly envisage more than creating for themselves some free space in society where they can breathe more freely and begin to stretch themselves. What is crucial at the moment is to create a base for joint action which is relatively free from control of the locally powerful. Wherever this has been achieved, people begin to move."

"What does expanding the space of freedom concretely mean? It can mean the ability to reduce maternal and child mortality, to double agricultural production by a scheme of dry farming, to get goats for all the families, to get rid of bondage to money-lenders."

These new experiences can convince the poor "of their capacities and new possibilities of collective action. By analyzing the obstacles they encounter in these endeavors they come to understand gradually the working of society and the deeper issues of a more just society."

Highly recommended.

Participatory Training for Development, book, 59 pages, by Kamla Bhasin, 1977, available from Freedom From Hunger Campaign/Action for Development Liaison Officer, Regional Office for Asia and the Far East, Phra Atit Road, Bangkok 2, Thailand.

This is the story of a training program for a new kind of "change agent" who can facilitate full community participation in solving local problems. Nine people took part in a traveling field level training process that included visits to 50 groups in 3 countries of Southeast Asia, to compare approaches and establish links. The program emphasized self-training and group learning through field visits and dialogue—all approaches that would be important later in the community work of the participants. "The best way to teach about 'bottom-up planning', people's participation and decentralization is by practicing these very ideas in a training program." The report describes some of their discussions, disagreements, and insights, and some of the participatory programs visited.

"Gradually it dawned on the change agents in the program that field workers like them should develop a capacity and knack to write about their work and their observations of village life and its problems...to feed the higher-ups and sympathizers with genuine accounts of issues related to the mobilization of people, cadre-building, rural power relationships, exploitation, injustice, etc... To reverse the flow of information, it is necessary that change agents function also as field level researchers. By learning simple techniques of research they can gather valuable information on the basis of which realistic and sensible decisions and policies can be made by the higher-ups."

"The participatory approach to development emphasizes the need for people to become aware of their own conditions, their socio-economic-political inter-relationships with others (like money lenders, landlords, government officials, etc.). It also emphasizes the need for people to be able to analyze their own situation and to take action to attain self-defined goals. In this approach the responsibility to direct change lies with the people, and not with outside agencies...Change agents must encourage the people to work in groups, because it is only through group action that the poor stand a chance of increasing their bargaining power."

This book provides a useful variety of down-to-earth views on participation, with good ideas on roles for change agents in village-level development efforts.

Highly recommended.

Assessing Rural Needs: A Manual for Practitioners, book, 127 pages, by Jeffrey Ashe, ACCION/AITEC, 1978, $3.95 from VITA.

Don't be surprised to find that this is a very different kind of publication than you would expect from the title. A more accurate title might be: "A Quick General Survey Method for International and Government Agencies". The booklet describes a low cost method, used for 2 years in Costa Rica, for collecting general information on what the authors claim are "priority needs" among the communities in a region. A single interview is held with a group of knowledgeable people in each community, cutting data collection time to **3 hours per community**. "We estimate that a well selected group can give accurate responses for a community as large as 1500 to 2000 inhabitants." This interview process is supposed to provide "a well structured opportunity for villagers and small farmers to clearly articulate their needs to the govern-ment." Information from the interviews is then compiled into reports for use by the government agencies working in the region.

We wonder: what person or group would use or act upon information which outlines community life on the basis of three hours of conversation? And why don't they have the time to get to know the community any better? This is a particularly disturbing offering from VITA at a time when appropriate tech-nology efforts of large aid agencies are tending to incorporate interventionist problem-solving approaches that pay lip service to "people's participation" in projects, but do not include them in any meaningful way in the problem identification process. In the literature of the appropriate technology move-ment, the term "practitioner" has come to mean people who do "hands-on", close-to-the-people work in experimentation, extension, and community organizing, involving technologies for basic needs. Not so in this book: "practitioners" are simply the people who carry out the survey.

No doubt low cost methods of keeping government agencies informed of the general patterns of migration, the condition of water supply systems, and so forth are vital to allocating resources where they are most needed. One would expect, however, that the relevant agencies would already have the informa-tion on most of the topics actually included in this survey. A public works department, for example, should know where it has already installed electricity and where the roads are paved.

In any needs-assessment survey, methods and format are important. The manual offers some useful suggestions on how to word and try out questions, train interviewers, and compile results. The specific topics in the sample questionnaire include migration patterns, agricultural production and problems, employment, water systems, sanitation, credit, electricity, roads and land ownership. Readers are expected to adapt the content and approach to their own information needs. This will have to be done carefully. For example, the method for determining priority needs for water systems is based on a point system, in which it is possible to give a "high priority" rating to the need in a community that has no interest at all in contributing to the water system. (Experienced observers suggest that this is a sure recipe for failure.)

The difficulty with any survey technique is that one tends to find only what one is looking for. A more open-ended process involving more of the community is likely to turn up "felt needs" that are not already predicted by the survey's designers.

Community, Culture and Care, book, 297 pages, by Ann Templeton Brownlee, 1978, free to Peace Corps volunteers from the Office of Multilateral and Special Programs, Peace Corps, Washington D.C.; others $10.95 from the C.V. Mosby Company, 11830 Westline Industrial Drive, St. Louis, Missouri 63141, USA.

" 'Once upon a time a monkey and a fish were caught up in a great flood. The monkey, agile and experienced, had the good fortune to scramble up a tree to safety. As he looked down into the raging waters, he saw a fish struggling against the swift current. Filled with a humanitarian desire to help his less fortunate fellow, he reached down and scooped the fish from the water. To the monkey's surprise, the fish was not very grateful for this aid...' " (from the Introduction).

For a cross cultural health program to be successful, the outsiders involved must understand what is needed and what is possible. This book will help an outsider to develop that understanding by investigating the various aspects of community life which influence health care. The book consists of a series of questions about the community, and strategies for answering them. The questions are grouped into topics for inquiry (e.g. income, attitudes toward work) and broader subject areas (e.g. economics). Quotes from health workers in the southwestern U.S. and Africa are included to illustrate important points. Concerning the Muslim fast (Ramadan) in North Africa: "When community people had their Ramadan, they wouldn't come to the dispensary. They wouldn't eat during the whole day, and for the same religious reasons they didn't want any ointment in their eyes, they didn't want anything in their bodies..."

This book does not give answers. Instead, it emphasizes the best ways of asking good questions. The author stresses the importance of "low-profile" roles for health care planners and workers, and the value of learning by living close to the people. Thus the book is a kind of cross-cultural study guide useful to a wide variety of people (such as extension workers and volunteers).

While it is essential that outsiders approach communities with sensitivity and a keen desire to learn, this should be only part of a process in which local people are supported in their own efforts to address their own problems.

Papers Describing A Training Program for Engineers for Rural Welfare, available from Fundacion para la aplicacion y ensenanza de las ciencias (FUNDAEC), Apartado Aereo 5555, Cali, Colombia.

FUNDAEC is a practical Colombian educational institute which trains agriculturalists, engineers, and other professionals to work in villages and rural areas. In recent years they have created an innovative program to train a new kind of professional, the "engineer for rural welfare". Students are drawn from the rural areas, and provided with a practical training program to develop a wide range of technical skills that will be important in helping their communities solve their own technical problems. This approach is similar to the "barefoot doctor" strategy now being used worldwide in health care systems.

"During the past few decades, it has become increasingly more evident that education on a worldwide basis has reached a level of unprecedented crisis. Among the diverse reasons, one can point to the fact that educational programs have remained marginally effective in providing the means to solve the social problems that confront humanity...New programs must be created to train new professionals who could participate more effectively in the process of development."

Once trained, the engineer for rural welfare would "explore technological possibilities for adapting existing tools from many parts of the world, and when necessary, for innovating those implements which would directly contribute to the welfare of the small farmer. The investigation of these possibilities ought to take into account the socio-economic conditions in the given region and include participation of the small farmer...This process will contribute to both the development of technical knowledge and the creative capacities of the small farmer in the region. The engineer for rural welfare who is being trained is an essential factor to assure effective mechanisms for this participation."

This FUNDAEC program is a pioneering effort of potentially major significance to the way technical problems of small communities may be solved in the future. The ongoing results of the program deserve careful attention from A.T. groups and educators around the world.

Training for Village Renewal, book, 200 pages, by Murray Culshaw, 1975, £1.00 or $2.00 from Lutheran World Service, c/o Murray Culshaw, Burton Cottage, Marehill, Pulborough, Sussex, United Kingdom.

"Rural development is very complex, there are many aspects to it. This handbook is concerned with one aspect; that of training youth in technical skills required to improve social and economic conditions in rural areas...Because people and conditions in the world are so varied no single idea can ever solve the problem of everyone, but today there are many ideas which are definitely working. Many groups, both government and private, have established patterns of training growing directly out of experience in rural areas, and it is these which we believe offer directions for the future."

This book is a directory of groups (most of them private voluntary organizations) working on educational and training programs for village development. These programs emphasize the use of local material and human resources in solving local problems. Most of the book is devoted to short description of hundreds of small innovative agricultural and vocational training projects, and nonformal or "out-of-school" educational programs. Most of the programs are in Asia and Africa, with only a few in Latin America. Also included are short sections on educational research institutions and functional literacy projects.

There is an interesting set of "guidelines for identifying rural skills" relying on the twin concepts "technological needs and the problems which arise in the ordinary life of the villagers" and "work opportunities".

This manual is not comprehensive, and some of its contents are undoubtedly out of date in 1980. It does not present programs in sufficient detail to serve as a "blueprint" for program design. It does illustrate the variety of innovation in local-level rural self-reliance schemes. It is also a low-cost "networking" tool, intended to help small organizations contact other organizations with similar interests.

World Education Reports, magazine published three times a year, about 24 pages per issue, free to serious groups, $5.00 per year to others, from World Education, attn. Martha Keehn, Director of Publications, 251 Park Avenue South, New York, New York 10010, USA.

A thoughtful, clearly-written journal, aiming "...to report on developments in the field, to share findings from innovative programs, and to stimulate thinking on emerging issues" in nonformal education. Articles document small NFE pilot projects (most of them located in the rural Third World) emphasizing: building "horizontal" working relationships in groups; materials development and holding workshops; the potential contribution of NFE to local self-reliance; and the limited roles foreigners can legitimately play in "development from below". Each issue includes an editorial section with letters from readers, and reviews of publications.

Highly recommended.

ADDITIONAL REFERENCES ON NONFORMAL EDUCATION AND TRAINING

A Solar Water Heater Workshop Manual describes a weekend training course given to a community group; see review on page 639.

Rural Small Scale Industries in the People's Republic of China notes that these industries have contributed to the "scientification" of the rural population; see review on page 390.

SMALL ENTERPRISES
AND COOPERATIVES

Small Enterprises and Cooperatives

The difficulties faced by the small business enterprise have always been great, and these problems have become more acute the world over with increasing consolidation of market control by transnational corporations. In the Third World, small producers, manufacturers, and consumers must overcome the high cost of capital (interest rates of 10-20% per **month** are common in the rural areas) and must survive market fluctuations caused by influences beyond their control. Penetration of local markets by cheap goods manufactured in the modern industrial sector, inconsistent governmental tax and import/export quotas and restrictions, and seasonal price cycles in agricultural produce, are only a few of these difficulties.

But because a viable local economy depends on preventing the drain of resources out of the community, small manufacturing and other commercial units seem to be an especially "appropriate" form of business organization. Small businesses channel investments which improve local capital stocks; they develop local skills and increase job opportunities while using local materials; and they allow the diversity which is crucial to a healthy, stable local economy. Significantly, locally-owned small businesses are unlikely to abandon a community in search of lower wages elsewhere.

A relative neglect of institutional support for small businesses, worker-owned enterprises, and cooperatives is found in both the United States and the Third World. On the whole small enterprises give a better return on investment (risks are higher, gains are higher) and create more jobs per unit of capital than large enterprises. Yet small enterprises have great difficulty in obtaining capital, due to the poor match between their capital needs and the operating rules of the capital markets. Compounding the problem is the fact that small enterprises in rich and poor countries are often failing to make best use of what capital and human resources they do have, and they face substantial difficulties in obtaining technical assistance to change this situation. **Small Enterprises in Developing Countries** documents the extent of this problem in the Third World through case studies, and makes program and policy recommendations. **Consultancy for Small Businesses** presents an innovative means for improving management skills in Third World small enterprises through a low-cost system of consulting. In rich and poor countries, a variety of governmental and non-governmental efforts are being initiated to stimulate the growth of small businesses and other enterprises.

One particularly promising type of small enterprise is the cooperative. By pooling resources and functioning as a unit, a group of producers or consumers can operate at a more efficient scale and share the benefits. They may be able to buy in quantity, or store and ship produce to more profitable markets, for example. The cooperative also has great potential as a mechanism for capital formation in the Third World. A group of individuals pooling small monthly surplusses may be able to finance a community improvement project, or establish a credit fund. Clearly cooperatives have great potential as tools to help break vicious circles of poverty and lack of opportunity, yet they have seen only limited success. Especially in the poor countries, cooperatives fall prey to unskilled and corrupt management, domination by ruling local interests, and manipulation by governments intent on using them for political purposes. Cooperatives have tended to be unable to help those most in need. For example, often farmers must own land to qualify for membership in Third World agricultural coops; this excludes tenants and laborers.

Like any tool or technique, the cooperative will be an appropriate organizational form only when it grows out of the aspirations of the people who will be members. The great mass of the literature, however, treats the coop as a predetermined organizational structure that extension agents should convince people to adopt on the assumption it will be good for them. Case studies in the **Handbook for Cooperative Field Workers in Developing Countries** show how this approach is bound to fail. There are, however, a flourishing variety of informal cooperative organizations in the Third World. The "arisan" is an Indonesian capital-formation club in which each individual contributes the same small amount to a monthly "pot", all of which goes to a different member each month. Other Indonesian villagers form savings associations to provide revolving loan opportunities to members. Interest is collected on small loans and paid on savings, at moderate rates. This is, in effect, an alternative to the commercial banks which are unable to respond to most village credit needs. Many experts view effective government support (in the form of training, facilities, consulting, etc.) as essential to a cooperative's success. Yet these Indonesian associations are informally organized, and avoid registration with the government ministry for cooperatives.

In the promotion of cooperatives and small business enterprises, as in so many other activities, it appears that government agencies and NGOs have been too quick to supply answers and promote adoption of fixed organizational structures, and have failed to support local initiatives or strengthen existing organizational forms. Certainly a process may be set in motion by carrying out a dialog with a community about the many forms a cooperative might take. But it appears that best use is made of training and other "inputs" when these are provided only in response to requests from businesses or cooperatives which have already defined their own activities. While traditional extension programs do not operate in this way, there are certainly other, more participatory approaches. Nonformal education, in particular, puts increasing emphasis on supporting group-defined educational and vocational goals. Small scale nonformal education programs might provide one of the best mechanisms for successful support of cooperatives and other small enterprises in the Third World.

In Volume I:

Small Enterprises in Developing Countries: Case Studies and Conclusions, book, 115 pages, by Malcolm Harper and Tan Thiam Soon, 1979, ₤2.95 from ITDG.

"The A&B Soap Manufacturing Enterprise was established by two brothers in 1973...A small factory which produces candy is located in Marcato, Addis Ababa...Mr. Qhoqhome manufactures leather belts at his home in Katlehong Village...During 1971 a Mr. Zabuli, who was a very experienced businessman, applied to the Ministry of Planning to set up a factory to produce Coca Cola, Fanta and Sprite..." These are the opening phrases from some of the case studies in this intriguing little book. Each of these short summaries provides background details and describes problems faced by a particular small business. Each case study is then followed by "a brief note summarizing some of the issues which were raised...including some suggestions as to what a small business promotion agency ought to do in the situation described."

"It is perhaps surprising, in view of its importance, that small-scale business, and ways in which governments can help or hinder it, have not been more intensively studied. A great deal of information is available about various types of appropriate technology, and some work has been done on systems of financing, treated from the point of view of a bank, a trainer, or perhaps a technologist, but very little information has been published on the ways in which the various methods of assistance actually impinge on their 'target'..." The authors note that variations in economic circumstances make it impossible to lay down absolute rules for promoting indigenous enterprise. Still, the cases illustrate certain truths and key factors. Entrepreneurial talent is a human resource found everywhere. On the other hand, small-scale business promotion officers, whose task is to help businesses become profitable, are often faced by businessmen content to operate at little profit or even at a loss. Small-scale enterprises can be established with relatively little capital investment; yet because of their limited resources they are especially vulnerable to supply and demand cycles, shifting import restrictions, price competition from larger enterprises, and innovations which make production techniques obsolete. In a section devoted to analysis of the cases, the authors discuss ways in which governments might help: "The first priority should be...to examine carefully the ways in which all its activities impinge on small business, and to make whatever changes are possible in order to avoid the unintentional discrimination which often arises when regulations, which are basically conceived for larger businesses or transferred from industrialized economies, are applied to small enterprises."

"The case studies demonstrate the need for well-planned assistance programmes...Staff responsible for implementing the programmes must appreciate that the manager of a small business is usually responsible for all the details of finance, production, marketing and personnel, as well as the long-term direction of the enterprise itself. The manager may not distinguish between these functions in his own mind, but any outside intervention is bound to affect the business as a whole, even if it is nominally concerned only with credit, markets or technology."

The book concludes with a short synopsis of how intermediate technologies are a vital part of strategies for small-scale industrialization, and a review of policy suggestions to encourage indigenous enterprise.

This is an excellent companion volume to **Appropriate Technology: Problems and Promises** (see review on page 31). Highly recommended.

Consultancy for Small Businesses, book, 254 pages, by Malcolm Harper, 1976, Ł7.50 from ITDG.

"It is generally realized that small businesses have a particularly important role to play in the development of employment opportunities and economic progress because they are in a far better position than large organizations to make use of 'intermediate technology'. It is not enough, however, for business people to be encouraged to use appropriate technology; they must learn how to decide what is right for their particular business, how to calculate costs and selling prices, how to sell their products and generally how to operate a successful and profitable enterprise. The system described in this manual is, in a sense, a way of conveying 'intermediate management' to small enterprises—the method itself is also labour intensive and may therefore be considered as an example of 'appropriate training'."

"Small businesses are widely scattered...competitive with one another... different from one another...vulnerable. Small business owners are usually not well educated...speak a variety of different languages and dialects...busy, often running their business singlehandedly." Because assistance to business-people involves teaching analysis skills and affecting attitudes and behavior, advice should be available on an individual consultancy basis. Yet "the cost and scarcity of suitable candidates usually means that individual advisory services are impossible...There is clearly a need, therefore, for some system which would enable less qualified and quite inexperienced staff to provide useful advice for small business people."

Part one of this manual explores the potential for low-cost small business consultancy in developing countries, and outlines a service in consultancy training which could be provided by government and/or development agencies, banks, or voluntary organizations. Topics covered include selection of consultant trainees (six or seven is probably a suitable number to start out with), the training period (approximately 21 days of instruction), supervision, administration, financing and evaluating the service. "One of the most common problems of small business people is that they think they need more money but are in fact using the money that they do have in the wrong way. The first object of any diagnosis must therefore be to discover how the businessman is using his capital."

Part two of the manual is the training course itself, nearly 200 pages of detailed outlines of lectures and discussions, role-playing exercises, field assignments, and tests. Hand-out exercises for each training session are also included.

Here is a carefully-assembled learning resource aimed at sharing and developing important skills on a decentralized basis. Highly recommended.

Size, Efficiency and Community Enterprise, book, 129 pages, by Barry Stein, 1974, $5.00 from Center for Community Economic Development, 1320 19th Street, N.W., Mezzanine Level, Washington D.C. 20036, USA.

The purpose of Barry Stein's well-written book seems to be to make a case and sketch the economic context for community owned and operated enterprises. Realizing that his greatest dogmatic obstacle lies in the notion of economies of scale, he spends the bulk of the book reviewing the economies of scale literature, and discussing the factors that determine firm efficiency. Simply stated, his general conclusion is that economically optimal plant size for many industries is quite modest.

Proceeding from the economies of scale discussion, Stein examines the rationale and proper role for community enterprise. He points out that firms actively seek, through advertising and packaging, to differentiate their product from the competition, thus achieving a monopoly position. Stein maintains that price competition is not significant for mass-produced differentiated products. He further maintains that for many staples, competing products are fundamentally the same so that firms spend large amounts to promote product identity. Thus he argues that these products can be the outputs of successful community enterprises, particularly in poor areas. The rationale is that community enterprises would not need to spend on product differentiation.

Although there is little treatment of practical problems such as capital formation, Stein's book should be useful to community organizers and others concerned with community economic development within a capitalistic consumer economy. Although his review of the economies of scale literature would be enlightening, most residents of developing countries may find the rest of the book somewhat outside their interest.

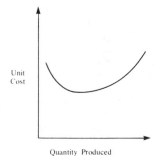

U-shaped curve:
Diseconomies of scale begin
to appear as plant size
continues to increase

Community Canning Centers: A Project Profile in Community Economic Development, report, 54 pages, by Stephen Klein, 1977, $2.50 from Center for Community Economic Development, Mezzanine Level, 1320 19th Street, N.W., Washington D.C. 20036, USA.

Small-scale community canning enterprises, many of them owned on a cooperative basis, have existed in the United States for most of this century. Canning (the preserving of foods in tightly sealed tins or jars) has long been a part of rural self-reliance, as farm families saved their own harvest-time surplus for consumption through the rest of the year. Relatively low-cost community-scale canning technology was developed in the 1930's, and thousands of government-subsidized canning centers were established in the effort to increase food supplies during World War II.

This report is a comparative survey of 16 community canning centers, most of which are cooperatively owned, producing from 7,000 to 12,000 quarts of food per year. The centers use glass jars and/or tin cans, and most of the equipment is hand-operated. Users are involved in the canning process, and locally-grown produce is processed for local consumption.

Whether or not a community canning center is a viable proposition in the U.S. depends on local conditions and initiative, as well as cost and availability of different types of equipment. Key choices for any center include production for personal use or for commercial sales, use of tins or jars, and self-service or staff-service food processing. The author discusses the different combinations of these variables and finds a surprising variety of strategies. "The combination (jars, self-service processing, commercial sale) occurs in upper New

England at the Gardens for All Community Canning Center in Shelburne, Vermont. Small farmers utilized a noncommercial, self-service canning center to process products for sale at their roadside stands, taking advantage of the center to can specialty items. Through direct marketing at their stands, they were able to charge a price that was sufficient to cover costs and still leave a fair profit."

Charts of projected monetary costs and savings for hypothetical canning centers are included, along with appendices on regulatory and technical considerations, and how to calculate project costs (at 1977 prices).

Most community canning centers are unable to cover their investment and overhead costs with proceeds from processing and sales. Government and other agencies often provide subsidies, and membership fees are charged. "In reviewing the costs and benefits of community canning we find ourselves asking why it is that towns, counties, states, and various funding agencies continue to build and support community canneries in increasing numbers despite the need for subsidization...(but) those whose support sustains community canning centers understand that in community economics, profits involve more than a direct dollar inflow. The benefits of community interaction, increased self-reliance, better quality food, and skill-building, plus monetary savings for families and added stability for area growers, are vital enough social reaons to far outweigh the costs of the initial investment and the ongoing subsidization."

Due to the costs of processing equipment and glass or tin containers for the food, community canning centers are likely to be feasible and appropriate only in industrialized countries.

A Handbook for Cooperative Fieldworkers in Developing Nations, series of seven booklets, 408 pages total, edited by Mark S. Ogden, 1978, free to Peace Corps volunteers and development organizations in developing countries, from Information Collection and Exchange, Peace Corps, 806 Connecticut Avenue N.W., Washington D.C. 20525, USA.

Produced to supplement the training of Peace Corps volunteers, this handbook is a compilation of excerpts from publications of international cooperative development groups, university cooperatives extension services, and Peace Corps volunteers (PCV) reports. "The resulting anthology is by no means 'the complete guide' for PCVs...individual variances from country to country and from program to program, make it difficult to arrive at...all-inclusive guidelines." The seven parts of the handbook include an introductory section using case studies of cooperative projects in Bangladesh and Peru, and a directory of organizations active in cooperative work in developing countries. Another section on cooperative organization uses a case history from a Guatemalan regional cooperative scheme to illustrate how indigenous ideas about debt and membership are important to a successful cooperative. A chapter entitled "Determining the Economic and Social Feasibility of a Cooperative" lists key questions for evaluating the potential of marketing, purchasing, and service cooperatives. "Cooperative Education and Training" provides interesting material on participatory group learning approaches (e.g. study circles, role-playing case histories of hypothetical cooperatives). Other useful sections discuss simplified accounting methods, cooperative forms of group credit, and a case history of a Nigerian handicraft marketing coopera-tive.

Some parts of this handbook are taken from books written about cooperatives in the United States; these sections are less directly relevant to conditions in developing countries. Still, the handbook is the best compendium we've seen of on-the-ground experience and insights as to why cooperatives do and don't work. Frequent quotes from PCV field reports show how difficult it is for a foreigner to sensitively enter a community and introduce a new organizational technology: "Recently a volunteer terminated and returned to school. He was well experienced in agricultural technology and a bona fide expert in hog production. He like all PCVs who are well trained, knew that Guatemalan farmers need more money, more protein for their diets, and product diversification. He deduced that his hog knowledge was well suited to meeting the needs of the people. One month after he left, the hog co-op with which he had worked for two years held a meeting to consider what it could do without their expert-in-residence. The result was an immediate dissolution of the co-op, distribution of the 30 hogs to the members and a sale of the assets (purchase price with AID assistance: $5,000) for $500. Each member received more money, ate a little pork, and briefly experienced diversification of production. Is this developing Guatemala?...Another volunteer, dedicated to patterning Guatemalan co-ops after his father's group in Iowa, arranged the purchase of a tractor for his co-op and happily left knowing he had effected progress. Today the co-op has a $200 debt and the implement company has the tractor. The members aren't sure why they couldn't make enough with the tractor to meet payments but are certain that they want no further heavy equipment. They wish the volunteer had not pushed the tractor purchase because now it is difficult to get new members interested in joining a cooperative with a substantial debt."

A useful document for cooperatives extension, non-formal education, and other community development workers.

Cooperative Organization: An Introduction, booklet, 34 pages, by the Co-operatives Panel, Intermediate Technology Development Group, 1977, Ł1.25 from ITDG.

A short, clear summary of cooperative principles emphasizing their potential application in developing countries. Cooperative organization allows individuals to combine their resources for greater economic strength, but experience shows that this pooling of resources cannot be done haphazardly. Democratic control, open membership, limited ownership of shares, and commitment to the cooperative through realization of common economic interest, are generally key factors in the success or failure of cooperative groups.

The advantages of each of the most important types of cooperatives (agricultural and farming, credit, supply and marketing, industrial production, housing) are outlined. Multi-purpose and linked cooperatives are discussed briefly.

The cooperative, as a source of credit and framework for education and innovation, often plays an important role in development schemes. "Comprehensive development projects are becoming more and more common...a typical development project provides a comprehensive package including roads and communications, agricultural extension, credit, supply and marketing services. Sometimes, all the services and control are provided by the project. Inputs are supplied direct, on loan, from the project to the farmer. Increasingly, however, it has come to be realized that this is an expensive way of doing things. Furthermore, there is often no provision for continuity after the project itself has come to an end, and the experts have gone home. Cooperatives are, therefore, being introduced into these projects..."

The booklet concludes with an appraisal of the most common of problems facing cooperatives in developing countries: "The ordinary members know little if anything about the way the cooperative is supposed to work...comparatively large sums of money are handled by people whose own income is relatively small...(domination by) a few wealthy or influential people, who direct the affairs of the cooperative in their own interest."

A brief but well-rounded overview of cooperatives.

Cooperative Organization Papers, booklet, 18 pages, by D.A. Huntington, 1975, $2.00 from META.

"Written as an informational resource for groups (in the United States) that wish to conduct a business under that form of legal organization known as a Cooperative...a special kind of corporation in which members usually contribute the same amount of investment for their shares, always have an equal voice in the direction of the business, and are limited in their liability...by the monetary value of their shares...if you are devoted to a one person-one vote-one monetary share philosophy, then a cooperative corporation is the way to go."

The required legal procedures for incorporation vary somewhat from state to state in the U.S., but often the process is simple enough to do by yourself. The author discusses pros and cons of incorporating on a for-profit or non-profit basis, and how to obtain necessary information on tax status and legal forms. Also included are sample articles of incorporation and by-laws. "Articles of Incorporation...are the papers that must be registered and filed with the powers that be. They give life to that legal individual called the Corporation... Your job in writing the Articles for your co-op is to adapt the general format accepted by the law to the needs and goals of your group. It pays here to be as farsighted as possible. Don't limit your co-op's future activities or possible areas of future interest by being too specific in the Articles, at least in the sections dealing with goals and powers. The best Articles are those that can be lived with no matter what direction your group turns in later years."

The Work of a Cooperative Committee, book, 87 pages, by the Co-operatives Panel, Intermediate Technology Development Group, 1978, £2.25 from ITDG.

"Committee:...the team of people elected by the members to direct their co-operative society." This is a self-guided learning text, in simple English, for members and prospective members of most kinds of primary cooperatives. Simple explanations of the function of the committee within the cooperative organization are followed by thought questions; you turn the page to see the correct answer. While this question-answer format almost seems simplistic, the sequence focusses effectively on the essential role of the committee as leader, protector of the members' interests, and builder of consensus. Lists of questions for study group discussion follow each of the book's eight sections.

An Initial Course in Tropical Agriculture for the Staff of Cooperatives, book, 54 pages, by Peter Yeo, 1976, £1.95 from ITDG.

This, like **The Work of a Co-Operative Committee** (see review in this section), is a self-guided learning text presenting very basic information on soils, fertilizers and plant nourishment, controlling infestations, and animal husbandry. Each part is followed by a progress test.

"If you are a Co-operative officer or otherwise concerned with rural development programmes in a tropical country...(you) ought to know in outline the conditions under which the products co-operatives market for farmers can flourish...The idea that 'agriculture is the job of another department and, therefore, does not concern the cooperative officer' is indefensible."

"The course should help you to cooperate better with people who know more about agriculture than you do...It certainly doesn't justify any feeling of superiority over those who make their living by using traditional methods of agriculture."

The fact that there is a need for a book like this is itself an alarming commentary on the way in which cooperatives are often structured. It seems clear that those people responsible for administration of agricultural cooperatives should have intimate knowledge of the local agricultural situation; certainly the manager of a farming cooperative should be a farmer. Unfortunately, the people who implement cooperatives programs often have had more experience in bureaucratic work than in the productive activities of the cooperatives they supervise. It is partly because of this that cooperatives programs can be seen by farmers as an intervention from outside, willed upon them by a government agency.

Business Arithmetic for Cooperatives and Other Small Businesses, book, 87 pages, by the Co-operatives Panel of Intermediate Technology Development Group, 1977, Ł1.95 from ITDG.

"This manual deals with...business calculations. A good book-keeping system is no good unless the calculations on which it is based are accurate, and it is hoped that this manual will help those many of us whose arithmetic was not very advanced or has gone rusty." In fact, most of the book is arithmetic; a section on calculating interest, mark-up and margin, and gross and net profit, accompanies the basic arithmetic review.

None of the mathematics presented goes beyond what primary school students learn; undoubtedly there are lower-cost publications that can be used to review this material almost anywhere. The explanations and exercises using business calculations might be adapted for use in cooperatives education programs.

Co-operative Book-Keeping: 1—Marketing Co-operatives, 2—Consumer Co-operatives, 3—Savings and Credit Co-operatives, and **4—Industrial Co-operatives**, set of four large booklets, 27 to 51 pages each, by the Co-operative Education Materials Advisory Service, Ł1.50 each from ITDG.

"It is a common complaint that, in many primary cooperatives, the standard of book-keeping is poor. The need has long been recognized for a basic, simplified system of book-keeping, for use in...developing countries, in order to help improve that situation." This system, originally introduced in Botswana, uses a double-entry method: both a credit and a debit are written down for every transaction. Thus, when the accounts of the cooperative are accurate, the sum of credits equals the sum of debits and the accounts "balance". Each of these booklets includes examples and explanations of each type of ledger, deposit, and slip used in the cooperatives' transactions, as well as sample budgets and monthly reports. Exercises require the reader to transform lists of incomes and expenditures into a main ledger.

The importance of accurate, understandable accounts to the success of new cooperatives cannot be overemphasized. "The members are the owners of the society (cooperative). They need to know how their business is doing and how their funds are being used...the book-keeping system must therefore show:

a. How much the society **owes**. These are its **liabilities** and indicate the source of the funds in use in the society.

b. How much the society **owns**. These are its **assets** and show the use being made of these funds.

c. Whether the society has **financial stability** and is able to pay its debts as they arise.

d. Whether the society is **operating efficiently**, covering its costs and providing a net surplus.''

Clearly and thoughtfully written, these materials could be adapted for use by a wide variety of cooperative and small business education programs.

A Single-Entry Bookkeeping System for Small-Scale Manufacturing Businesses, booklet, 54 pages, by Derry Caye, 1977, $3.25 from VITA.

''VITA publishes this manual in the belief that it can help support efforts to increase local self-reliance and to create job opportunities in developing areas by serving as a valuable tool for use by small business managers and advisors...This manual describes a bookkeeping system which is contained in a kit, or carrying case, of some kind. There are suggestions in this manual for building one kind of kit, but the system could be used as well in different containers. The important thing is to package the system in some way so that important records are all in one place.'' The uses of entry books, record-keeping, and files are explained step-by-step. The booklet includes an annex on business letters and a glossary of business terms, as well as a section on how to use inventory and monthly summaries to pinpoint strengths and weaknesses of a business.

The recordkeeping forms explained in this booklet could be stenciled or mimeographed and bound at very low cost. In this way this bookkeeping system would be simplified or adapted to the needs of cooperatives, learning groups, and other group enterprises.

LOCAL COMMUNICATIONS

Local Communications

Historically, one part of the field of development communications has been concerned with dissemination of information from centralized institutions to distant rural recipients. The theory and techniques surrounding "diffusion of innovations" strategies have been based on the assumption that central agencies can successfully identify the information that rural people need (usually without consulting them) and get it to them. What has developed is essentially a marketing approach to dissemination of information and technical innovations, in which attempts are made to persuade the target population to adopt new technologies or otherwise change their behavior. This kind of strategy assumes that the recipient population has no significant and useful traditional technologies, a poor understanding of their own problems, and no capability or interest in problem-solving. (In fact, the term "innovator" has been used to mean not a creative, inventive person, but merely someone who accepts a technology or behavior change recommended by the extension agency.) Programs developed in this style have often promoted irrelevant, unrealistic, or simply incorrect recommendations to a rural population that has become increasingly resistant, denied the opportunity to set their own priorities and seek help tailored to their own perceived needs.

As a consequence, there has been a failure to foster communication systems that respond to information requests. A villager trying to solve a technical problem has a very difficult time obtaining any relevant help from the formal extension agencies, unless he happens to be asking something that is on their list of what they think he needs to know.

Appropriate communications technologies, on the other hand, imply a self-reliant, community-based development strategy. All over the world people rely on local informal networks for technical information. When building a house, for example, a family will tap the knowledge and skills of other members of the community. For many kinds of information, these personal sources appear to be more accessible and responsive than the official channels; in this way, tried-and-true experiences (and useful new ideas) are shared and circulated naturally. Thus a communication strategy for a grassroots-based appropriate technology group might emphasize finding ways to make the technical skills of the group available to those existing informal networks, without simply attempting to tell people what to do. **A Solar Water Heater Workshop Manual** (reviewed on page 639) describes an interesting skills-development and technology extension effort that consciously worked with existing networks of associated people, at their invitation. These people were later able to aid each other in completing their projects.

Grassroots A.T. groups and other community organizations wishing to share their successful ideas with a larger audience will be faced with the problem of how to make this information available in a low-cost, understandable form.

Even in a horizontal information exchange strategy, published documentation has a role to play, when distance or the technical complexity of an innovation demands it. The books reviewed in this chapter should be helpful in choosing ways to present information effectively and at low cost. **Experiences in Visual Thinking** shows how most people can use sketching as a tool for developing ideas in the problem-solving process. **Visual Literacy in Communication** and **Communicating with Pictures** are concerned with finding drawings and pictures, especially those based on common local images, that can be effectively used to communicate ideas from one community to another and among illiterates. **Visual Communication Handbook** and **Screen Printing** offer many ideas on low-cost visual communications media that can be produced in small communities. An example from another section of this book is the Peruvian natural wool dyeing manual **Tintes Naturales**, reviewed on page 772. This innovative publication includes packaged native plant samples and tufts of dyed wood that clearly indicate the colors achieved. Such a format would not have been possible in a high-volume printing operation. **Grass Roots Radio** provides the basic outlines for the low-cost production of radio programs in rural communities, for broadcast throughout a region. Here is an approach for two-way communication, in which rural people can seek technical help and discuss their own successful technological innovations.

Clearly media and techniques like those described in this section are but a few of the possible tools of a grassroots-based horizontal communications strategy. Such an approach will also involve close collaboration between researchers and beneficiaries, and mechanisms to ensure that technical support for problem-solving can be provided when requested by villagers. Visits by individuals from one community to another, technical databanks responsive to rural requests, and low-cost catalogs covering a broad range of topics (like the **Liklik Buk** and **Appropriate Technology Sourcebook**), also have a place.

In Volume I:

Rural Mimeo Newspapers, by Robert de T. Lawrence, 1965, page 262.
Print: How You Can Do It Yourself, by J. Zeitlyn, 1975, page 263.
How to do Leaflets, Newsletters, and Newspapers, by Nancy Brigham, 1976, page 263.
Basic Bookbinding, by A. Lewis, 1957, page 263.
The Organization of the Small Public Library, by I. Heintze, 1963, page 263.
Visual Aids Tracing Manual, by World Neighbors, page 49.
Use of Radio in Family Planning, by World Neighbors, page 50.

Experiences in Visual Thinking, book, 171 pages, by Robert McKim, 1972, $9.95 from Brooks/Cole Publishing Company, 10 Davis Drive, Belmont, California 94002, USA.

'' 'Visual Thinking' is used to describe the interaction of seeing, imagining and idea-sketching.'' This is a book on thinking about design, and how sketches and other representations of design ideas can be a great aid in releasing the creativity of the reader as a designer. The reader is taken through a series of small problems to develop an understanding of many different ways in which 'visual thinking' can aid in design work. The author has drawn from a wide literature on creativity, mental processes, and the history of great inventions.

"Unlearn the stereotype that places drawing in the category of Art... Drawing, most of all, stimulates seeing...Almost everyone learns to read and write in our society; almost everyone can also learn to draw."

"Graphic ideation is not to be confused with graphic communication. The former is concerned with conceiving and nurturing ideas; the latter is concerned with presenting fully formed ideas to others. Graphic ideation is visually talking to oneself; graphic communication is visually talking to others...The graphic ideator...can sketch freehand, quickly and spontaneously, leaving out details that he already understands...he feels free to fail many times on the way to obtain a solution."

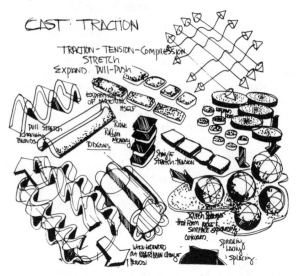

Visually working out ideas with drawings

This is a valuable tool for strengthening the design and problem-solving abilities of individuals and groups. It could be used for a short course for members of an appropriate technology unit.

Highly recommended.

Visual Literacy in Communication: Designing for Development, book, 144 pages, by Anne Zimmer and Fred Zimmer, 1978, from the International Institute for Adult Literacy Methods, P.O. Box 1555, Tehran, Iran.

This is about how to communicate effectively with drawings and pictures. It is intended for use by artists and most development workers (good English language ability required). The author presents a systematic strategy for improving the effectiveness of drawings, with a lot of examples.

Many posters and drawings fail to communicate what is intended. Partly this is because "most communication theory as we know it today was developed in the industrialized West...it takes little notice of the kinds of communication—visual and otherwise—that have been important in spreading and maintaining traditional...cultures." Foreign methods of visual presentation can be as hard to understand as a foreign language.

"The first job of the visual communicator is not to draw pictures. It is to find out what visual communication is already going on among the people he wants to reach, and to get the other information he needs in order to design materials that communicate properly. To do this, he makes a collection, called a 'visual inventory'. Instead of putting together elegant designs from all over the world, he samples the visual communication his intended audience already sees. Then he finds out how—and whether—these examples communicate by asking question..."

"The message is: read your own culture and understand your own visual language as you design visual messages for use in your particular cultural setting."

Improving realistic drawings. Before: viewing angle does not make the important action clear. Details are confusing and contrast is weak.

After: new composition, stronger contrast, and details all focus on the important activity.

Communicating with Pictures, booklet, 24 pages, by UNICEF-Nepal, 1975, available on request from UNICEF, P.O. Box 1187, Kathmandu, Nepal.

This booklet is a summary of a full report also obtainable from UNICEF in Nepal. The booklet describes the results of a study that was undertaken to discover how effectively pictures could be used in communication.

"Is it possible to communicate ideas and information to villagers by using pictures only? Probably not. In the course of the study, over 20 pictures intended to convey ideas (rather than just to represent objects) were shown to villagers. Many (but not all) of the villagers could recognize the objects shown in the pictures. But the ideas behind the pictures were almost never conveyed to the villagers. For example, one picture was intended to convey the idea that people who drink polluted water are likely to get diarrhea. It was shown to 89 villagers, and only one of them understood the message behind the picture."

The reasons for the failure of pictures to convey ideas are thoroughly discussed. Many different types of pictures (illustrations, sketches, photos, and other graphics) were used: the disadvantages and advantages of each type are covered. The effects of colors are mentioned too.

"People are interested and attracted to pictures, even though they may need help to interpret them...During the study one picture was taken to six villages and shown to over 100 people. In five villages, none of the villagers who saw the picture could understand it. But in the sixth village, many villagers could explain exactly what the picture meant. They could understand it because five months before, some health workers had visited their village and talked about

TB, and had shown them this picture.''

Anyone attempting to use pictures to communicate ideas and information would find this booklet useful. Highly recommended.

Lifting big pumpkins is a trial of strength in some parts of Nepal. The artist intended to portray "strength" and "weakness" with these two pictures. However, less than 5% of the villagers who were shown these pictures said that one man was strong and the other weak. 24% understood the pumpkins to be flowers.

Visual Communication Handbook, book, 127 pages, by Denys J. Saunders, 1974, Ł2.75 from TALC.

The subtitle of this book is ''Teaching and Learning Using Simple Visual Materials.'' The author does an excellent job of explaining how to use a great variety of inexpensive visual aids, including paper pictures, posters, flannel boards, and puppets. This book would be useful to anyone trying to carry out an education program as cheaply as possible.

A simple device (a pantograph) is shown which can be used ''to make enlargements up to eight times the size of the original. By means of a screw you fix the pantograph to the table or the drawing board. With the pointer you trace the lines of the original picture and a pencil draws the enlarged picture on another sheet of paper.''

A very useful book.

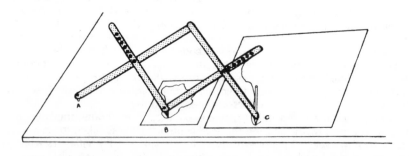

A pantograph can make accurate enlargements easily

57 How-to-do-it Charts on Materials & Equipment & Techniques for Screen Printing, book, 63 pages, by Harry L. Hiett, 1959, $3.00 from META.

Screen printing (also called 'silk screen printing') is an excellent way to print pictures and words on leaflets, posters, clothing and other materials. The methods of screen printing are very thoroughly described in this book. There are a very large number of illustrations, which will be very helpful to those people who don't read English easily.

Highly recommended to anyone looking for information on screen printing as a means of cheaply and efficiently producing high quality graphics.

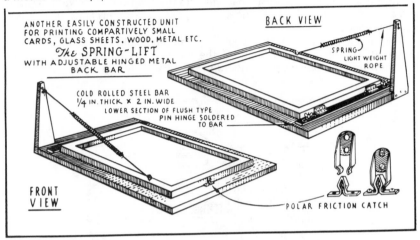

A spring lift on a silk screen printer

Grass Roots Radio: A Manual for Fieldworkers in Family Planning and Other Areas of Social and Economic Development, book, 66 pages, by Rex Keating, 1977, $6.50 (contact if currency difficulties prevent you from paying for this), from Distribution Department, International Planned Parenthood Federation Publications, 18-20, Lower Regent Street, London SW1Y 4PW, England.

"Rural broadcasting, as practiced in most developing countries, is a one-way line of communication, the specialist or government official instructing farmers and other members of the rural community...But any form of adult education yields its best results when the communication is two-way and this is the secret of the Farm Forum success..." By 1974, most villages in Senegal were listening to farm forum broadcasts, many of them holding organized discussion groups. Members of the broadcasting team "are always on the move, systematically covering the countryside, and in each village they hand over their microphone to anyone willing to use it. The program's producers insist that 3/4 of the time on the air, in the three weekly programs, is devoted to what the villagers have to say...The broadcasts embrace all aspects of the rural scene, from animal husbandry and crop production to public health and prevailing market conditions. (This broadcast) has brought about a better mutual understanding between farmers and the officials who run the technical services of the countryside."

This manual is intended to introduce the techniques of successful low cost production in rural areas of taped interviews and scripts for broadcasts. The reader is expected to be a development field worker who through the use of this book will be able to produce good quality tapes on topics in his or her area of activity. Written primarily as a guide for use by family planning workers, it has a bias towards information dissemination from a central authority rather than grass-roots information sharing.

Central to the production of low cost grass-roots radio programs is the use of cassette recorders, which are now available at reasonable prices and capable of excellent performance. Steps for the operation of these recorders for best results are presented.

The author offers some ideas that will help the fieldworker make interviews and scripts appealing to the listeners. He suggests how to organize discussions and news shows.

The language used in this book is sometimes difficult, and will pose problems to field workers. Some of the suggestions for script writing and interviewing are relevant primarily in the English language.

Cassette recorder for making grass roots radio programs

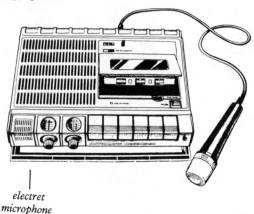

electret microphone

ADDITIONAL REFERENCE ON LOCAL COMMUNICATIONS
Small Technical Libraries has some ideas that may help in organizing a small appropriate technology library; see review on page 382.

BEEKEEPING

Beekeeping

Beekeeping—the controlled raising of bees in hives to obtain honey—has a very long history. In the industrialized countries, beekeeping is a solid income-generating venture for many people, and a hobby for many others, both in rural and semi-urban areas. The basic piece of equipment in these countries is the rectangular wooden hive with interchangeable parts, including movable frames into which commercially-made wax honeycomb foundation can be inserted, to control and accelerate honey production. **The Beekeeper's Handbook** provides a practical summary and guide to the tools and techniques of beekeeping as it is practiced in the rich countries.

Tree trunks, hanging logs, baskets, and jars are among the simpler hives traditionally used by beekeepers in the Third World. Beekeeping could play a greater role in supplementing rural incomes in these countries. Yet almost all of the commercially published beekeeping literature is on techniques and equipment (such as rectangular wooden hives) used in temperate climate industrialized countries (see pages 265-268). **Beekeeping in Rural Development** and **Apiculture in Tropical Climates** are efforts by apiculturists and rural development agents to counter this trend, and initiate sharing of knowledge about many different traditional beekeeping systems. Improved "hybrid" methods should result, some of them "intermediate" between indigenous and manufactured technologies.

An excellent example of a promising "hybrid" is a modification of the Tanzanian top bar hive. **A Beekeeping Handbook** provides step-by-step instructions for making a low-cost cowdung and cardboard version of this simple hive now being promoted in Botswana. Several other simple, low-technology hive designs are presented in **A Beekeeping Guide. Home Honey Production** is included because it presents a technique for making wax honeycomb foundation, which should be of interest in tropical countries where distance, cost and high temperatures make it difficult to get commercially-made wax foundation in good condition.

Beekeeping in Rural Development, large paperback book, 196 pages, Ł2.50 from Commonwealth Secretariat Publications, Marlborough House, London SW1Y5HX, U.K.

This collection of 20 papers reviews beekeeping practices and potential in developing countries of the Commonwealth. Though there is some overlap in content with **Apiculture in Tropical Climates** (see review in this section), this book deals specifically and extensively with traditional practices and the introduction of adapted or new methods. Indigenous techniques and current development programs are discussed for 9 African nations, India, Sri Lanka, the Guianas, Belize, Panama, and the Pacific islands. Photographs of Kenyan log hives and Tanzanian pegged-bark hives are included.

An introductory article presents a valuable summary of geographical distribution of colony-forming honeybee species, honey production and trade, and traditional vs. modern equipment and methods. "Traditional hives are simple containers made of whatever material is used locally for other containers; hollowed logs, bark, woven twigs or reeds, coiled straw, baked or unbaked clay, plant stems and leaves, or fruits such as gourds. In the tropics and subtropics almost all these hives lie or hang horizontally. In the most primitive form of beekeeping the bees are killed or driven out once or twice a year when the honey and wax are taken, the colony being destroyed in the process...At the other end of the scale are the movable-frame hives used in modern apiaries throughout the world, which consist of a tier of accurately manufactured wooden boxes...Between these two extremes—each irreplaceable in its appropriate context—there are various 'intermediate' hives that provide some of the benefits of movable-frame beekeeping with a much reduced need for precision...In movable-comb frameless hives, used successfully in development programs in East Africa...the rectangular frame fitted with foundation wax is replaced by a top-bar only, rounded on the under side and smeared with wax (or perhaps supplied with a narrow strip of wax). The top-bars must be at the correct distance apart to give the bees' natural intercomb distance (beespace) but that is the only precision measurement."

A useful overview of beekeeping's potential as a low-cost, appropriate technology for supplementing rural incomes in many parts of the developing world.

Apiculture in Tropical Climates, large paperback book, 208 pages, edited by Eva Crane, 1976, Ł15 ($28) from International Bee Research Association, Hill House, Gerrards Cross, Bucks., SL9 0NR, England.

This large volume is the full report of the First Conference on Apiculture (beekeeping) in Tropical Climates. The twenty-six papers presented to the conference describe apiculture popularization and development programs in several Asian, African, and Latin American countries.

"The Senegalese authorities began to take an interest in beekeeping development and rationalization in 1962...A transitional stage was undertaken to get beekeepers accustomed to intermediate hives (with frameless movable combs). The Rivka hive, of wood, has rope slings so that it can be hung in trees like a traditional hive, The David hive is made of straw, bamboo and reeds...These two hives are not so very different from the traditional hives but make better management possible: examination of bees and combs; honey harvesting without destroying colony; and a real increase in production."

Other papers cover traditional bee management practices (using log hives) in Africa, honey and wax quality and processing, the various species of honeybees in the tropics and subtropics, and crop pollination.

The editor notes: ''In reading the papers...several points have struck me especially. One is the immense variety of conditions in the tropical regions where beekeeping could be extended, usefully on an economic basis, and with

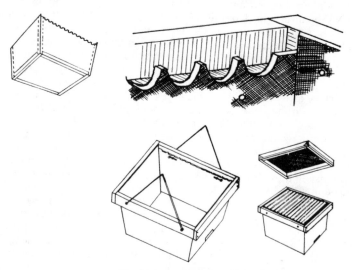

Bar hive now being marketed

reasonable safety on an environmental one. A second is the immense size of the tropical regions...they must represent at least 40% of the earth's total land area where beekeeping would be viable. The movable-frame hive was first developed and promulgated in 1851, and...an intensely inventive phase of beekeeping came in the fifty years that followed. I think that in the tropics beekeeping is now at the threshold of a similar phase of innovation and expansion.''

A Beekeeping Handbook, 65 pages, by B. Clauss and L. Tiernan, price unknown, write to Ephraim Kilon, Beekeeper, KRDA, P/Bag 7, Molepolole, Botswana.

Here is an excellent combination: a primer on honeybees and a manual for setting up and keeping colonies using simple low-cost equipment. ''On a small scale the prospects of beekeeping in Botswana are good...(it) can be completely home based; the hives are made in Kanye and Molepolole or the individual can try constructing his own, from a cardboard box and cowdung.'' Both the simple manufactured hive and the cowdung hive (a cardboard box strengthened and protected with a plaster of cowdung and sand) are of the top bar type, and do not require frames or commercial comb foundation. A smoker made from a tin can and a feather (for brushing bees off combs) are the key accessories. The handbook gives detailed instructions on starting a colony from a swarm or capturing an existing wild colony. Appendices discuss problems and pests, costs of hives and materials, and honey production as a source of income.

Photographs show children doing all of the handling operations. Clear, convincing; a welcome document on low-cost beekeeping methods. Highly recommended.

Hiving a Swarm

cut away excess leaves, branches, flowers, etc.

cut main branch

spray bees with medicated syrup or water

shake swarm in front of prepared hive

The Beekeeper's Handbook, large paperback, 131 pages, 1978, by Diana Sammataro and Alphonse Avitabile, $7.95 from Peach Mountain Press, Box 126, Dexter, Michigan 48130, USA.

"There are hundreds of beekeeping books, but there is an almost universal complaint that beginners' books are not sufficiently explicit...(this book) will not only give you good understanding of the life history and behavior of bees, but it also tells you how to manage bees, how to control their diseases, how to remove and process honey, and many other 'how-to-do-it' aspects." Especially useful for its simple, clear discussions of bee behavior and various methods of locating, starting, feeding and maintaining hives. The authors assume that beekeepers will buy commercial hive parts, but line drawings and text may provide enough information to improvise some equipment.

A clear, comprehensive introduction to beekeeping.

Home Honey Production, book, 72 pages, 1977, by W.B. Bielby, £1.45 from EP Publishing Ltd., East Ardsley, Wakefield, West Yorkshire WF32JN, United Kingdom; write for addresses of local distributors in India, Philippines, Kenya, Malaysia, Singapore, and West Indies.

A "do-it-yourself" manual for the beginning beekeeper, less complete than **The Beekeeper's Handbook** (see review in this section). This book, however, includes drawings and instructions for making hives, plaster molds for wax foundations (the patterned surface on which bees build honeycombs), a solar wax extractor, and candles. Dimensional drawings and a list of materials are included for the catenary hive, which is shaped so that bees will build their combs on a home-made foundation hanging from a single strip of wood. This design eliminates the rectangular frames and beespaces of a conventional hive.

One step in making a plaster mold for wax foundations

A Beekeeping Guide, booklet, 34 pages, by Harlan Attfield, illustrated by Marina Maspero, Technical Bulletin No. 9, $3.00 from VITA.

This booklet presents construction details for beehives made of wood, tree trunks, clay jars, woven bamboo/reeds/straw, and empty kerosene tins. A

Bellows smoker

honey extractor and several smokers are also included, along with guidelines for selecting sites, caring for hives, and choosing proper clothing. This booklet was originally published in Bangladesh by the Appropriate Agricultural Technology Cell.

Making and Using a Solar Wax Melter, leaflet no. 2788, plans, 3 pages, 1975, one copy free from Publications, University of California, Division of Agricultural Sciences, 1422 Harbour Way South, Richmond, California 94804, USA.

MATERIALS: wood, sheet metal, 2 pieces of glass, nails or woodscrews, 4 small metal or leather hinges, black paint
PRODUCTION: hand tools

This glass-covered box uses solar heat, collected in a black metal pan, to melt and recover beeswax from old combs and hive scrapings. Drawings showing construction details are very clear. A metal pan measuring 24'' by 36'' by 6'' can recover wax from up to 60 hives. The size can be varied. This melter can be closed during operation, protecting the wax from robber bees.

Solar wax melter

SOME HOME INDUSTRIES

Some Home Industries

Cottage-scale industries which produce goods for local consumption play an important role in a healthy rural economy. (See the chapter on SMALL ENTERPRISES AND COOPERATIVES for a discussion of how small businesses might contribute to local economic self-reliance.) Many of the tools and equipment found in other chapters could be made or used in cottage industries. This chapter contains reviews of publications on a few basic processes: dyeing fibers, weaving cloth, sewing machine maintenance, and soapmaking. Plans for a small food texturizer are also reviewed.

Make Your Own Soap, booklet, 16 pages, by Department of Social Welfare and Community Development (Ghana)/African Bureau of the German Adult Education Association, from German Adult Education Association, Africa Bureau, P. O. Box 9298, Accra, Ghana.

This book is valuable for two reasons: First, it shows you how to make several types of soap, using materials available in Ghana and almost everywhere else. Secondly, it's an interesting example of a booklet in which words and multi-colored screen printed illustrations are mixed to produce an inexpensive effective product.

Soap making, based on indigenous technology, used to be a very prominent local industry, particularly in the rural areas of the country...The practice is being revived as a major contribution to the Self-Reliance Programme...this soap making campaign is particularly geared towards the younger folks who could take this up as a trade instead of roaming the cities and towns, desperately looking for jobs which are not there.''

Also describes how to teach soap making by conducting public soap making demonstrations. Highly recommended.

Making Homemade Soaps and Candles, book, 46 pages, by Phyllis Hobson, 1974, $2.95 from Garden Way Publishing, Charlotte, Vermont 05445, USA.

Lots of recipes for making soap with animal fat and leftover kitchen grease, and for making candles out of animal fat, wax or paraffin. Includes instructions for making lye out of wood ashes (needed in soap making).

*Cutting soap
with a wire tool*

Basic Sewing Machine Repair, book, 63 pages, by K. Kiri and S. Kalmakoff, March 1979, $2.30 in South Pacific region, $3.50 in Asia, Africa, Latin America, $5.00 in Australia, Europe, and North America, from South Pacific Appropriate Technology Foundation, P.O. Box 6937, Boroko, Papua New Guinea.

A well-illustrated book on the proper adjustment and care of several common varieties of sewing machines. Oiling the machine, adjusting and fixing the stitch regulator, replacing broken springs, and adjusting needle timing are among the topics presented. A trouble-shooting chart helps in identifying the likely source of specific problems. Very simple language is used along with the 200 drawings.

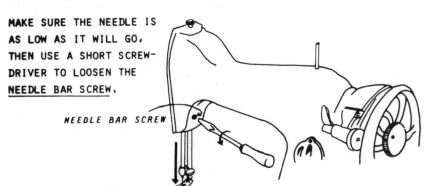

MAKE SURE THE NEEDLE IS AS LOW AS IT WILL GO, THEN USE A SHORT SCREW-DRIVER TO LOOSEN THE NEEDLE BAR SCREW.

NEEDLE BAR SCREW

Tintes Naturales, book, 90 pages, by Hugo Zumbuhl, 1979, $8.00 from S.E.P.A.S., Apartado Postal 53, Huancayo, Peru, or $12.00 from Swiss Center for Appropriate Technology (SKAT), Varnbuelstrasse 14, CH-9000 St. Gall, Switzerland.

Written in Spanish, this well-illustrated natural wood dyeing manual is a remarkable attempt to better communicate with the campesinos of the Peruvian Andes. Samples are included of chemicals used to make the plant dyes more permanent; there are also drawings and real samples of native plants and insects along with small tufts of the dyed wood that give a clear indication of the colors achieved. The quantity of plant material and the recommended method of dyeing are given for each plant/color combination. Wool preparation before dyeing is also discussed.

Instructions for one natural dye

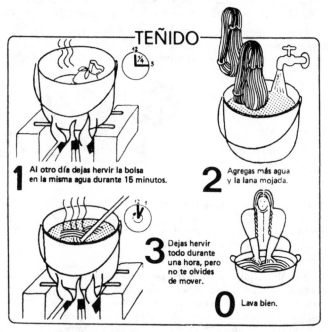

TEÑIDO

1 Al otro día dejas hervir la bolsa en la misma agua durante 15 minutos.

2 Agregas más agua y la lana mojada.

3 Dejas hervir todo durante una hora, pero no te olvides de mover.

0 Lava bien.

Vegetable Dyeing: 151 Color Recipes for Dyeing Yarns and Fabrics with Natural Materials, book, 146 pages, by Alma Lesch, 1970, $9.95 from META.

This book is about dyes from vegetable and other natural sources (such as clay and insects). A large number of the dyes are from tropical and sub-tropical plants in addition to temperate zone plants. The simple recipes and techniques can be used by a beginner. Sources for dyeing equipment and materials are listed. General principles of dyeing are covered, along with specific instructions for particular dyes.

The author notes that different readers will produce slightly different shades when following the same recipe, due to water composition, timing, temperature and other factors. "Time of year when the dyestuff is collected perhaps most influences the final color. The amount of moisture during a season, the number of daylight hours, and the type of soil where the plant grows are also factors that will affect its dye properties. Generally, parts of the plant above ground need a lot of sunshine to produce strong dyes. Barks may be an exception.''

A dye substance information chart lists the common name (but not the Latin name) of each plant, the part of the plant required, the time of year for harvesting (in northern hemisphere temperate zones), and methods of preservation. A color information chart lists the proper cloth to use, the color, the proper mordant (dye preservative to prevent fading), and the relative performance of the dye. There is a bibliography and an index.

Handloom Construction: A Practical Guide for the Non-Expert, looseleaf manual, 163 pages, by Joan Koster, 1979, $6.95 from VITA.

MATERIALS: boards, sticks, or tree limbs up to 10 cm. diameter; dowels, wood glue and screws.

PRODUCTION: drill, saw, hammer and a few other woodworking handtools.

''With inexpensive machine-made cloth increasingly available almost everywhere, it seems likely that fewer and fewer people will be interested in producing their own cloth...Yet weaving can be done in one's spare time using free or inexpensive fibers available locally, and simple, efficient looms can be built from local materials at little cost. Therefore, as long as the loom and fibers cost little, the finished cloth requires an investment in time rather than money...Because people all over the world have been weaving since the very earliest times, there are many styles and varieties of looms. This is a book about building and using some of these. Three types of looms, including two variations of a foot-powered loom, are presented here. The book gives 1) detailed directions for building each kind of loom, 2) the advantages and disadvantages of each, and 3) instructions for weaving.''

Large, clear line drawings show materials, construction sequences, and weaving techniques for frame looms, pit and freestanding footpowered looms, and the Inkle loom (a small loom for rapid weaving of strong strips of cloth). All the looms are made from low-cost, commonly available materials. The choice of loom will depend upon the types of fibers available and the kind and quantity of articles to be woven. Tables show fiber and product types and their suitability for the various loom styles. Planning weaves and patterns, finishing fabrics, and use of colors are also discussed. A well-written reference.

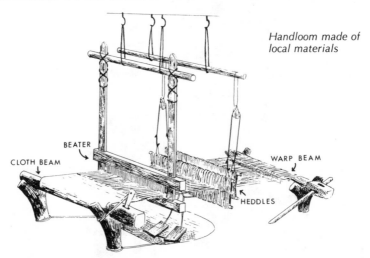

Handloom made of local materials

BEATER

CLOTH BEAM

WARP BEAM

HEDDLES

The Village Texturizer, booklet, 76 pages, Meals for Millions Foundation, 1977, $3.95 from VITA.

MATERIALS: scrap motor vehicle parts, mild steel (flat, plate, rod, bar stock), nuts and bolts, wood

PRODUCTION: metal hole drilling, cutting, shaping and metal turning

This hand-operated device was adapted from a Korean design used by street vendors to make snacks from sweet potato pellets. This modified version creates texturized food products from high-protein, low-fat flours (from legumes such as soy, peanut, or chickpea), seeds, and dried vegetables. The products do not spoil quickly and are easily digested, especially by children. A variety of foods with different protein and calorie levels and suitable flavorings can be produced.

There are good construction drawings and detailed sections on operational costs and nutritional composition of raw materials and end products.

"The machine described in this manual is an excellent example of an intermediate technology: construction costs are low (it can be built with pieces of metal and old auto parts for roughly $50); operation is labor-intensive; it requires no special knowledge (only experience) to operate and a minimum of maintenance; it can produce a wide variety of products which are both highly nutritious and tasty; and it can be used in a variety of situations—from home to small business."

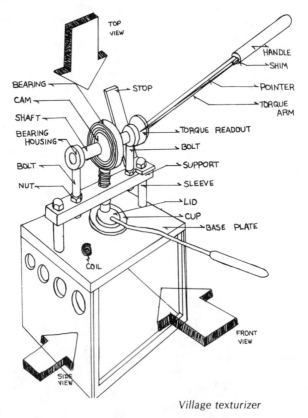

Village texturizer

PERIODICALS

Periodicals

Magazines, journals, and newsletters are important tools for documenting appropriate technology developments and discussing other issues related to community self-reliance. Most of the periodicals listed in volume one (see reviews on pages 280-284) are edited in the industrialized countries; in this section we also present a number of publications produced in the Third World which provide more locally relevant (and more specialized) perspectives. Some of these items are published using few sophisticated processes or equipment, and as such are interesting attempts to disseminate information regionally at low cost. Other periodicals that focus on topics within one of the chapter divisions can be found in the relevant chapters.

In Volume I:

Mazingira, quarterly journal, published in English, French and Spanish, about 100 pages average length, subscriptions US$6.00 per year ($3.00 per year for individuals in Africa, Asia and Latin America), from Subscriptions Fulfilment Manager, Pergamon Press Ltd., Headington Hill Hall, Oxford OX30BW, U.K.

Mazingira, which means "environment" in Swahili, is a forum on environmental issues in development, supported by the United Nations Environment Programme. The focus is on problems in the Third World, but the context is global; the linkage of all development issues to the planet's carrying capacity is a central theme. Articles by contributors from all over the world are some of the most thoughtful in all the development literature related to appropriate technology. Each issue contains a short section surveying potentially important applications of appropriate technologies.

Available at low cost. Highly recommended.

The Ecologist, magazine, six to ten issues per year, average length about 60 pages, 80p per issue, subscriptions from Maria Parsons, Managing Editor, 73 Molesworth Street, Wadebridge, Cornwall, United Kingdom.

Subtitled "journal of the post-industrial age", this magazine presents broadranging articles on humankind's changing place in the environment. Contributions are from scholars, technologists, scientists, and authors from around the world. The contents reflect a belief that an understanding of worldwide ecological problems should affect choices made by individuals.

Includes editorials, letters, and a section of book reviews. Highly recommended.

Ecodevelopment News, journal published monthly in English and French, average length about 80 pages, subscriptions free from C.I.R.E.D., 54, Boulevard Raspail, Room 309, 75270 Paris CEDEX 06, France.

Like **Mazingira**, this is published with the support of the United Nations Environment Programme. Articles discuss the emerging concept of eco-development and examine village development efforts in light of the concept. An "Ecodevelopment at Work" section in each issue reviews projects and literature on promising tools and techniques for environmentally sound development. Also included are brief reports of international conferences and seminars on the topic.

A good networking tool; may be difficult reading for non-native speakers of English or French.

Impact, magazine, monthly, 30 pages average length, approximately $6.00 per year to subscribers in Asia; $10.00 in Australia, PNG, New Zealand, Africa and Latin America; $12.00 in North America and Europe. For nearest distribution outlet write IMPACT, P.O. Box 2950, Manila, Philippines.

Subtitled "a monthly Asian magazine for human development", Impact presents a combination of Asian current events and feature articles on the lot of Asia's rural and urban poor. With correspondents and outlets in many of the countries in the region, this publication may reach more development workers at the grass roots in Asia than any other of its kind.

An effective link among voluntary and humanitarian individuals and organizations, available at low-cost. Highly recommended.

New Internationalist, magazine, monthly, 30 pages average length, about US$20.00 per year depending on location; for Australia: P.O. Box 82, Fitzroy 3065, Victoria; for Papua New Guinea: Wanfolk Publications Inc., P.O. Box 1982, Boroko; for India: Central News Agency Private Ltd., Delhi 110001; for U.S and Canada: New Internationalist, 113 Atlantic Avenue, Brooklyn, New York 11201; for U.K. and rest of the world: New Internationalist, Montagu House, High Street, Huntingdon, PE18 6EP, Cambridgeshire, U.K.

"The New Internationalist exists to report on the issue of world poverty; to focus attention on the unjust relationship between rich and poor worlds; to debate and campaign for the radical changes necessary within and between nations if the basic needs of all are to be met; and to bring to life the people, the ideas and the action in the fight for world development." Each issue discusses a particular topic or theme (e.g. "The Struggle for Control of Third World Farming", "The Rich World's Poor", "Women and World Development"). Articles relate individual human poverty and oppression to exploitative national and international structures.

A good educational resource with a liberal perspective on a broad range of development topics. Includes a section of book reviews.

Asian Action: Newsletter of the Asian Cultural Forum on Development, bimonthly, 20 pages average length, $12.00 per year from ACFOD, Room 201, 399/1 Soi Siri off Silom Road, Bangkok-5, Thailand.

"**Asian Action**...attempts to reflect Asian initiatives in development; to provide information on regional trends and to pin-point the effects...of current development strategies being followed by many governments and institutions." The newsletter reports on the plight of Asia's rural poor, and discusses the kinds of helping initiatives that might be undertaken by non-governmental organizations (NGOs). Also included are reports of ACFOD's conferences, seminars, and training sessions for development workers aimed at building cooperation among such organizations in the region.

Reading Rural Development Communications Bulletin, newsletter, 25 pages average length, published quarterly, available from University of Reading Agricultural Extension and Rural Development Centre, London Road, Reading RG1 5AQ, U.K.

Not a technical bulletin, this magazine contains in-depth critical essays and case studies exploring the philosophy and politics of Third World rural development and appropriate technology. Up to half of the bulletin is devoted to long and thoughtful book reviews, including information to help the more isolated reader obtain hard-to-find volumes.

Ap-Tech Newsletter, published 3 or 4 times a year, average length 14 pages, subscriptions $2.00 per year plus postage from ATDA, Post Box 311, Gandhi Bhawan, Lucknow 226 001, India.

This is a collection of "concepts, views, news, research and development reports, practices, and techniques related to appropriate technology", published by the Indian partner and affiliate of ITDG, the Appropriate Technology Development Association. Most items are short summaries of Indian training courses and projects, although some articles deal with developments in other countries. Also includes a section of annotated book reviews. Recommended.

ADAB News, newsletter, monthly, 25 pages average length, annual subscription TK 35 in Bangladesh, $5.00 or TK 75 in other countries, $10.00 or TK 150 airmail to other countries; from ADAB, 79 Road 11A, Dhanmondi, Dacca-9, Bangladesh.

Information for and about appropriate agricultural development in Bangladesh, of interest throughout the humid tropics. Issues contain articles and case studies within one general topic area, such as rural women's organizations, livestock, integrated pest management, or grain storage. There are also articles on specific pests or crops, notably tree crops. Some issues are devoted entirely to one subject, offering opposing viewpoints.

R.E.D., bimonthly newsletter published in English, French and Spanish, 12 pages average length, yearly subscriptions $3.00 in Guatemala, $4.00 in

Central America, Panama and North America, $4.50 in South America, $6.50 in Europe, Africa, Asia, Australia, Oceania, and $7.00 in the Philippines and Hongkong; from CEMAT/R.E.D., Apartado Postal 1160, Guatemala City, Guatemala, Central America.

Reports compiled by the Mesoamerican Center for Studies of Appropriate Technology (CEMAT) on experiments in agriculture/nutrition, non-conventional sources of energy, rural health, alternative construction, and non-formal education. "The newsletter's main function is to be informative and a link between groups that are working in rural development with appropriate technologies in the Mesoamerican region...For the preparation of the newsletter we are mainly interested in experiences of campesino groups that have experimented directly with an appropriate technology, that has solved some problem in the community and has generated a certain amount of approval by the groups that have known of its existence. We would like to take advantage of this opportunity to invite appropriate technology groups of the Mesoamerican region to share with us their experiences and accomplishments."

Articles are short and non-technical.

Yumi Kirapim, newsletter, average length 12 pages, bimonthly, free to serious groups from ODV-SPATF, P.O. Box 6937, Boroko, Papua New Guinea.

Reports on the activities and research carried out by the South Pacific Appropriate Technology Foundation, and the services offered by the government's Office of Village Development. (These are the groups that produced **Basic Sewing Machine Repair** and **A Blacksmith's Bellows**). Food processing techniques, loans, and advice for establishing small scale businesses, sources and methods of water supply are topics of recent issues. Printed in English and Pidgin.

An interesting experiment in presenting locally relevant technical and extension service information, at a level understandable to virtually any literate person.

Tarik, small magazine, published once or twice a year, average length about 30 pages, write to Publications Unit, Project Dian Desa, P.O. Box 19, Bulaksumur, Yogyakarta, Indonesia.

A journal aimed at documenting and sharing information on appropriate technologies being developed by Indonesian community development groups. Each of the first two issues has described a technique in enough detail for readers to attempt construction themselves. (These include building a Lorena stove and fabricating 9-cubic-meter ferro-cement and bamboo-cement water storage tanks for use with rooftop catchments.) Photos and drawings are mostly self-explanatory for those unfamiliar with Indonesian or Malay languages.

A local example of a low-cost way of documenting and disseminating important ideas. Those wanting copies of **Tarik** should offer their publication in exchange, or submit money for postage.

Practical Self-Sufficiency, magazine, bimonthly, average length 40 pages, from Broad Leys Publishing Company, Widdington, Saffron Walden, Essex CB11 3SP, U.K.

A journal of the "back-to-the-land" movement in England, aimed at an audience interested in small-scale farming and reducing dependency on government institutions. Articles are mostly of the how-to-do-it variety: poultry keeping, beginning beekeeping, cutting and laying hedges, and so forth. There are active reader correspondence and classified advertising sections. This magazine is relevant mainly in the U.K. and other industrial nations.

International Foundation for Development Alternatives Dossier, journal, published bimonthly, available from IFDA Secretariat, 2, Place du Marche, CH-1260 Nyon, Switzerland.

This publication presents papers in French or English with abstracts in both languages, on a wide variety of general subjects. Some titles from the Jan.-Feb. 1980 issue: "Third World Commodity Policy at the Crossroads: Some Fundamental Issues", "Another Development for Japan". Language is scholarly, and may be difficult for non-native speakers of French or English.

ADDITIONAL PERIODICALS
IFOAM Bulletin, page 431.
Quarterly Review of the Soil Association, page 431.
Jojoba Happenings, page 441.
Agroforestry Review, page 506.
Bamidgeh (on fish farming), page 521.
Soft Energy Notes, page 565.
Steam Power Quarterly, page 574.
Live Steam Magazine, page 577.
Nepal Biogas Newsletter, page 654.
World Education Reports, page 741.
Medical Self Care, page 716.
Salubritas: Health Information Exchange, page 717.
Appropriate Technology for Health, page 717.

a priori—at the beginning; given.

abdomen—the part of the body containing the stomach and intestines.

abolition—the banning of something.

abundant—plentiful; available in large quantities.

accessible—easily reached or obtained.

accommodate—to adjust.

accumulated—gathered or saved (up).

acupuncture—the ancient Chinese practice of piercing parts of the body with needles to treat disease or relieve pain.

acute—severe, serious.

adept—skillful.

adobe—unburnt sundried brick.

adverse—unfavorable.

adze—a metal cutting tool like an axe, but with a blade at right angles at the handle.

aeration—the process of mixing with air or oxygen.

aerial ropeway system—a transport system using a permanent set of ropes or cables to carry goods over rough terrain.

aerodynamics—the study of the motion of air and the forces acting on bodies in motion (such as windmill blades).

aesthetic—related to taste or beauty.

afforestation—the process of planting trees in an area that does not have them.

agile—quick; able to move quickly.

agronomist—a person who is trained in the science and economics of crop production and the management of farm land.

ailment—illness, health problem.

algae bloom—the uncontrolled growth of algae in a pond.

alien—foreign.

alienating—causing loss of sense of purpose.

alleviate—to make less hard to bear; to lighten or relieve.

allied—joined together for a common purpose.

all-inclusive—complete; covering all parts or aspects.

allocation—the amount of something set aside for a particular purpose.

all-terrain vehicle—a heavy-duty vehicle especially designed to operate in rough and wet terrain (including hills, swamps, and creeks).

amateur—a person who is not a professional.

ambient—surrounding, on all sides.

ambitious—demanding great effort, skill or enterprise.

ambivalent—undecided.

amenities—comforts.

ammeter—an instrument which measures the strength of an electric current in the form of amperes (amps).

amortize—to gradually pay off a debt.

ample—plenty.

anaerobic fermentation—fermentation in the absence of air or oxygen.

analysis—breaking a problem or question down into parts.

ancestral—of anything regarded as prior to a later thing.

anecdote—a short account of some happening.

anemia—a condition in which there is a reduction in the number of red blood cells or of the total amount of hemoglobin in the blood stream, resulting in weakness.

anemometer—a simple device that is used to measure wind speed.

animal husbandry—a branch of agriculture concerned with the raising of animals.

animal power gear—a gear that converts the power of a horse or other animal walking in a circle into the high-speed motion of a drive shaft, used to operate equipment (such as a thresher).

anodize—to coat with a protective film using electric current.

antecedent—any happening or thing prior to another.

anvil—a heavy steel block on which metal is pounded for shaping (blacksmith's tool).

aperture—the opening in a camera or telescope through which light passes into the lens.

apiary—a place where bees are kept.

aquaculture—the raising of fish and other marine organisms.

aquarium—a glass-walled container for fish and other animals and plants, that allows careful observation of their behavior.

aquatic—having to do with water (ponds, streams, oceans).

arable land—land which can be farmed.

arbitrary (arbitrarily)—without reason.

armature—the iron core with wire wound around it, in a generator, alternator, or electric motor.

array—a regular arrangement or series.

artisan—craftsman, artist.

aspirations—hopes, desires for the future.

aspirator—a device for moving air or fluids by suction.

assimilation—the process of becoming part of something.

astute—accurate; showing a clever mind.

attached greenhouse—a solar greenhouse attached to a house, where it helps in heating by acting as a solar collector.

attributable to—due to, caused by.

auger—a tool for boring holes.

authoritarian—characterized by unquestioning obedience to authority, as that of a dictator, rather than individual freedom of action.

auxiliary—extra, reserve.

axial-flow turbine—a turbine in which water flows parallel to the axis.

backlash—a strong political reaction resulting from fear or resentment of a movement.

backslide—to slide backwards, failing to fully implement a political promise.

backward—from earlier times, not modern.

bacteriological—related to the study of tiny life forms present in all organic matter.

band saw—a saw that has a long narrow continuous band for a blade; the band travels in one direction only, rotating around several wheels.

banish—to send away permanently.

ballyhoo—noise and hollering.

barometer—an instrument for measuring atmospheric pressure; anything that indicates change.

baseboard heating—a space heating system which radiates heat from panels on the wall near the floor.

beaker—a glass container used in scientific experiments.

bearing—any part of a machine on which another part revolves.

becak—pedicab, three-wheeled taxi (Indonesia).

belt sander—a machine with a long abrasive belt that travels around two or more rotating cylinders; the belt is used for sanding and smoothing rough pieces of wood.

bemo—small transport vehicle in Indonesia.

benign—not dangerous; not causing damage or hurt.

bevel gear—a gear wheel meshed with another so that their shafts are at an angle of less than 180 degrees.

biased—unfairly affected or directed; not fair, prejudiced.

bikeway—path or lane reserved for bicycle use only.

bilharzia—see schistosomiasis.

bi-metal strip—a device made of two strips of different metals that expand at different rates when heated; the strip bends or curls when heated.

biodegradable—capable of being decomposed by bacterial action.

biogas plant—see methane digester.

biomass—the total amount of living organisms in a particular area or volume.

biotic—of live, or caused by living organisms.

bloat—to swell.

block and tackle—a set of pulleys and ropes for hauling and lifting.

blueprint—a large set of detailed plans.

board feet—a unit of measure of lumber equal to a board one foot in length on two sides and one inch thick.

bona fide—real; made in good faith.

borehole—hole drilled in the earth to make a well.

borne out—proved to be true or accurate.

botany—a branch of biology that deals

with plants.

brace—a support; also a tool into which a drill bit or auger is inserted for drilling.

brazing—to bond two pieces of metal using a metal rod with a lower melting temperature than either of the pieces being connected; usually uses copper wire, and can be done with a small propane torch.

bridle—a head harness for guiding a horse.

brittle—easily broken.

brunt—the major portion of negative consequences.

bulk—greatest part.

bungalow—a small house with a porch.

byproduct—other or additional product.

cable plow—a plowing system in which a lightweight plow is pulled across a field by cable instead of by a tractor or draft animal.

cadre—a local level leader and motivator.

calculus—calculation; estimation.

calibrated—carefully and correctly adjusted.

cam—a bump on a turning shaft which lifts or pushes.

campesinos—rural people (Spanish).

canning—the preservation of foods in tightly sealed cans or jars.

capital formation—gathering resources and buying or making tools, equipment or buildings to be used in production.

carburetor—that part of an engine in which air and fuel are mixed.

cardiovascular—of the heart and the blood vessels.

carnivorous—meat eating.

carrying capacity—amount of life or activity that the ecosystem can support.

caseharden—to harden the outer layer of a piece of metal.

cash crop—a crop that is produced for sale rather than for consumption by the farm family.

casting—the process of making products from a mold, usually using hot molten (liquid) metal.

catalyze—to act as a stimulus in causing something.

caulking compound—a filling material

used to make a boat or other object "water-tight" so that water cannot enter or escape.

cellulose—the bulky or fibrous part of plants, consisting of natural sugars.

centralize—to concentrate the power or authority of a central organization; to gather together; to focus on a center.

centrifuge—a spinning machine used to separate particles of different density.

certified—having a license issued by an authority, proving the ability to do something.

charitable—kind and generous in giving money or help to those in need.

chassis—the part of a motor vehicle that includes the frame, suspension system, wheels, steering mechanism and so forth but not the body or engine.

chemical coagulation—bringing together suspended particles in water, by adding a chemical.

chisel—a tool for cutting grooves or shavings from wood or metal.

chlorination—purifying water by adding tiny amounts of the chemical chlorine to it.

chromatography—the process of separating the elements in a mixture by having a solution flow through a column of absorbent material on which the different substances are separated into distinct bands.

chronic—happening again and again.

circuit board—an electrical system laid out on a board for use in teaching.

circuitry—the elements of an electrical system.

cited—noted, identified.

clarify—make clear.

clear-cutting—cutting down all of the trees and plants.

climatology—the study of weather patterns.

clogged—blocked or stopped flow.

closed loop—in a solar energy system, using water or another liquid to move heat from a collector to a storage area, and then returning the same liquid to the collector.

coalition—a group of organizations that agree to cooperate.

coercive—based on the use of force.

coherent—fitting together well; making sense.

collaboration—working together.

collaborative—from working together.

collateral—something of value owned by a borrower, such as a house or land, used as a guarantee to a lender that a debt will be paid; if the debt is not paid, the lender takes the collateral as payment.

colleague—a fellow worker in the same profession.

combat—fight.

combustion—the process of burning.

commencing—beginning.

commend—to praise; to favorably point to.

commutator—in a generator or electric motor, a revolving part that collects the electric current from, or distributes it to, the brushes.

companion planting—a strategy used in intensive gardening in which different plants are raised next to each other to take advantage of nitrogen-fixation, insect repelling properties, shade, etc.

compatible—going together well; fitting together well.

compelling—convincing, persuasive.

compensated—paid.

compendium—collection, compilation, summary.

compost toilet—a waste disposal system in which wastes break down to become fertilizer.

compounding—adding to.

comprehensive—including all aspects.

compression—being pushed or squeezed together.

compulsory—required.

computing—figuring out using numbers.

concave—curved inward, like a bowl.

concerted—concentrated, deliberate, vigorous.

concientizacion—a group discussion process aimed at creating an expanded awareness of the factors that keep people poor, and stimulating action for change.

condensation—the process whereby water vapor or another gas changes into a liquid as its temperature drops.

condenser—a device for converting a gas into a liquid.

configuration—arrangement of parts.

congealed—become solid or firm.

conical—shaped like a cone.

conscientious—very careful and consistent.

consensus—decision-making by a group in which all members participate and are satisfied with the outcome.

constraints—limits; problems.

construed—understood, interpreted.

containment—where animals are held inside.

contamination—dirtying or poisoning.

continuity—the act of proceeding smoothly over time; ongoing.

contour—an imaginary line around the side of a hill that maintains the same elevation.

convergence—combining; coming or flowing together.

convey—communicate.

conveying—communicating, showing.

cope—to deal with problems effectively.

coppicing—the controlled production of small trees repeatedly from the same stumps (root systems).

corollary—a proposition related to one that has been proven correct.

corrode—to eat into or wear away gradually, as by rusting or the action of chemicals.

corrugated—having parallel grooves and ridges.

corrupt—dishonest in handling money; using influence unfairly.

counter-sink—a tool used to drive a nail or screw below the surface of a piece of wood.

crannies—small cracks.

creativity—the ability to use the imagination and invent.

creosote—unburned gas from a wood fire, that has condensed to form a sticky, dark substance.

crop rotation—a system of growing successive crops that have different nutrient requirements, thereby preventing soil depletion, and breaking disease cycles.

crop-lien system—a system in which a future crop is sold at a low price to store owners or other middlemen, in order to acquire credit for essential purchases by a farm family.

cross fertilization—stimulation and improvement through exchange (of ideas).

cross-flow turbine—a wheel with curved

vanes driven by the pressure of water flowing through it, and in which the water acts on the vanes twice, once while entering and once while leaving the turbine.

crucible—a container used to hold metal while it is being melted.

cube (**math**)—the product of multiplying a number by itself three times. The cube of the number 2 is 8 (2 x 2 x 2 = 8).

culmination—the highest point, the climax.

cultivator—an implement used to loosen the soil and remove weeds while crops are growing.

curricula—plural of curriculum.

curriculum—the set of concepts being taught in a class.

currier—a worker who treats leather.

cutical (**insect**)—skin or covering of an insect.

cutlery—knives; tools used for eating.

cycle rickshaw—pedicab, three-wheeled taxi.

cyclist—a person riding a bicycle.

cynical—antisocial; believing that all people's actions are based on selfishness, and thus basing one's own actions on selfishness as well.

cyst—a growth in the skin.

damper—a piece of metal used to control the flow of air and hot gases in a stove.

data bank—a place where information is collected and stored for later use.

dawn on—become clear to.

debilitating—making weak.

debit—amount to be subtracted.

debris—rough, broken bits of material left after a war or other disaster.

debt servicing—interest paid on a loan.

decentralization—a shift in the patterns of decision-making and production so that these activities go on in many more places than before.

decentralize—to break up a concentration of governmental decision making, industry or population, and distribute it more evenly.

decorative—a plant of interest due to its appearance only.

deduced—realized; understood.

deep litter bedding—straw, leaves, or wood shavings used in a deep layer to cover the bottom of a chicken coop.

deficit—the amount by which a sum of money is less than the required amount.

deflector—a device that can be used to change the direction of a flow of water in a turbine, to reduce the power produced.

deforestation—the destruction of forests.

degradation—making worse; becoming less usable.

dehydration—the draining of fluids from the body through diarrhea or perspiration; dangerous if the fluids are not replaced.

demoralizing—discouraging.

depletion—using up.

deplore—to regard as unfortunate.

derive—obtain, get.

desertification—the creation of deserts.

destitute—very poor.

deterioration—the process of becoming worse.

determinant—cause.

detract—undermine, reduce, subtract.

devastated—having suffered great destruction.

diagnosis—the process of deciding the nature of a diseased condition, by examination of the symptoms.

dialects—different forms of a language; local languages.

dialogue—conversation; talking between two people or groups.

diaphragm pump—a pump which moves water through the alternating expansion and contraction of a chamber.

diarrhea—excessive looseness and frequency of bowel movements.

diatribe—a bitter, abusive criticism.

diesel set—an electric generator driven by a diesel engine.

dietary—related to what a person eats.

differential—an arrangement of gears connecting two axles in the same line and dividing the driving force between them, but allowing one axle to turn faster than the other; used in the rear axles of cars and rear-drive tricycles to permit wheels to turn at different rates when turning corners, to avoid tipping over.

differentiate—show difference among or between; separate.

diffraction—the breaking up of a ray of light into the colors of the spectrum.

diffusionist—an approach to technological change in which new techniques chosen by central agencies are spread, concentrating on community leaders.

digression—a wandering from the main subject.

dilemma—problem for which a solution is not evident.

diligence—hard-working, responsible.

direct gain—solar energy that enters a building, without the use of collectors.

discharge—release.

discredit—to show reasons for disbelief.

disinfectant—a substance which cleans and kills disease-causing organisms.

dismantle—take apart.

dispel—remove, clear away.

dispersed—spread out.

disposable income—that portion of an income which can be spent.

dissecting tool—a tool used in separating the parts of an animal or plant.

dissemination—spread.

dissolution—breaking apart; breakdown.

distill—carefully select the essential elements of; evaporate and condense.

diversified—having many different activities or components.

divert—to move water or resources away from their normal channels.

dogmatic—closely following the rules; unwilling to listen to other ideas.

donor—a group that provides funds.

dosage—the exact amount of a medicine to be given at one time.

double acting pump—a pump designed so that water is lifted during both the up and down strokes of a piston or diaphragm.

double-digging—a technique used in intensive gardening, in which the topsoil is removed, the subsoil is loosened, and the topsoil is then replaced.

dowel—a round length of wood used to join two other pieces of wood.

draught—British spelling of draft.

draw-knife—a two-handled knife used in making precise cuts in wood.

drill press—a machine used to drill straight holes in metal or wood.

dropper—a glass or plastic tube used to pick up and transfer drops of liquid.

drought—an abnormally long period of time with lower than normal annual rainfall.

drudgery—hard, boring work.

dry steam—high-temperature steam which contains little moisture.

dubious—doubtful; uncertain.

duplicate—to copy; to do again.

dynamic—moving, changing.

dynapod—a basic pedal power unit that can be attached to small machines.

dwindles—gets smaller quickly.

earthen—made of earth.

ecologically-sound—something that does not affect the natural balance of the environment and ecosystem.

economies of scale—savings that come with increasing size of a business or activity.

ecosystem—a system made up of a community of people, animals, plants and bacteria, and the physical and chemical environment with which it is connected.

edible—that which can be eaten.

effluent—material or waste flowing out.

eke out—scrape together.

electrical conduit pipe—lightweight metal tubing used to protect electrical wires.

electric grid—system of electric lines which distribute and regulate electricity in a community.

electrolysis—the process of changing an electrolyte by passing an electric current through it.

electrolyte—a liquid or solution which conducts electricity and deposits a metal coating; used in electroplating.

electromagnet—a core of material that becomes a magnet when electricity is passed through a coil of wire around it.

electronic governor—a device which switches part of the electric current produced by a turbine away from the main line (for example, to heat water) when the electric demand falls; this allows the turbine to be run at a constant speed, avoiding the need for an expensive governor to regulate the amount of water flowing through the turbine as electric demand changes.

electroplate—to coat with metal using electricity passed through a solution.

elicit—to draw out (a response).

emery stone—a stone for grinding the

edges of tools to sharpen them.

empirical—based on practical experience and observation rather than theory.

endeavors—efforts, projects.

endorse—recommend.

enhance—improve, make better.

enteric pathogens—organisms causing disease in the intestine.

entrepreneur—someone who sees an opportunity to start a new enterprise or activity; businessman.

entrepreneurial—related to undertaking the risks and management of a new enterprise or activity.

environmentally-sound—see ecologically-sound.

envisage—imagine.

epidemic—a disease that is spreading rapidly among many individuals in a community at the same time.

epoxy—liquid material which hardens in the air, used in glues.

equate—to consider the same.

equitable—fair, equal for all.

escalating—rising, increasing.

ethanol—alcohol made from grain or other vegetable.

euphemistically called—given a nice name.

exacerbate—make worse.

excerpt—a piece taken from a longer article or book.

exclusion—the leaving out of something.

existential—involving awareness of being a free individual.

exotic—highly unusual; not part of daily life.

expediting—to speed up.

explicit—directly, obviously.

exponential—rapidly increasing.

extraction—the process of taking something out.

extractive—something that is drawn out or removed.

extrapolation—a conclusion reached by estimating beyond a known range.

extruded—to be forced out.

fabricate—make, construct.

facilitate—enable, help to happen.

fad—a temporarily popular activity.

fall prey—become a victim; be taken advantage of.

fallible—possibly wrong; capable of making mistakes.

fallow land—land not planted in a crop for a growing season, to allow improvement in soil fertility.

feasible—possible, practical.

fecal coliform bacteria—microscopic (tiny) organisms in human waste which can cause sickness.

fecal matter—solid human waste; shit.

fencerow—a row of bushes forming a fence.

fermentation—the breakdown of complex molecules in an organic material, caused by a bacteria; action of yeast making vinegar or alcohol.

ferrocement—a cement-sand shell reinforced by wire mesh.

fiber—any substance that can be separated into threads for spinning, weaving.

fiberglass insulation—insulation made of small fibers (similar to hairs).

field wash—soil erosion caused by the flow of water.

firebreak—a strip of land on which trees and other plants have been removed, to prevent the spread of forest fires.

firebrick—special brick that will not break at high temperatures.

flagstone—a hard stone that splits into flat pieces.

flannel board—a board on which scenes and processes can be illustrated for an audience; the flannel holds the movable pieces in place.

flap valve pump—a simple low-lift hand pump with a valve on top but no piston; same as inertia pump.

flat plate collector—a glass or plastic-covered metal panel which traps the solar energy that falls on it; this heat is then transferred by a water or air system for hot water heating or home heating.

flaws—mistakes.

flow regulator—a device that controls the amount of water flowing through a turbine, to match the power needed at any moment.

fluctuations—variations.

focal point—the central point at which activities are directed and effects are felt.

fodder—plant food for animals, such as leaves and straw.

foliage—plant growth.

forage—food for domestic animals; to search for this food.

foreign exchange—money in the form of foreign currency that can be used to buy things from outside the country.

forerunner—one which came before.

forge—a blacksmith's furnace for heating iron or steel hot enough so that it can be shaped by pounding.

format—general arrangement.

formulation—a theory or plan.

forum—a place where discussion and exchange of ideas can take place.

fossil fuel—coal, oil, natural gas.

foundryman—a person who operates a workshop in which iron is melted and poured into molds to make tools.

foyer—entryway, entry room.

fragile—delicate, easily broken.

fragments—breaks apart; small pieces.

fringe areas—margin; edges.

frugal—economical; not wasteful.

fry—young fish.

fungicide—a substance used to kill fungus.

galvanized—coated with zinc for protection from rust and corrosion.

galvanometer—an instrument for detecting and measuring a small electric current.

gasogen—a stove-like device carried by a vehicle, producing gas through the partial burning of charcoal or wood.

gauze—a very thin, loosely woven piece of cotton or silk.

genetics—the branch of biology that deals with heredity and variation in similar or related plants and animals.

germination—the process of starting to grow or sprout.

germplasm—the portion of the reproductive cells of an organism involved in heredity.

gestation—a development, as of a plan in the mind.

glazing—the plastic or glass covering on a solar device.

gleaned—picked out of.

gouge—a tool like a chisel used to remove chunks of wood.

governor—a device that controls the amount of water flowing through a turbine, to match the power needed at any moment.

graphic ideation—the use of drawings to express and develop ideas.

grassroots—local communities; where people live and work.

green manure—a crop that is plowed back into the soil while still green, for its beneficial nitrogen-adding effect.

greywater—waste water from sinks or washing machines.

gutted—destroyed by fire.

gypsum board—a thin board formed of layers of gypsum plaster and paper, used on interior walls of buildings.

halter—a rope or strap for tying or leading an animal.

hamper—to make difficult; to hinder.

hands-on—practical.

haphazardly—in a disorganized way; carelessly.

hard-pressed—faced with a very difficult task.

harnessing—using to advantage.

harrowing—using an agricultural implement with spikes or discs to break up and level plowed ground.

hatchery—place where fish are raised from eggs to small but viable size before being released to feed and grow larger.

have-nots—those who don't have enough wealth and income to live at an acceptable standard of living.

haves—those who have enough wealth and income to live relatively comfortably.

head—the total usable height water falls when used in a waterwheel, turbine or hydraulic ram pump; or the distance water is lifed by a pump.

hedgerow—a row of bushes forming a boundary or fence.

helicak—motorized three-wheeled taxi (Indonesia).

helical—winding or circling around a center or pole while getting smaller and smaller; spiral.

hence—therefore, thus.

herbicide—a chemical substance used to control weeds or other undesired plants.

herbivores—animals that eat only plants.

heritage—something handed down from one's ancestors or the past.

hierarchical—having people arranged in order of rank.

high carbon steel—steel that has a relatively high carbon content and can be hardened for this reason.

hitch—a connecting device.

hock, (in)—in debt, with house or land or other asset as collateral.

holdover—something staying on from an earlier period.

honey extractor—a device that removes honey from honeycomb, usually by spinning.

hookworm—any of a number of small parasitic roundworms with hooks around the mouth, that infest the small intestine of humans, especially in tropical areas.

horticulturalist—one who works with the science or art of growing flowers, fruits, vegetables or shrubs, especially in gardens or orchards.

host—a plant or animal that has a parasite living on or in it.

humidity—dampness or wetness in the air.

humility—humble attitude or approach; the opposite of proud or arrogant.

humus—black or brown decomposed organic matter.

hurdle—a portable frame made of branches, used as a temporary fence or enclosure.

hybrid—a new variety created by breeding, often producing higher yields but sterile (the crop cannot be used for seed).

hydraulic ram pump—a device used to pump water, using only the energy in the falling water itself to pump a small portion of the water to a much higher level.

hydrology—the study of where water is and how it behaves.

hydropower—energy generated by falling water.

hypothesis—an unproved theory or proposition.

hypothetical—for example; imaginary.

ideologically-tainted—associated with an ideology and therefore appearing biased.

illiterate—unable to read or write.

illiterates—people who cannot read and/or write a language.

immunology—the branch of medicine dealing with immunity to disease and biological reactions such as allergies.

impaled—pierced through with something pointed.

impenetrable—cannot be entered.

imperative—the evidence that some action must be taken; an urge; necessity.

impinge—affect.

implicit—suggested or understood though not plainly stated.

imposition—hardship or burden forced from outside.

imprinting—a learning mechanism operating very early in the life of an animal in which a stimulus creates a behavior pattern that is remembered.

improvisation—something made with the tools and materials at hand to fill an immediate need.

improvise—to solve a problem using what is available.

inadequacy—not enough; not good enough.

incandescent light bulb—a light bulb that glows due to intense heat caused by electricity passing through a special wire coil.

incentive—something that stimulates one to take action or work harder.

income disparity—the difference or gap between high and low incomes.

income stratification—the division of a community or nation into several very different income levels.

incremental—involving small changes or improvements.

incubator—a special compartment used to keep chicken eggs at a warm temperature.

indigenous—native; originally found in local area.

indispensable—something that cannot be left out.

inertia pump—see flap valve pump.

infestation—attack by insects or other pests causing damage to crops.

infuse—to fill with something.

ingenuity—creative ability.

inherent(ly)—by itself; existing in someone or something as a natural quality.

inhibition—a mental process that restrains or suppresses an action.

injurious—harmful.

in-kind—with goods or food, not money.

innumerable—too many to count.

inoculant bacteria—nitrogen-fixing bacteria that are spread on seeds to aid in later plant growth.

inoculation—the spreading of bacteria or other life forms into soil or water for beneficial growth; the injection of a disease agent into an animal or plant to build up an immunity to it.

insecticides—chemicals used to kill insects and therefore protect crops.

insolation—the amount of solar energy falling on an area, usually measured in BTU's per unit of area.

integrate—to mix together, to combine.

intercropping—planting two or more crops together.

intergranular spaces—the air spaces between the kernels in a pile of grain.

interlocking—linked together.

intermittent—off and on; now and again.

internal combustion engine—an engine in which the fuel is burned inside the chambers in which expansion takes place and moves the pistons.

inter-row cultivator—a tool that is used to remove weeds in several rows at once.

intervening—entering and altering the normal flow of activities in a community.

intervention—a project begun by an outside agent or agency.

intertwining—interconnecting, linked.

intolerable—unbearable; too painful to be endured.

intravenous—directly into a vein.

inventive—skilled in devising a process or mechanism for the first time.

inventory—a detailed list.

invertebrates—spineless organisms.

invoked—referred to with reverence.

jargon—special words used and understood only within a particular field of activity.

jig—a guide for a tool that allows the repeated production of the same cut or part.

judicious—careful; well-placed.

juvenile—young.

keel—the primary timber of piece of steel that extends along the length of the bottom of a boat or ship.

keen—eager, strong.

kernel—a grain or seed, as of corn, wheat or peanut.

kiln—a structure for the high-temperature treatment of bricks or pottery for hardening; or for the production of lime or cement; or for the reduction of wood to charcoal; or for the drying of wood.

kinetic energy—the energy of a body that results from its motion.

kink—a short twist, curl or bend in a rope, wire or chain.

knack—talent, ability.

Ku Klux Klan—a secret society of white men created in the United States following the Civil War in 1865, to re-establish and maintain white supremacy.

kymography—the study of wavelike motions or variations.

landfill—area that has been filled in with a mixture of soil and solid waste.

landmark—something that marks an important place.

land reform—the redistribution of agricultural land by breaking up large landholdings and spreading them among all of the rural population.

larvae—the young worm-like form of an animal that changes structurally when it becomes an adult (e.g., caterpillars to butterflies).

laterite soil—a red soil formed by the decomposition of many kinds of rocks, and found especially in tropical rain forests.

lathe—a machine which turns wood or metal while it is being cut by a tool.

latitude—a distance measured in degrees, north or south of the equator.

leaching—the draining away of important soil nutrients by water action.

ledger—list of amounts of money.

legume—any of a group of plants that add nitrogen to the soil, such as the soybean or any other bean.

leguminous—of the family of plants that produce pods, to which peas and beans belong; legumes.

leprosy—a chronic and infectious disease caused by a bacterium that attacks the skin, flesh, nerves, etc; characterized by white scaly scabs and wasting of body parts.

lever—a bar for prying or lifting.

liabilities—debts.

lift—the height water is raised by a pump.

linkage—connecting mechanism.

lorry—truck.

Luddites—a group of workers in England (1811-1816) who smashed new labor-saving textile equipment in protest against reduced wages and unemployment.

lunar—having to do with the moon.

machete—a large knife, used for chopping brush and other heavy cutting.

machining—shaping metal by turning, milling, or planing.

magnifying lens—a hand lens used to enlarge an image for closer inspection.

magnitude—size; amount.

malnutrition—inadequate nutrition.

manipulative—affecting events or other people without consulting them.

manometer—an instrument for measuring the pressure of gases.

manure spreader—a specially equipped wagon used to spread barnyard manure around the fields.

marine borers—small animals that live in sea water and eat holes in the hulls of ships and in the posts of docks.

mash—mixture of grain or other vegetable, yeast, and water.

masonry—the fitting together of blocks, bricks, or stones, in construction.

master—completely understand.

mattock—a tool for loosening the soil.

meager—very small.

media—various means of communication.

mediate—to act as a communication channel and intermediary between two people or groups.

medicated—containing drugs.

metabolizable energy—food energy which can be converted for use.

metal spinning—the technique of bending and shaping metal by pressing on it while it is turned on a lathe.

metaphorically—using words to create an image in the mind to illustrate an idea.

metazoa—any of a group of very small animals that have cells for different functions; many of these are parasites.

methane digester—a sealed tank in which manure and vegetable waste decompose to form a gas for fuel (methane) and liquid fertilizer.

methane plant—see methane digester.

methanol—alcohol made from wood.

methodology—a system of methods.

microbe—a microscopic organism.

microhydroelectric turbines—small power systems with generators that use falling water to produce electricity; usually in the range of 1-40 kw of power.

microorganism—any very tiny organism that can only be viewed through a microscope.

midwife—a woman who assists women in childbirth.

migrant—a farm worker who moves from place to place with the agricultural calendar.

millenia—thousands of years.

milling machine—a metal cutting machine in which the surface of the work is shaped by being moved past revolving toothed cutters.

mimeod—printed on a low cost machine called a "mimeograph" machine.

minimum tillage—an agricultural strategy in which plowing and cultivating is kep to a minimum to reduce soil erosion and encourage soil microorganisms.

misconception—mistaken idea.

mitigate—reduce the negative effects of; make less bad or serious.

mocking—showing scorn, contempt, or defiance.

mode—method, way of operating.

modular—made in small units which can be combined as needed.

molecular model—a visible model of the structure of molecules, used in teaching.

momentum—the quantity of motion of a moving object, equal to mass times velocity.

monetary—involving money.

monitor—carefully observe.

monoculture—the raising of only one crop.

monolithic—solid, the same throughout; all of one kind.

moped—a motor-assisted bicycle.

moratorium—temporary halt.

morbidity—the incidence of disease among a population; how many people are sick.

mordant—dye preservative to prevent fading.

mortar and pestle—a very hard bowl in which softer substances are ground or pounded to a powder with a hard tool.

motive power—power to move something, as an engine.

motocross bikes—a heavy duty bicycle recently popularized among children in the United States, in imitation of motorcycles used in racing over hilly terrain on dirt tracks.

mowing machine—a farm machine with a reciprocating blade that cuts standing grain or grass.

mulch—top covering of the soil consisting of organic material (grass, compost, straw) that serves to keep moisture in the soil and reduce weed growth.

multiple cropping—involving more than one crop.

mundane—common, unexciting, normal.

mutual—involving both or all sides.

mystification—the process of making something deliberately hard to understand.

natural calamity—a disaster such as a hurricane, flood, or forest fire.

natural phenomena—processes and events that occur normally in nature.

naught—nothing.

needs assessment—technique for deciding what people need.

niche—place.

night soil—human excreta.

nitrogenous fertilizer—fertilizer that contains nitrogen.

nodule—a small knot on a root that contains nitrogen-fixing bacteria.

nonconformist—one who does not do things in the way in which they are normally done.

nooks—small hidden places.

nonviability—unable to be sustained.

novice—beginner.

nozzle—a device at the end of a hose or pipe with which a stream of water can be controlled and directed.

nuisance—a person or thing causing trouble or inconvenience.

nurture—the process of raising or promoting the development of.

nutrient cycling—the process of moving nutrients through the agricultural system, from fodder to manure to fertilizer to additional plant growth, for maximum production and continued fertility.

nutritive—promoting health through a balanced diet.

obscure—relatively unknown.

obsolescence—the process of going out of use.

obtainable—available.

oilseed—any of a number of plants grown for the oil contained within their seeds.

oligarchy—a form of government in which the ruling power belongs to a few people, families, or groups.

opelet—small transport vehicle in Indonesia.

open loop—in solar energy system, heating water or air by passing it through or over a collecting surface and then moving it to where it is needed.

opt—choose.

optical—having to do with the sense of sight.

optimize—to obtain the most efficient or maximum use of.

optimum—best.

organic gardening—a form of gardening using only natural materials.

organic solvent—liquids distilled from vegetable matter than can be used in cleaning.

outmoded—out of date; no longer useful.

outstripping—surpassing, increasing at a greater rate.

ova—eggs.

overhead functions—tasks performed which enable other more basic activities to go on.

overriding—extremely important.

overuse—too much use.

oxy-acetylene welding—a welding system that uses compressed oxygen and acetylene gas to supply heat.

paddy weeder—a hand tool used to remove weeds between the rows of rice plants in a paddy.

papier-mache—a material made of paper and glue or flour that is easily shaped when wet, but dries hard.

paradigmatic—showing a model or pattern.

paramedical—of auxiliary medical personnel.

parasite—a plant or animal that lives on or in an organism of another species (the host) while usually doing harm.

parboiling—a preliminary cooking process which serves to seal the outer surface of a grain such as rice.

parcelization—dividing up into small pieces.

pare—to cut away the outer covering of something.

particle board—board made of small pieces of wood or other material compressed together.

passive solar—any solar technology that uses natural energy flows in the materials and orientation of a building for heating, without the use of special collectors, pipes, and pumps.

pathological growth—growth or increase in size which is unhealthy, or which is not good for people.

peak power—the highest level of power that can be provided at any time.

pedicab—a three-wheeled pedal powered taxi.

peer—people who do the same work, or are of the same social status or age.

pendulum—a weight hung from a fixed point so as to swing freely under the combined forces of gravity and momentum.

perennial—any plant that produces from the same root structure year after year; important in soil conservation.

perforated—having holes.

perseverance—continued, patient effort.

pertinent—significant, important.

pesticide persistence—the tendency for pesticides to remain in the soil and water supply after use.

pesticides—chemical substances used to kill plant pests.

pharmaceuticals—drugs.

pharmacological—having to do with the science of study of the effects of drugs on living organisms.

phenomenally—amazingly.

philanthropic—showing a desire to help mankind by gifts to charitable institutions.

photovoltaic array—a set of photovoltaic cells.

photovoltaic cells—solar energy devices that directly convert solar energy into electricity.

physiologically—having to do with the functions and vital processes of living things.

pickling—a process for canning or bottling vegetables using vinegar.

pictorial—using pictures.

piecemeal—bit by bit; not organized very well.

pioneers—early workers in a new field.

pise—rammed earth; a construction technique in which earth is pounded inside movable forms to make walls.

pit latrine—a toilet in which human waste accumulates and is buried in a hole in the ground.

pivotal—most important.

plankton—microscopic animal and plant life found in water, used by fish for food.

plateau—a high flat or level place.

plight—a sad or dangerous situation.

pneumatic tires—air-filled tires.

pollination—the act of transferring pollen between the parts of a flower, important in the production of fruits and vegetables, carried out by bees and other insects.

polyethylene—a plastic material used in sheets and waterpipes.

poly-phase electric motor—a motor driven by more than one alternating electric current.

polythene—British spelling of polyethylene, a plastic used in sheeting.

polyvinylchloride—a plastic material used in water pipes.

porridge—a soft food made of cereal grains boiled in water or milk until thick.

potable water—safe drinking water.

power drive—the system of spinning shafts and gears used in transmitting mechanical energy.

power tiller—a small engine-driven machine for plowing or breaking the soil; usually has two wheels.

pragmatic—practical, taking into account organizational constraints and capabilities, for example, when deciding a path of action; taken too far, being ''pragmatic'' can mean taking the easier path such that fundamental problems are never addressed.

precarious—dangerous, insecure.

predators—animals that eat other animals.

prejudice—an opinion held in disregard of facts that contradict it.

preliminary—introductory; coming before or leading up to the main action.

producer-gas engine—an engine which runs on gas produced in a charcoal-making process.

product differentiation—marketing technique of making products appear to be different from other similar products with the same function.

productivity—the amount of product created or work accomplished per unit of something (usually labor time) invested.

production version—the final form of a product to be made in large quantities for the market.

proliferation—spread, increase.

prolific—producing a large amount.

propagandistic—involving the uncritical promotion of particular ideas and doctrines.

propagation—the reproduction or multiplication of a plant or animal.

propeller—a device with two or more twisted blades that rotate with the hub in which they are mounted.

prototype—an experimental version for testing.

protozoa—any of a group of microscopic animals made up of a single cell or group of identical cells; many of these are parasites.

protracted—extended.

prudent—wise.

pulp—a mixture of ground-up wood from which paper is made.

pulse—any member of the legume family (peas, beans, lentils, etc.).

punch—a tool for making holes.

punitive legislation—laws that declare certain activities illegal and create punishments for these activities.

qualitative—not numerical; involving kind or type.

quantitative—numerical; involving numbers or quantities.

quarried stones—pieces of rock cut out from under the earth.

quasi-—somewhat; to a certain extent.

radial-flow turbine—a turbine in which water flows around the axis.

rammed earth—a technique of building construction in which earth is pounded inside movable forms to make walls.

rapacious—taking by force, greedy.

rarity—a very unusual thing or event.

rasp—coarse file for removing wood or metal.

rationale—reasons used to support a decision or conclusion.

rattan—a long slender tough stem that comes from a climbing palm and is used in making furniture.

reallocation—spending in a different place, or for a different purpose.

realm—area.

reamer—a tool used to smooth out or enlarge the inside of a pipe.

reap—to harvest.

reaping machine—a machine that cuts grain in the fields.

rearing—raising.

reclamation—a reclaiming; especially the recovery of wasteland or desert by irrigation, drainage, replanting, etc.

recoup—make up for something lost; recover.

rectifier—a device that converts alternating current into direct current.

recurring—happening again and again.

reforestation—the process of planting trees in an area that once had them.

refraction—the bending of a ray of light as it passes from one medium to another.

refractory cement—cement that will survive high temperatures.

refuge—a place of protection against storms, etc.

rehabilitation—repair or rebuilding.

rehydration—the process of restoring the body to its natural balance of fluids.

rejuvenate—to revive.

remote—far away.

render—make.

renewable energy—energy from sunlight, wind, falling water or biological sources that is continually recharged by the sun.

renovation—the process of repairing and rebuilding.

repatriated—sent back to the country from which it came.

repudiate—to deny; to refuse to accept

or support.

reputable —respectable.

reservoir —a body of water held behind a dam.

resin —a solid or honey-like substance from plants.

resistor —a coil of wire used in an electrical system to provide resistance and thus heat.

respiratory system —the system of organs, including the lungs, involved in the exchange of oxygen and carbon dioxide with the environment.

restoration —a putting back into a former, normal condition.

retort —a container in which a material is heated to extract gases.

retrogression —a return to a less complex or worse condition.

revegetation —the process of replanting in an area that has lost most of its plant life.

rhetorical —having unnecessary, exaggerated language or style in making a point.

riddled —filled; filled with holes.

rigorous —very strict, thoroughly accurate.

ripple tank apparatus —a device made of glass and filled with water, used to show wave motion.

rivals —is similar to.

rivet —a metal pin used to bind two pieces of metal together.

roadbed —the foundation of a road.

romantic —without a basis in fact.

roof pond —a shallow reservoir of water on a roof used to collect, store, and release energy to heat or cool a building.

rote —(to memorize) mechanically and unthinkingly.

rototiller —a small motorized hand-tractor or cultivator usually with rotating blades.

routine —normal, regular, common.

rubble —damaged building material.

rudimentary —primitive, simple.

salinity —level of salt content.

sandcrete blocks —building blocks made of sand and cement.

sanitation —the use of hygienic measures such as the drainage and disposal of sewage.

savannah region —a treeless plain found in tropical and subtropical regions; a transition zone between rain forest and desert.

scanty —too little, not enough.

schematic drawings —drawings that show the complete layout of a system with all of its connecting parts.

schistosome —a parasite that causes schistosomiasis when it enters and lives in the human body.

schistosomiasis —a tropical disease that involves problems in the liver, nervous system, urinary bladder, or lungs; spread by a parasite that spends part of its life in the body of a snail.

scholarly —showing much knowledge, accuracy, and critical ability; presented in standard form acceptable to professors and other academics.

screen printing —a method of printing through a piece of silk or other fine cloth on which all parts not to be printed have been coated with film that prevents ink from passing through.

scrutiny —careful examination.

scythe —a hand tool with a long handle and metal blade, used to cut grain or grass.

seam —the line formed by sewing together two pieces of material.

seasonal —taking place only during certain seasons of the year.

sedimentation —accumulation of mud, sand, and gravel carried by water.

seedbed —the earth in which seeds are planted in a garden.

seed dressing drum —a rotating drum in which seeds are mixed with fertilizers or pesticides.

seed drill —an implement used to place seeds into the ground, usually dropping them through a tube.

segregate —keep divided.

self-actualizing —helping people to know themselves better.

semantic —related to meaning in language, the relationship between words and the concepts they represent.

sentimental —romantic, unrealistic, emotional.

septic tanks —large tanks for settling and decomposing human waste.

shearing force —a force tending to cause

a piece of metal tubing, for example, to separate in a direction perpendicular to the tubing.

shears—heavy scissors for cutting sheet metal.

shelterbelt—a barrier zone of trees or shrubs planted to protect crops and soil from strong winds and erosion.

shoot—a new growth, sprout or twig.

sickle—a hand tool with a curved metal blade for cutting grain or grass.

sieve—a tool with holes in it that allows water to pass through but traps anything else.

silhouette—outline against a light.

silk screen printing—see screen printing.

silt—small particles of soil intermediate in size between sand and clay.

simultaneous—at the same time.

siting—choosing the location for.

skewed—distorted in one direction.

slang—popular words not found in dictionaries.

slit—a narrow cut or crack through which light can pass.

sludge—the outflow of a digester or sewage treatment plant.

soakway—a place for waste water to sink into the ground.

social account—the net economic effects of an investment or other action, measured as they affect an entire community or nation; which investments and policy measures appear wise may be different in social account than when only individual investment-profit effects are considered.

sod houses—houses with roofs or walls made from strips of soil still containing the roots of grass.

soil amendment—a substance that aids plant growth indirectly by improving the condition of the soil.

soil cement—a mixture of soil and a small amount of cement, used in making blocks without sand, for construction purposes.

solar distillation—the use of solar energy to purify water by forming water vapor and then condensing it to form water, leaving behind impurities.

solar greenhouse—a greenhouse that depends primarily on solar energy for heating; differs from a conventional greenhouse that uses fossil fuel energy to control the inside temperature.

solar radiation—energy from the sun.

soldering—a technique of lightweight metal bonding, melting a soft metal.

Solomon—a very wise King and judge in the Bible.

solvent—a liquid substance capable of dissolving or dispersing another substance.

spar—long pole which supports a sail.

sparingly—in small quantity.

sparse—not common, few.

spawning—producing or depositing eggs, sperm or young.

species diversity—the number of different species in an area.

specific gravity—the ratio of the weight or mass of a given volume of a substance to that of an equal volume of water (liquids and solids) or air (gases).

spectrum—range.

spinoff—a secondary development.

spontaneous—happening suddenly or without an obvious cause.

sporadic—occasional, irregular.

sprung frame—a frame mounted on springs.

squatters—people who have built shelters and houses on land they do not own.

stabilized soil—soil that has emulsified asphalt, cement or other material added to make it resist erosion when made into blocks for construction purposes.

stagnant—unchanging, lifeless.

staples—basic foods such as grains, beans, and tubers.

state of the art—latest, most current, most technically advanced.

static—unmoving.

stator—the fixed outside part of an electric motor or generator.

stereotype—a fixed concept about a group of people or an idea.

sterile—free from living germs.

stewardship—taking care of in a responsible manner to preserve quality and benefits over the long term.

stratification—the division into groups of different rank, status, income.

stricken—affected by something painful or sickening.

striving—attempting; trying hard.

stroboscope—a revolving disc with holes around the edge which allows flashes of light to pass through it at regular intervals.

stucco—a coating for the outside walls of buildings, applied like plaster.

subscribe (to an opinion)—believe.

subservient—humbly submissive.

subsidy—a grant of money.

subtropics—regions bordering on the tropics, having a nearly tropical climate.

sundial—a clock that indicates the time of day using the shadows caused by the sun.

super helicak—small taxi (Indonesia).

superimpose—to place on top of.

superphosphate—a chemical fertilizer made by treating bone or phosphate rock with sulfuric acid.

surging—rapidly increasing.

surpass—to be better than.

suspension bridge—a bridge that is hung from cables.

sustainable—that which can be continued indefinitely into the future.

swab—to clean with a small piece of cotton.

swell—increase greatly.

swine—pigs.

symptom—a condition that serves as an aid in diagnosing an illness.

synchronous inverter—device for changing direct current.

synergy—parts or ideas working together.

syringe—a medical device used with a needle to give injections.

tabular—arranged in a table.

tandem—a two-wheeled bicycle that has seats and pedals for two riders.

tannery—a place where animal hides and skins are preserved.

tapered—becoming smaller at one end.

tap-root—a large main root found on many plants.

technical fix—an attempt to solve a social, political or economic problem through a purely technical change, which may simply postpone the problem.

technocratic—of government in which all economic resources and the social system are controlled by scientists and engineers.

technological determinism—a theoretical point of view which holds that technological change is the primary cause of political and social change.

tempeh—an Indonesian food made from soybeans inoculated with bacteria.

temper—harden.

temperate zone—either of two zones of the earth between the tropics and the polar regions.

tenant—someone who pays rent to use a house or piece of land.

tension—being pulled apart.

tenure—the act of holding property.

tenurial—related to holding property.

terminology—words from a particular field of activity.

terracing—the building of flat areas along contour lines of a hillside, to prevent soil erosion while allowing productive use of the land.

terrain—the surface of the land.

terrarium—a glass-walled container for small animals that allows careful observation of their behavior.

therapeutic—serving to cure, heal, or improve health.

thermal pollution—heat from a power plant or other source that can disturb the ecological balance.

thermal storage—heat storage during the warm or sunny parts of the day, for use during the colder parts of the night.

thermosiphon—natural circulation based on the principle that hot air and hot water rise, while cold air and cold water fall.

thresher—a machine used to knock grain or beans away from the unwanted straw or other plant material.

thriving—growing very well.

tiebars—strips of metal for securely fastening roofing pieces.

tier—level.

tillage—the plowing of land.

timescale—period of time during which an activity is planned or expected to take place.

tolerance—the amount of variation in the dimensions of parts that is acceptable when constructing a machine.

toolbar—a frame to which different tools can be attached for various land preparation activities, such as plowing or cultivating.

torque—the force that acts to produce rotation.

torrid—very hot.

torsion—being turned or twisted.

totalitarian regime—a government in which one group maintains complete control under a dictatorship.

tragedy of the commons—the phenomenon that land, water, air and other things owned in common are frequently abused, polluted, or otherwise damaged, to the disadvantage of all.

transaction—exchange of something for something else.

transfer pipette—a tube of glass or plastic used to move liquid from one container to another.

transformer—a device containing two or more coils of insulated wire that is used to change the voltage and amperage levels of an electric current.

transit level—an instrument for identifying a horizontal line or plane.

transmission loss—the amount of power lost between a turbine and the machinery it is operating; or electricity lost between the generator and the point of use.

treadle—a foot-powered mechanism that converts an up and down motion of the foot on a board into a rotating motion on a machine; commonly seen on sewing machines.

trip-hammer—a heavy mechanical hammer that is regularly lifted and dropped; used in blacksmithing.

Trombe wall—a space heating system that involves a wall covered with glass that traps solar energy and circulates the heat through vents into the building.

tropics—the region of the earth lying between the Tropic of Cancer and the Tropic of Capricorn, marking the limits of the apparent north and south journey of the sun.

troubleshooting—seeking the source of a problem in a piece of equipment.

truss plates—metal plates attached to the beams that support a roof.

T-square—an instrument for drawing or cutting 90-degree angles.

tuberculosis—an infectious disease that affects the lungs.

tuft—a small piece of wool.

turbidity—cloudiness.

turbine—a wheel with curved vanes on a shaft, driven by the pressure of water.

typhoon—hurricane.

u-bolt—a bolt shaped like the letter "u".

undermining—removing the justification for.

unduly—unnecessarily.

unfettered—without restrictions.

unforeseen—unexpected.

unhampered—not restricted.

unicycle—a pedalled cycle having only one wheel.

unsubstantiated—undocumented, unproven.

urine—liquid human waste.

vaccinate—to give an injection that produces immunity to a specific disease by causing the formation of antibodies.

validity—truth, accuracy.

vaporizing—converting from a liquid to a gas by heating.

veil—to hide.

velocity—speed.

ventilation—a flow of fresh air.

verbiage—an excess of words beyond those needed to say what is meant.

verging on—bordering, approaching.

versatile—can function or be used in many different ways.

vertical shaft kilns—relatively small kilns used in the production of cement, having a vertical main shaft, differing from the more capital-intensive rotary kilns.

viable—successful, possible, practical.

virus—any of a group of extremely small infective agents that cause disease in animals, people and plants.

vogue—popularity or fashion.

volatile gas—unburned, energy-containing gases in the smoke of a fire.

volumeter—an instrument used to measure the volume of liquids and gases directly, and of solids by the amount of liquid they displace.

vulnerability—the state of being open to attack or damage.

vulnerable—easily hurt.

water-borne—carried by water.

water catchment—an apparatus for collecting and storing water.

water seal privy—a human waste disposal system that has a passageway filled with water which prevents odors, gases, and disease organisms from returning through the passageway.

weaning—the process of causing a young child to eat other foods than mother's milk.

weatherization—the process of sealing air leaks and insulating a house to reduce heat loss.

wed—join.

welding—high temperature heavy duty metal bonding, joining two pieces of metal using very high temperatures.

well-tuned—well matched.

wheelwright—a craftsman whose job is the production or repair of wheels and wheeled vehicles.

whole heartedly—enthusiastically.

winch—a device for hauling, pulling, or raising another object, that allows the operator to slowly move something that he would normally not be able to move at all.

winch plow—a plow pulled across the field by a cable attached to a winch.

windbreak—a hedge, fence, or row of trees that serves as protection from wind.

winnower—a machine used to separate grain from hulls or straw.

wither—to dry up.

wobbling—unsteady; unstable.

woodlot—land planted with trees grown for fuelwood.

yurt—a traditional Mongolian dwelling.

Index

Publications of the Appropriate Technology Project

1. **Appropriate Technology Sourcebook, Volume One**, by Ken Darrow and Rick Pam, second edition, November 1976 (updated January 1981), 320 pages. $5.50 (plus $0.83 postage). Special rate of $2.75 (plus $0.83 postage) only for local groups in developing countries. Bulk discounts.

This guide to practical books and plans for village and small community technology has more than 30,000 copies in print. It is being used in more than 100 countries to find a wide range of published technical information that can be directly used by individuals and small groups. Reviewed are selected publications on renewable sources of energy, farm implements, workshop tools and equipment, agriculture, low cost housing, health care, water supply systems and related subjects. Small-scale systems using local skills and resources are emphasized. Publications were chosen that provide enough information to be of significant help in understanding principles and in actually building the designs presented and using the techniques described. Entries were selected for low price, clear presentation, and unique subject matter. Reviews 385 publications from U.S. and international sources. Price and address are given for each publication. 250 illustrations. Extensive index. Includes a 15-page introduction to the principles of appropriate technology.

2. **Appropriate Technology: Problems and Promises**, Part One: The Major Policy Issues, by Nicolas Jequier, OECD Development Centre. Reprinted with permission, 1977. 88 pages. Distribution restricted by agreement to the United States and the developing countries only. $2.50 (plus $0.83 postage) in U.S. Special rate of $1.00 (plus $0.83 postage) to local groups in developing countries.

This is the most important conceptual book on appropriate technology since the late E.F. Schumacher's **Small is Beautiful**. Through special arrangement we are able to offer a low cost edition. The author examines the roles of craftsmen, aid agencies, governments, research and development centers, universities, and domestic development agencies in the promotion of appropriate technology. He offers a thorough treatment of the policy issues facing appropriate technology advocates.

3. **Lorena Owner-Built Stoves**, by Ianto Evans, January 1979, revised and expanded edition January 1981, 100+ pages. $4.00 (plus $0.83 postage). Special rate of $3.00 for local groups in developing countries.

The Lorena stove is a low cost, relatively efficient cookstove made with a mixture of sand and clay. Almost anyone can build it, without special tools, with only this book or a few days' training. In illustrated step-by-step fashion, this manual explains Lorena stove construction: how to test for suitable sand-clay mixtures, design the stove, and build and carve out the sand-clay block. Cooking methods and possible design modifications are suggested. A final section describes research on use of the stoves by the people of highland Guatemala. New material added in 1981 shows how to reduce construction time by using a dry mix, smaller one- and two-burner stoves made of lorena and other materials in Senegal and Java, efficient ceramic stoves, and stove efficiency testing techniques that can be used with any design. An appendix describes and illustrates rice hull burning stoves from Java.

All available from: Appropriate Technology Project, Volunteers in Asia, Box 4543, Stanford, California 94305, USA. (We accept UNESCO coupons.)